Recent Trends in Catalysis for Syngas Production and Conversion

Recent Trends in Catalysis for Syngas Production and Conversion

Editors

Fanhui Meng
Wenlong Mo

Basel • Beijing • Wuhan • Barcelona • Belgrade • Novi Sad • Cluj • Manchester

Editors

Fanhui Meng
State Key Laboratory of Clean and
Efficient Coal Utilization, College of
Chemical Engineering and Technology
Taiyuan University of Technology
Taiyuan
China

Wenlong Mo
State Key Laboratory of
Chemistry and Utilization of
Carbon-Based Energy Resources
Xinjiang University
Urumqi
China

Editorial Office
MDPI
St. Alban-Anlage 66
4052 Basel, Switzerland

This is a reprint of articles from the Special Issue published online in the open access journal *Catalysts* (ISSN 2073-4344) (available at: https://www.mdpi.com/journal/catalysts/special_issues/catalysis_for_syngas_production_and_conversion).

For citation purposes, cite each article independently as indicated on the article page online and as indicated below:

Lastname, A.A.; Lastname, B.B. Article Title. *Journal Name* **Year**, *Volume Number*, Page Range.

ISBN 978-3-0365-9008-0 (Hbk)
ISBN 978-3-0365-9009-7 (PDF)
doi.org/10.3390/books978-3-0365-9009-7

© 2023 by the authors. Articles in this book are Open Access and distributed under the Creative Commons Attribution (CC BY) license. The book as a whole is distributed by MDPI under the terms and conditions of the Creative Commons Attribution-NonCommercial-NoDerivs (CC BY-NC-ND) license.

Contents

Fanhui Meng and Wenlong Mo
Recent Trends in Catalysis for Syngas Production and Conversion
Reprinted from: *Catalysts* 2023, 13, 1284, doi:10.3390/catal13091284 1

Lina Wang, Fanhui Meng, Baozhen Li, Jinghao Zhang and Zhong Li
A Dual-Bed Strategy for Direct Conversion of Syngas to Light Paraffins
Reprinted from: *Catalysts* 2022, 12, 967, doi:10.3390/catal12090967 5

Guoqiang Zhang, Jinyu Qin, Yuan Zhou, Huayan Zheng and Fanhui Meng
Catalytic Performance for CO Methanation over Ni/MCM-41 Catalyst in a Slurry-Bed Reactor
Reprinted from: *Catalysts* 2023, 13, 598, doi:10.3390/catal13030598 17

Yuan Ren, Ya-Ya Ma, Wen-Long Mo, Jing Guo, Qing Liu, Xing Fan, et al.
Research Progress of Carbon Deposition on Ni-Based Catalyst for CO_2-CH_4 Reforming
Reprinted from: *Catalysts* 2023, 13, 647, doi:10.3390/catal13040647 31

Manel Hallassi, Rafik Benrabaa, Nawal Fodil Cherif, Djahida Lerari, Redouane Chebout, Khaldoun Bachari, et al.
Characterization and Syngas Production at Low Temperature via Dry Reforming of Methane over Ni-M (M = Fe, Cr) Catalysts Tailored from LDH Structure
Reprinted from: *Catalysts* 2022, 12, 1507, doi:10.3390/catal12121507 57

Zahra Gholami, Zdeněk Tišler, Eliška Svobodová, Ivana Hradecká, Nikita Sharkov and Fatemeh Gholami
Catalytic Performance of Alumina-Supported Cobalt Carbide Catalysts for Low-Temperature Fischer–Tropsch Synthesis
Reprinted from: *Catalysts* 2022, 12, 1222, doi:10.3390/catal12101222 75

Xu Wu, Heqin Guo, Litao Jia, Yong Xiao, Bo Hou and Debao Li
Effect of MnO_2 Crystal Type on the Oxidation of Furfural to Furoic Acid
Reprinted from: *Catalysts* 2023, 13, 663, doi:10.3390/catal13040663 93

Yuan Zhou, Guoqiang Zhang, Ya Song, Shirui Yu, Jingjing Zhao and Huayan Zheng
DFT Investigations of the Reaction Mechanism of Dimethyl Carbonate Synthesis from Methanol and CO on Various Cu Species in Y Zeolites
Reprinted from: *Catalysts* 2023, 13, 477, doi:10.3390/catal13030477 109

Yanqiu Li, Yuan Su, Yunfeng Yang, Ping Liu, Kan Zhang and Keming Ji
Study on the Formaldehyde Oxidation Reaction of Acid-Treated Manganese Dioxide Nanorod Catalysts
Reprinted from: *Catalysts* 2022, 12, 1667, doi:10.3390/catal12121667 123

Ting Liu, Yincui Li, Yifan Zhou, Shengnan Deng and Huawei Zhang
Efficient Pyrolysis of Low-Density Polyethylene for Regulatable Oil and Gas Products by ZSM-5, HY and MCM-41 Catalysts
Reprinted from: *Catalysts* 2023, 13, 382, doi:10.3390/catal13020382 139

Ley Boon Sim, Kek Seong Kim, Jile Fu and Binghui Chen
The Promoting Effect of Ti on the Catalytic Performance of V-Ti-HMS Catalysts in the Selective Oxidation of Methanol
Reprinted from: *Catalysts* 2022, 12, 869, doi:10.3390/catal12080869 153

Editorial

Recent Trends in Catalysis for Syngas Production and Conversion

Fanhui Meng [1],* and Wenlong Mo [2]

1 State Key Laboratory of Clean and Efficient Coal Utilization, College of Chemical Engineering and Technology, Taiyuan University of Technology, Taiyuan 030024, China
2 State Key Laboratory of Chemistry and Utilization of Carbon-Based Energy Resources, Xinjiang University, Urumqi 830046, China
* Correspondence: mengfanhui@tyut.edu.cn

Synthesis gas (or syngas) is a mixture of CO and H_2 that can be produced from fossil fuels or biomass. Syngas is one of the crucial platform chemicals for the production of a variety of high-value compounds, such as synthetic hydrocarbons and oxygenated fuels. More syngas will be required to meet industrial demand. This Special Issue contains articles that contribute to syngas production, syngas conversion or application, and various methods of the catalyst synthesis. Whether it is strategies for synthesis/conversion of syngas or catalyst synthesis processes and reactions, these provide new ideas for syngas applications as well as in the field of catalysis.

This Special Issue contains ten articles, of which nine are research articles and one is a review article. In the review article [1], the thermodynamics, kinetics and reaction mechanism of the CO_2-CH_4 reforming reaction (CRM reaction) are reviewed. Since Ni-based catalysts exhibit high activity but have the problem of easy deactivation of carbon deposition, this paper further summarizes the research situation regarding carbon deposition on Ni-based catalysts, including the types of carbon deposition, the amount of carbon deposition, and the elimination of carbon deposition. As for how to improve the anti-carbon deposition ability of the Ni-based catalyst and how to eliminate carbon deposition, this paper focuses on two aspects: one is the resistance of carbon deposition from the perspective of catalyst optimization; the other is the elimination of carbon deposition from the perspective of process condition adjustment. The authors also present the perspectives on how to inhibit carbon deposition to improve the activity and stability of the CRM catalyst.

In [2], bimetallic layered double oxide (LDO) NiM (M = Cr, Fe) catalysts with nominal compositions of Ni/M = 2 or 3 were tailored from layered double hydroxides (LDH) using co-precipitation method to investigate the effects of trivalent metal (Cr or Fe) and the amount of Ni species on the structural, textural, reducibility and catalytic properties for CH_4/CO_2 reforming at low reaction temperatures (400–650 °C). The influences of the molar ratio and cationic composition in the preparation of the LDH precursors on the physicochemical properties of the target catalysts and on their performance for the dry reforming of methane were also evaluated.

Syngas to hydrocarbons via the Fischer–Tropsch (F-T) synthesis pathway is a valuable way to achieve syngas utilization. In [3], a series of cobalt carbide (Co_2C) catalysts were synthesized by exposure of Co/Al_2O_3 catalyst to CH_4 at different temperatures from 300 °C to 800 °C for F-T synthesis to hydrocarbons. The result shows that by increasing the carbidation temperature, the Co_2C content decreased, and the metallic cobalt content increased, which resulted in higher catalytic activity. For the catalysts prepared at higher temperatures, the presence of less Co_2C, which is transformed into hcp cobalt during the reduction with hydrogen, and the presence of less metallic fcc cobalt resulted in lower CO conversion and less heavy hydrocarbons.

In addition to the F-T synthesis mentioned above, the Special Issue also includes research using a new strategy for syngas to hydrocarbons. In [4], a dual-bed strategy

Citation: Meng, F.; Mo, W. Recent Trends in Catalysis for Syngas Production and Conversion. *Catalysts* **2023**, *13*, 1284. https://doi.org/10.3390/catal13091284

Received: 4 September 2023
Accepted: 6 September 2023
Published: 7 September 2023

Copyright: © 2023 by the authors. Licensee MDPI, Basel, Switzerland. This article is an open access article distributed under the terms and conditions of the Creative Commons Attribution (CC BY) license (https:// creativecommons.org/licenses/by/ 4.0/).

was adopted to directly convert syngas to light paraffins, which includes an STD catalyst (CZA+Al_2O_3(C)) in the upper bed and methanol/DME conversion catalyst SAPO-34 in the lower bed. This dual-bed strategy allows the different catalysts to be operated at different temperatures in the reaction process, which can save energy, and the synthesis of methanol at low temperature is beneficial to extend the catalyst life and also provides a potential route for syngas conversion to valuable chemicals.

In [5], MCM-41 is selected as support to prepare xNi/MCM-41 catalysts with various Ni contents and the catalytic performance for CO methanation in a slurry-bed reactor is investigated under different reaction conditions. The reason for catalyst deactivation after reaction was analyzed. The aim is to clarify the relationship between the structures and the catalytic methanation performance.

The participation of CO in the oxidative carbonylation of methanol to dimethyl carbonate (DMC) is also a pathway for CO conversion. In [6], the reaction mechanisms governing oxidative carbonylation of methanol to DMC with Cu^+, Cu^{2+}, Cu_2O and CuO species in Y zeolites using density functional theory (DFT) were studied. The results are expected to guide the selection and preparation of CuY catalysts with the best catalytic activity for DMC synthesis.

Methanol and formaldehyde are important building blocks in the production from syngas, and the conversion of these products is worthy of attention. In [7], Ti-HMS supported vanadium oxide catalysts exhibited higher activities in the selective oxidation of methanol to dimethoxymethane, and the enhanced activity of the V-Ti-HMS catalyst is attributed to the improved dispersion and reducibility of vanadium oxides. In [8], a series of MnO_2 catalysts were obtained for formaldehyde oxidation using acid treatment, and the structure and properties of the acid-treated catalysts were investigated and the effect of acid treatment on the catalytic oxidation activity of formaldehyde was explained.

In [9], catalytic pyrolysis of LDPE was performed in a fixed-bed reactor using ZSM-5, HY and MCM-41 catalysts to obtain the three-phase products. The effect of pyrolysis temperature and catalyst type on product yield was explored. In [10], four MnO_2 with different crystalline structures were synthesized in the present study and the catalytic activity was evaluated in the oxidation of furfural to furoic acid and the factors affecting the catalytic performance over different MnO_2 are discussed.

In conclusion, the guest editors of the Special Issue "Recent Trends in Catalysis for Syngas Production and Conversion" would like to thank all the authors for their contributions, which demonstrate the importance of ongoing research in the field of syngas production and conversion. We also thank the reviewers for their hard work. In addition, we thank the journal *Catalysts* for the great opportunity to produce this Special Issue.

Acknowledgments: We are grateful to all the authors for submitting their impressive state-of-the-art research papers for this Special Issue and to the anonymous reviewers for their time and effort in reviewing the manuscripts.

Conflicts of Interest: The authors declare no conflict of interest.

References

1. Ren, Y.; Ma, Y.-Y.; Mo, W.-L.; Guo, J.; Liu, Q.; Fan, X.; Zhang, S.-P. Research Progress of Carbon Deposition on Ni-Based Catalyst for CO_2-CH_4 Reforming. *Catalysts* **2023**, *13*, 647. [CrossRef]
2. Hallassi, M.; Benrabaa, R.; Cherif, N.F.; Lerari, D.; Chebout, R.; Bachari, K.; Rubbens, A.; Roussel, P.; Vannier, R.-N.; Trentesaux, M.; et al. Characterization and Syngas Production at Low Temperature via Dry Reforming of Methane over Ni-M (M = Fe, Cr) Catalysts Tailored from LDH Structure. *Catalysts* **2022**, *12*, 1507. [CrossRef]
3. Gholami, Z.; Tišler, Z.; Svobodová, E.; Hradecká, I.; Sharkov, N.; Gholami, F. Catalytic Performance of Alumina-Supported Cobalt Carbide Catalysts for Low-Temperature Fischer-Tropsch Synthesis. *Catalysts* **2022**, *12*, 1222. [CrossRef]
4. Wang, L.; Meng, F.; Li, B.; Zhang, J.; Li, Z. A Dual-Bed Strategy for Direct Conversion of Syngas to Light Paraffins. *Catalysts* **2022**, *12*, 967. [CrossRef]
5. Zhang, G.; Qin, J.; Zhou, Y.; Zheng, H.; Meng, F. Catalytic Performance for CO Methanation over Ni/MCM-41 Catalyst in a Slurry-Bed Reactor. *Catalysts* **2023**, *13*, 598. [CrossRef]

6. Zhou, Y.; Zhang, G.; Song, Y.; Yu, S.; Zhao, J.; Zheng, H. DFT Investigations of the Reaction Mechanism of Dimethyl Carbonate Synthesis from Methanol and CO on Various Cu Species in Y Zeolites. *Catalysts* **2023**, *13*, 447. [CrossRef]
7. Sim, L.B.; Kim, K.S.; Fu, J.; Chen, B. The Promoting Effect of Ti on the Catalytic Performance of V-Ti-HMS Catalysts in the Selective Oxidation of Methanol. *Catalysts* **2022**, *12*, 869. [CrossRef]
8. Li, Y.; Su, Y.; Yang, Y.; Liu, P.; Zhang, K.; Ji, K. Study on the Formaldehyde Oxidation Reaction of Acid-Treated Manganese Dioxide Nanorod Catalysts. *Catalysts* **2022**, *12*, 1667. [CrossRef]
9. Liu, T.; Li, Y.; Zhou, Y.; Deng, S.; Zhang, H. Efficient Pyrolysis of Low-Density Polyethylene for Regulatable Oil and Gas Products by ZSM-5, HY and MCM-41 Catalysts. *Catalysts* **2023**, *13*, 382. [CrossRef]
10. Wu, X.; Guo, H.; Jia, L.; Xiao, Y.; Hou, B.; Li, D. Effect of MnO_2 Crystal Type on the Oxidation of Furfural to Furoic Acid. *Catalysts* **2023**, *13*, 663. [CrossRef]

Disclaimer/Publisher's Note: The statements, opinions and data contained in all publications are solely those of the individual author(s) and contributor(s) and not of MDPI and/or the editor(s). MDPI and/or the editor(s) disclaim responsibility for any injury to people or property resulting from any ideas, methods, instructions or products referred to in the content.

Article

A Dual-Bed Strategy for Direct Conversion of Syngas to Light Paraffins

Lina Wang, Fanhui Meng *, Baozhen Li, Jinghao Zhang and Zhong Li *

State Key Laboratory of Clean and Efficient Coal Utilization, Taiyuan University of Technology, Taiyuan 030024, China
* Correspondence: mengfanhui@tyut.edu.cn (F.M.); lizhong@tyut.edu.cn (Z.L.)

Abstract: The authors studied the direct conversion of syngas to light paraffins in a dual-bed fixed-bed reactor. A dual-bed catalyst composed of three catalysts, a physically mixed methanol synthesis catalyst (CZA), and a methanol dehydration to dimethyl ether (DME) catalyst (Al_2O_3(C)) were put in the upper bed for direct conversion of syngas to DME, while the SAPO-34 (SP34-C) zeolite was put in the lower bed for methanol and DME conversion. The effects of the mass ratio of CZA to Al_2O_3(C), the H_2/CO molar ratio, and the space velocity on catalytic performance of syngas to DME were studied in the upper bed. Moreover, a feed gas with a $CO/CO_2/DME/N_2/H_2$ molar ratio of 9/6/4/5 balanced with H_2 was simulated and studied in the lower bed over SP34-C; after optimizing the reaction conditions, the selectivity of light paraffins reached 90.8%, and the selectivity of propane was as high as 76.7%. For the direct conversion of syngas to light paraffins in a dual bed, 88.9% light paraffins selectivity in hydrocarbons was obtained at a CO conversion of 33.4%. This dual-bed strategy offers a potential route for the direct conversion of syngas to valuable chemicals.

Keywords: dual bed; syngas; methanol; dimethyl ether; light paraffins; propane

1. Introduction

The overexploitation of petroleum resources makes it highly essential to develop a nonpetroleum way of producing chemicals. Syngas is an important platform for the utilization of carbon resources such as coal, natural gas, and biomass; moreover, it can be converted into a variety of chemicals and fuels [1]. Fischer–Tropsch synthesis (FTS) [2] and oxide–zeolite (OX–ZEO) bifunctional catalyst [3] routines are commonly applied for syngas conversion. For FTS, hydrocarbon selectivity remains a challenge due to the Anderson–Schulz–Flory limitation rule [4]. Recently, OX–ZEO catalysts have been intensively studied for the direct conversion of syngas to light olefins (C_2–$C_4^=$) [5–9], aromatics [10,11], and gasoline [12], which was due to the breakthrough of hydrocarbon selectivity.

The OX–ZEO catalyst is composed of a methanol synthesis catalyst and a methanol conversion catalyst. It is well-known the syngas to methanol conversion (STM, $CO + 2H_2 = CH_3OH$) has been realized in industrial production under high pressure since 1923; most of the new plants have adopted a low-pressure and low-temperature process since the middle of 1970s. Right now, the commonly used methanol synthesis catalysts for STM are Cu-based catalysts [13], and great efforts have been devoted to improve the catalytic performance of Cu-based catalysts [14], which is due to the fact that methanol is a platform that can be used as a fuel and also converted to valuable chemicals [15]. The conversion of methanol to DME (MTD, $2CH_3OH = CH_3OCH_3 + H_2O$) has the same reaction temperature as that of STM, and the most widely used catalysts are the γ-Al_2O_3 [16] and ZSM-5 [17] zeolite. By combining the catalyst of STM and MTD, syngas can be directly converted to DME (STD, $2CO + 4H_2 = CH_3OCH_3 + H_2O$) [18], and reports have shown that CO equilibrium conversion can be significantly enhanced by the STD process.

For the conversion of syngas to light olefins with an OX–ZEO catalyst, a high reaction temperature (400 °C) to convert methanol to olefins (MTO, $CH_3OH(g) \rightarrow C_nH_{2n}(g) + H_2O(g)$)

is needed, which needs the methanol synthesis catalyst to be operated at around 400 °C to match the MTO reaction for C_2–$C_4^=$ synthesis. However, high temperature is not beneficial for the conversion of CO due to the thermodynamic limitation. Thus, a dual-bed fixed-bed reactor was designed for the conversion of syngas to DME to obtain high CO conversion at low temperatures, accompanied by the methanol/DME conversion performed at high temperatures [19]. In such a dual-bed fixed-bed reactor, the direct conversion of syngas to DME was performed in the upper catalyst bed, and the produced intermediate products were further passed through the lower catalyst bed. For the conversion of methanol or DME to olefins, the SAPO-34 zeolite is a commonly used catalyst [20], and the production of olefins has been commercialized in China since 2011. For the direct conversion of syngas, the presence of large amounts of unconverted H_2 could lead to the generation of olefins further hydrogenated to paraffins, which could result in the low ratio of light olefins to light paraffins (C_2–C_4^0) [21–23]. The results indicate that product distribution can be regulated by changing the reaction conditions. To the best of our knowledge, the direct conversion of syngas to light paraffins in a dual-bed fixed-bed reactor has been seldomly studied.

In this work, a dual-bed strategy was adopted to directly convert syngas to light paraffins, which includes STD catalyst CZA + Al_2O_3(C) in the upper bed and methanol/DME conversion catalyst SP34-C in the the lower bed. For the STD process, the mass ratio of CZA to Al_2O_3(C), the H_2/CO ratio, and the space velocity on catalytic performance were explored, while for the conversion of DME, a feed gas with a CO/CO_2/DME/N_2/H_2 molar ratio of 9/6/4/5 balanced with H_2 was simulated and studied in the lower bed over SP34-C. The adopted catalysts were characterized and the stability for syngas conversion to light paraffins was studied. This dual-bed strategy makes the different catalysts to be operated at different temperatures in the reaction process, which can save energy, and the synthesis of methanol at low temperature is beneficial to prolonging the catalyst life. Moreover, the dual-bed strategy provides a potential route for syngas conversion to valuable chemicals.

2. Results and Discussion

2.1. Catalyst Characterization Results

XRD patterns of the CZA and Al_2O_3(C) catalysts are displayed in Figure 1a. The characteristic diffraction peaks at 2θ of 35.5°, 38.7°, and 61.5° are attributed to CuO (PDF#45-0937), while the diffraction peaks at 2θ of 31.8°, 34.4°, 36.2°, and 47.5° correspond to ZnO (PDF#36-1451). A diffraction peak appeared at 2θ of 26.6° accounting for graphite, which was generally used to enhance catalyst mechanical strength and heat transfer efficiency [24]. The Al_2O_3(C) catalyst shows the characteristic diffraction peaks at 2θ of 19.6°, 31.3°, 37.6°, 39.5°, 45.8°, and 66.8°, confirming the existence of the γ-Al_2O_3 phase (PDF#29-0063). SP34-C in Figure 1b shows the characteristic diffraction peaks of SAPO-34 (PDF#47-0429) [25].

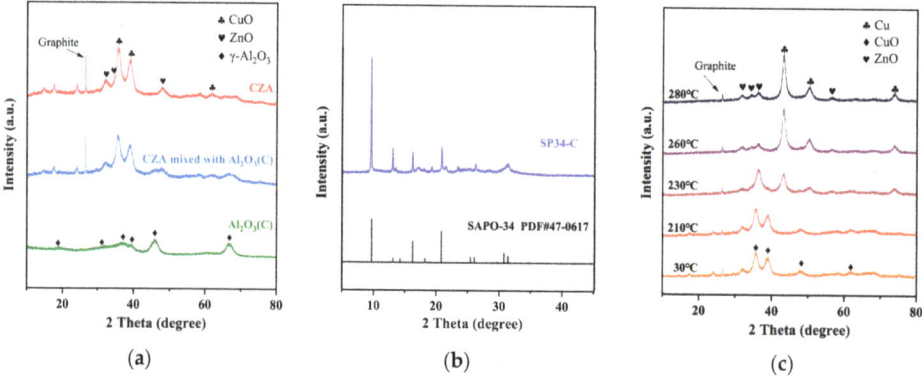

Figure 1. XRD patterns of the catalysts. (a) XRD patterns of CZA and Al_2O_3(C); (b) XRD patterns of SP34-C; (c) in-situ XRD patterns of the CZA catalyst.

To evaluate the structural evolution of the CZA catalyst during the reduction process, in-situ XRD patterns of CZA were performed in an H_2 atmosphere under various temperatures. The results are shown in Figure 1c. It is obvious that when the temperature was 210 °C or below, the crystalline phase was not changed, indicating that the CZA catalyst cannot be reduced at this temperature. As the temperature increased to 230 °C, the peak intensity of CuO slightly weakened, and the diffraction peaks attributing to Cu metal at 2θ of 43.3°, 50.4°, and 74.1° (PDF#85-1326) appeared. With a further increase in temperature to 260 °C, the diffraction peaks of CuO disappeared. The diffraction peaks at 280 °C were almost the same as at 260 °C, indicating that CZA was completely reduced.

H_2-TPR was performed to further study the reduction behavior of the CZA catalyst. The profile is displayed in Figure 2a. The ZnO species cannot be reduced at 400 °C or below [26]. Thus, the two peaks centered at 186 °C and 205 °C could be attributed to the reduction of CuO to Cu_2O and of Cu_2O to Cu, respectively [27], which was consistent with the in-situ XRD results. The profile of the CO adsorption ability is shown in Figure 2b. The amount of desorbed CO listed in Table 1 is 0.10 mmol/g. It is obvious that the CZA catalyst exhibited a large desorption peak for CO, which is beneficial to the conversion of CO.

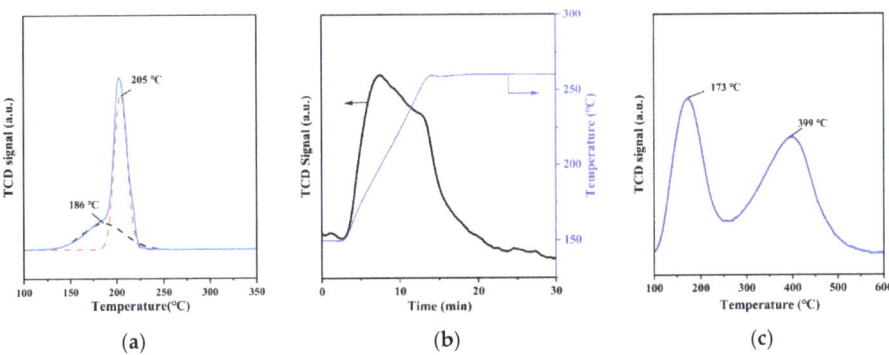

Figure 2. (a) H_2-TPR profile of the CZA catalyst; (b) CO-TPD profile of the CZA catalyst; (c) NH_3-TPD profile of the SP34-C catalyst.

Table 1. Textural properties of the catalysts and amounts of desorbed CO for CZA.

Samples	S_{total} [a] $(m^2 \cdot g^{-1})$	V_{total} [b] $(cm^3 \cdot g^{-1})$	D [c] (nm)	Amounts of Desorbed CO [d] (mmol/g)
CZA	68	0.19	9.4	0.10
Al_2O_3(C)	238	0.41	4.9	–
SP34-C	488	0.49	68.7	–

[a] S_{total} (total BET surface area) was calculated based on the multipoint BET equation. [b] Single-point desorption total pore volume of pores, $p/p_0 = 0.99$. [c] Barret–Joyner–Hallender (BJH) desorption average pore diameter. [d] Calculated from the CO-TPD profile.

The acidity of a SAPO-34 molecular sieve plays a vital role in determining the selectivity of light hydrocarbons [28]. NH_3-TPD profiling for SP34-C was performed, and the result is shown in Figure 2c. It can be seen that SP34-C exhibits two desorption peaks at 173 °C and 399 °C. The desorption peak at low temperature belongs to the physically adsorbed ammonia, whereas the high-temperature desorption peak is attributed to Si−OH−Al groups of the molecular sieve (Brønsted acids) [29]. It has been reported that during the MTO reaction, weak acid sites do not have a strong influence on the formation of light olefins, only the Brønsted acid sites play an important role [30].

N_2 adsorption–desorption was used to determine the textural properties of the samples. Figure 3 shows that the CZA catalyst exhibits a typical IV-type isotherm curve and an H1-type hysteresis loop, indicating that CZA is a mesoporous material. The Al_2O_3(C) catalyst also exhibits a IV-type isotherm with a type H3 hysteresis loop. The SP34-C catalyst exhibits a typical I-type isotherm characteristic, and when the relative pressure p/p_0 was lower than 0.01, the adsorption capacity increased sharply due to the filling of micropores, indicating the existence of large amounts of micropores [31]. The hysteresis loop appearing at p/p_0 of 0.98 was due to the stacking of SP34-C nanoparticles.

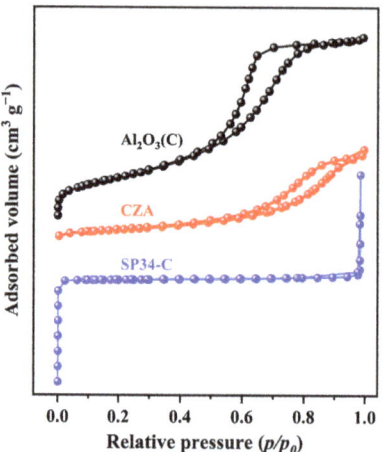

Figure 3. N$_2$ adsorption–desorption isotherms of the different catalysts.

The textural properties are summarized in Table 1. CZA possesses a BET surface area of 68 m$^2 \cdot$g^{-1} with an average pore diameter of 9.4 nm. The Al$_2$O$_3$(C) sample shows a high surface area of 238 m$^2 \cdot$g^{-1}, while for SP34-C, the BET surface area increased to 488 m$^2 \cdot$g^{-1}. Both the Al$_2$O$_3$(C) and SP34-C samples exhibit a much larger total pore volume than CZA. The large average pore diameter of SP34-C was mainly attributed to the stacking of SP34-C nanoparticles.

2.2. Catalytic Performance of Syngas to DME

The direct conversion of syngas to light paraffins contains three reactions, i.e., the syngas conversion to methanol, the methanol dehydration to DME, and the DME conversion reaction. In this part, the catalytic performance for the direct conversion of syngas to DME over a bifunctional catalyst composed of CZA as the methanol synthesis catalyst and Al$_2$O$_3$(C) as the methanol dehydration catalyst was studied. The results are displayed in Figure 4. Figure 4a shows the effect of the mass ratio of CZA to Al$_2$O$_3$(C) on catalytic performance. As the mass ratio of CZA to Al$_2$O$_3$(C) increased from 1:10 to 2:1, the CO conversion increased from 30.4% to 72.2%, and the selectivity of byproduct CO$_2$ gradually decreased from 36.0% to 28.8%. It is because the content of the methanol synthesis CZA catalyst increased, which promoted the conversion of CO. Subsequently, the generated methanol converted to DME over the Al$_2$O$_3$(C) catalyst and thus produced water. The increase in the mass ratio of CZA to Al$_2$O$_3$(C) indicates that the relatively low content of Al$_2$O$_3$(C) in such a bifunctional catalyst generated less water, which was not beneficial for the water gas shift reaction (WGSR, CO + H$_2$O = H$_2$ + CO$_2$) and thus decreased the selectivity of CO$_2$. It is obvious that when the mass ratio of CZA to Al$_2$O$_3$(C) achieved 2:1, a high CO conversion and a low CO$_2$ selectivity were obtained; thus, the CZA-to-Al$_2$O$_3$(C) ratio was set at 2:1 in the following studies.

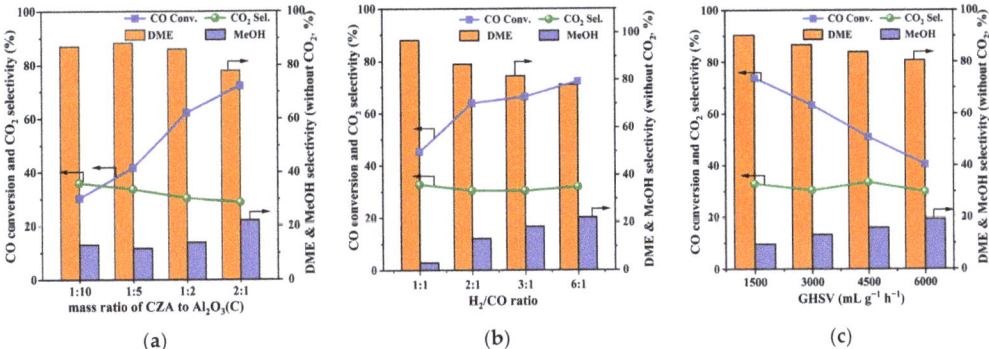

Figure 4. Effect of (**a**) the mass ratio of CZA to Al$_2$O$_3$(C), (**b**) the H$_2$/CO molar ratio, and (**c**) the GHSV on syngas conversion of DME. Reaction conditions: 260 °C and 2.0 MPa, (**a**) GHSV = 3000 mL·g^{-1}·h^{-1}, H$_2$/CO = 6:1; (**b**) GHSV = 3000 mL·g^{-1}·h^{-1}, m(CZA): m(Al$_2$O$_3$(C)) = 2:1; (**c**) H$_2$/CO = 2:1, m(CZA): m(Al$_2$O$_3$(C)) = 2:1.

The effect of the H$_2$/CO molar ratio on catalytic performance is shown in Figure 4b. It could be found that the CO conversion gradually increased from 45.4% to 66.3% as the H$_2$/CO ratio increased from 1:1 to 3:1 under the conditions of 260 °C, 2.0 MPa and 3000 mL g^{-1} h^{-1}, while the selectivity of DME decreased from 96.9% to 81.7%. With a further increase in the H$_2$/CO ratio to 6:1, the conversion of CO increased to 72.2%. The results indicate that a high H$_2$/CO ratio benefits the CO conversion and methanol selectivity, while it goes against the selectivity of DME.

Figure 4c shows the effect of gas hourly space velocity (GHSV) on catalytic performance. As the space velocity increased from 1500 mL g^{-1} h^{-1} to 6000 mL g^{-1} h^{-1}, the conversion of CO linearly decreased from 73.8% to 40.5%. The selectivity of MeOH gradually increased while the selectivity of DME gradually decreased. The reason was that a high GHSV reduced the residence time of methanol in the catalyst bed, which resulted in the methanol not being converted in time.

The CO equilibrium conversions of the STM and STD reactions were calculated using the HSC 6.0 software. The results are shown in Figure 5a. It is obvious that the CO equilibrium conversions of the STM and STD reactions both decreased as the temperature increased. The reason is that the STM and STD reactions are both thermodynamically exothermic: the increase in temperature is not conducive to the conversion of CO. For the STM reaction, the increase in the H$_2$/CO ratio gradually enhanced the CO equilibrium conversion, while for the STD reaction, the feed gas with the H$_2$/CO ratios of 3:1 and 6:1 exhibited almost the same CO conversions. It is because the presence of a large amount of H$_2$ inhibited the WGSR in the STD reaction, which thus showed a similar CO conversion.

Figure 5b shows the effects of the H$_2$/CO ratio on CO equilibrium conversions for the STD and STM reactions at 260 °C and 2.0 MPa. Moreover, the CO experimental conversion for the STD reaction at 2.0 MPa and 260 °C was selected for comparison. It could be found that the increase in the H$_2$/CO ratio enhanced the CO conversion, and the CO equilibrium conversion of STD was much higher than that of STM. The CO experimental conversion was lower than the CO equilibrium conversion for STD; however, it was much higher than the CO equilibrium conversion for STM, which is due to the thermodynamic driving force that promotes CO conversion.

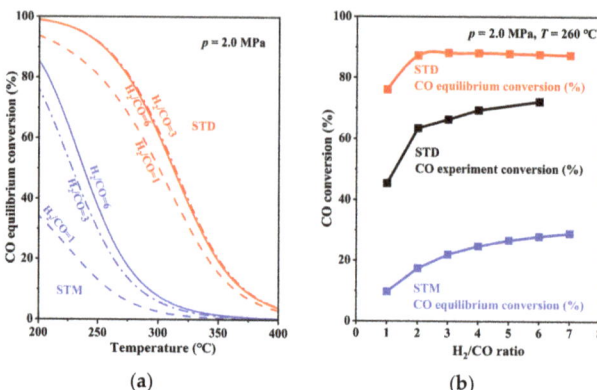

Figure 5. (a) Equilibrium conversion of the STD (red) and STM (blue) reactions with the H_2/CO ratios of 1/1, 3/1, and 6/1 at various temperatures, p = 2.0 MPa. (b) Comparison of the CO equilibrium conversion and the CO experimental conversion for STD and STM under the conditions of 260 °C and 2.0 MPa.

2.3. Catalytic Conversion of Reaction Intermediates

Under the reaction conditions of H_2/CO ratio = 3/1, GHSV = 3000 mL·g^{-1}·h^{-1}, and m(CZA)/m(Al$_2$O$_3$(C)) = 2:1, the component of the obtained products was close to the reaction intermediate with a molar ratio of $CO/CO_2/DME/N_2/H_2$ = 9/6/4/5 balanced with H_2. Thus, the reaction intermediate was used as a feed gas to study the catalytic conversion of reaction intermediates over SP34-C. The results are shown in Figure 6. Under the reaction conditions, no DME was detected in the product, indicating that DME was completely converted. Figure 6a shows the effect of space velocity on product selectivity under the conditions of 410 °C and 2.0 MPa. As the space velocity decreased from 16,000 mL g^{-1} h^{-1} to 6000 mL g^{-1} h^{-1}, the selectivity of light olefins decreased, while the selectivity of paraffins increased. The reason was due to the low space velocity possessing a long residence time in the catalyst bed for the generated products, which resulted in the excessive hydrogenation of the generated light olefins to paraffins. Specially, the ratio of light paraffins to light olefins increased from 0.9 to 24.5. Interestingly, it can be found that the selectivity of C_3H_8 increased from 36.9% to 76.7%. Thus, to enhance the selectivity of light paraffins, it is better to operate the reaction at a low space velocity.

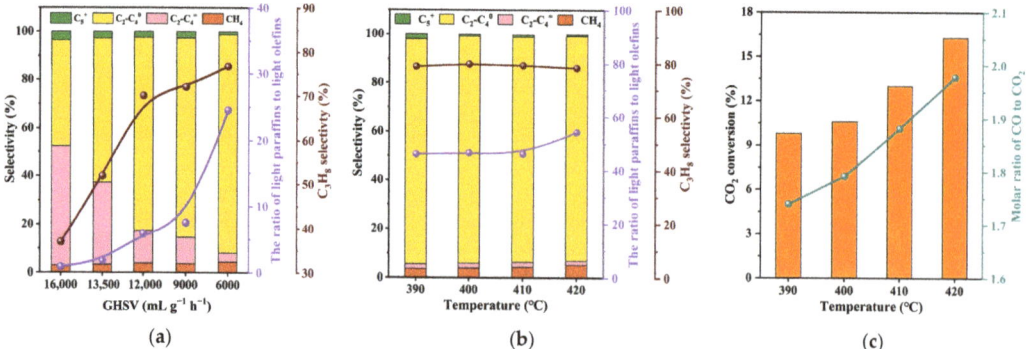

Figure 6. (a) Effect of space velocity on product selectivity. (b) Effect of reaction temperature on product selectivity. (c) Effect of reaction temperature on CO_2 conversion. Reaction conditions: m(SP34-C) = 0.2 g, $CO/CO_2/DME/N_2/H_2$ = 9/6/4/5 balanced with H_2, (a) T = 410 °C, p = 2.0 MPa; (b,c) p = 3.0 MPa, GHSV = 7600 mL g^{-1} h^{-1}.

The effect of reaction temperature on product selectivity is shown in Figure 6b. It could be found that the increase in reaction temperature only slightly changed the selectivity of hydrocarbons. The ratio of light paraffins to light olefins at 420 °C reached 54.4, which is higher than those at low temperatures. It is because the selectivity of light olefins (1.7%) at 420 °C was slightly lower than at other temperatures (2.0%). The results also indicate that the reaction temperature has little effect on the hydrocarbons selectivity.

The performance for the conversion of CO_2 at different temperatures is shown in Figure 6c. As the reaction temperature increased from 390 °C to 420 °C, the conversion of CO_2 increased from 9.8% to 16.3%. Moreover, it could be found that the molar ratio of CO to CO_2 increased from 1.74 to 1.98, which was due to the increase in CO_2 conversion at high temperatures. The results also indicate that high temperature intensified the reverse water gas shift reaction (reverse WGSR, $H_2 + CO_2 = CO + H_2O$).

To study the performance of a reverse WGSR over SP34-C, the reaction was carried out by using H_2 and CO_2 as feed gases. The results in Figure 7a show that the conversion of CO_2 is approximately 13% under the reaction conditions of 410 °C and 3.0 MPa, the main product is CO (99.2%), with a tiny selectivity of byproduct CH_4 (0.8%). The results suggest that a reverse WGSR could occur over the SP34-C catalyst, and the catalytic performance is relatively stable during the reaction.

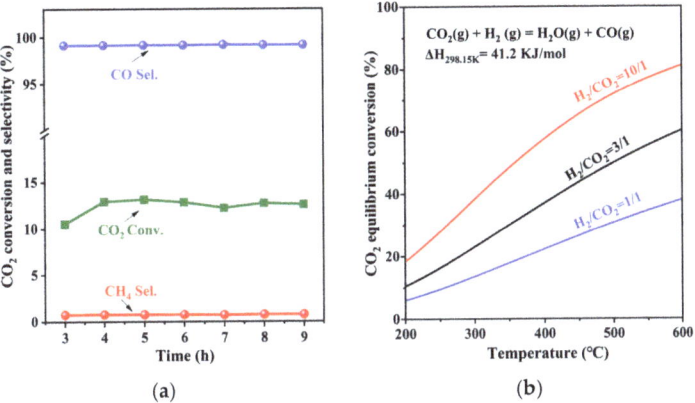

Figure 7. (a) Catalytic results of the reverse WGSR over SP34-C. Reaction conditions: m(SP34-C) = 0.4 g, H_2/CO_2 = 10/1, GHSV = 6000 mL g^{-1} h^{-1}, p = 2.0 MPa, T = 410 °C. (b) Effect of the H_2/CO_2 molar ratio and temperature on the CO_2 equilibrium conversion for the reverse WGSR.

Figure 7b shows the equilibrium conversion of CO_2 for the reverse WGSR calculated by the HSC 6.0 software. It could be found that increasing the temperature enhanced the CO_2 equilibrium conversion, which is attributed to the endothermic characteristics of the reverse WGSR (41.2 kJ/mol). Moreover, as the H_2/CO_2 ratio in the feed gas increased from 1:1 to 10:1, the CO_2 equilibrium conversion increased from 22.4% to 57.6% at 410 °C, attributing to the presence of large amounts of H_2 in the reaction.

2.4. Catalytic Performance of Syngas to Light Hydrocarbons

The direct conversion of syngas to light paraffins was carried out over a dual-bed catalyst. The schematic diagram of the dual-bed reactor is shown in Figure 8. For the upper catalyst bed, the catalyst was CZA + Al_2O_3(C), with the m(CZA):m(Al_2O_3(C)) ratio of 2:1, used at 3.0 MPa, 260 °C, and GHSV = 3000 mL g^{-1} h^{-1}, while for the lower catalyst bed, the catalyst was SP34-C and the reaction temperature was 410 °C. The catalytic performance results are shown in Figure 8 and Table 2. During the reaction process, almost no methanol and DME were detected (both less than 0.2%), indicating that the generated methanol and DME both converted under the reaction conditions. Table 2 shows that the conversion of CO was 33.4% and the selectivity of CO_2 was 32.6%, the selectivity of light paraffins reached 88.9%, which is much higher than the selectivity of light olefins (2.9%). Moreover, the selectivity of propane in hydrocarbons reached 75.4%. Figure 8 displays that the catalyst exhibited good stability after a reaction time for 1350 min.

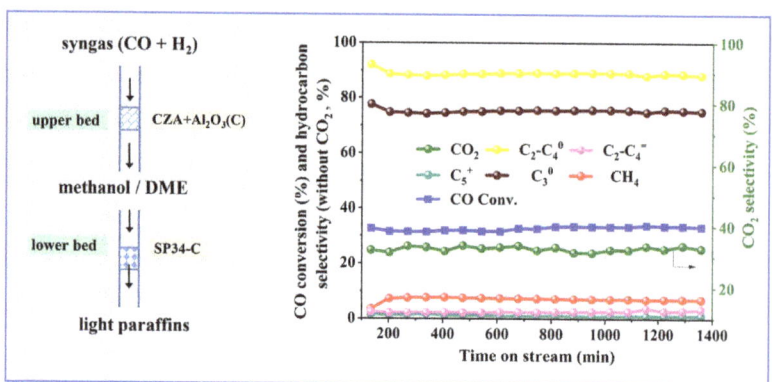

Figure 8. Schematic diagram of a dual-bed reactor and catalytic stability of syngas to light paraffins in a dual-bed reactor over the (CZA + Al_2O_3(C))/SP34-C catalysts. Reaction conditions: p = 3.0 MPa, T (upper bed) = 260 °C, m(CZA):m(Al_2O_3(C)) = 2:1, m(CZA + Al_2O_3(C)) = 0.6 g, GHSV = 3000 mL g^{-1} h^{-1}; T (lower bed) = 410 °C, m(SP34-C) = 0.3 g.

Table 2. Products selectivity in the direct conversion of syngas to light paraffins.

Product	CH_4	C_2–C_4^0			C_2–$C_4^=$			C_5^+	CO_2
		C_2^0	C_3^0	C_4^0	$C_4^=$	$C_4^=$	$C_4^=$		
Selectivity (%) [a]	7.2	7.3	75.4	6.2	1.1	1.0	0.8	1.0	32.6

[a] Average value from Figure 8.

3. Materials and Methods

3.1. Catalyst Preparation

CuZnAlO$_x$ (C302, denoted as CZA) was purchased from Southwest Chemical Industry Design and Research Institute (Chengdu, China); the alumina catalyst (Al_2O_3(C)) was purchased from TOAGOSEI Co., Ltd. (Tokyo, Japan); SAPO-34 (denoted as SP34-C) with a SiO_2/Al_2O_3 ratio of 0.5 was supplied by Tianjin Nanhua Catalyst Co., Ltd. (Tianjin, China).

CZA and the Al_2O_3(C) bifunctional catalyst: CZA powder and Al_2O_3(C) powder were mixed in an agate mortar and physically grinded for 10 min, then pressed into tablets and sieved to obtain 30–60 mesh particles. The SP34-C was also pressed and sieved to obtain 30–60 mesh particles.

3.2. Catalyst Characterization

Powder XRD of the catalysts was performed on a SmartLab SE (Rigaku, Tokyo, Japan), Cu Kα radiation was used as the radiation source (λ = 0.154056 nm) operated at 40 kV and 40 mA, with a scanning speed of 10° min^{-1}. In-situ XRD of the CZA measurement was carried out under a flow of 10% H_2 in Ar on a Max 2500 diffractometer equipped with an XRK-900 reaction chamber, the heating rate was 5 °C min^{-1}, and the gas flow was 20 mL/min.

H_2-temperature-programmed reduction (H_2-TPR) and temperature-programmed desorption of CO or NH_3 (CO-TPD, NH_3-TPD) were performed on an Autochem II 2920 (Micromeritics, Norcross, GA, USA) instrument.

For the H_2-TPR, 40 mg of the CZA catalyst were placed into a U-shaped tube and purged with N_2 at 300 °C for 30 min. After cooling to 50 °C, the N_2 was switched to a 10%

H_2/90% Ar gas mixture and heated to 600 °C at 10 °C/min. In this process, the signal of H_2 was detected by a thermal conductivity cell detector (TCD).

For the CO-TPD, 40 mg of the CZA catalyst were placed in a U-shaped tube and pretreated at 300 °C in a He atmosphere for 30 min. After the sample was cooled to 280 °C, the He was switched to 10% H_2/90% Ar and reduced for 30 min. Then, the sample was cooled to 50 °C in flowing He and allowed to fully adsorb 10% CO/90% He for 30 min; after that, the sample was purged with He for 1 h, the temperature increased to 150 °C and held for 10 min. Once the signal was stable, the temperature was further increased to 260 °C at 10 °C/min and maintained for 30 min. The amounts of desorbed CO were detected by the TCD.

For the NH_3-TPD, 100 mg of SP34-C were put in a U-shaped tube and pretreated in a He atmosphere at 300 °C for 1 h; after cooling down to 100 °C, the He was switched to 5% NH_3/95% He and held for 1 h; the TCD was purged with He until the baseline was stable, then the temperature was increased to 700 °C at 10 °C·min^{-1}, and the desorbed NH_3 was recorded by the TCD.

N_2 adsorption–desorption was carried out on a 3H-2000PS2 instrument (Beishide, Beijing, China). Before the test, 100 mg of the sample were weighed and degassed under vacuum and 250 °C for 4 h; the Brunauer–Emmett–Teller (BET) equation was used to calculate the specific surface area of the sample; the Barrett–Joyner–Halenda (BJH) method was used to calculate the pore size and pore volume.

3.3. Catalyst Performance Evaluation

The performance test of syngas to DME was carried out in a fixed-bed reactor furnished with a quartz tube. Typically, the physical mixed CZA and Al_2O_3© bifunctional catalyst was reduced at 280 °C for 4 h by using a mixture of 25%H_2/75%N_2. Then, the reduction gas was switched to feed gas with different CO/H_2 molar ratios, Ar was used an internal standard gas, and the reaction was performed at 260 °C. During the reaction process, the products were analyzed by an on-line gas chromatography Haixin GC-950, the equipped Agilent DB-624UI column (30m × 0.32m × 1.80μm) was used to analyze methanol and DME, and the packed column TDX-01 was used to analyze Ar, CO_2, and CO.

The performance of simulated reaction intermediates (obtained from the component of syngas to DME) and the performance of reverse WGSR were carried out in a fixed-bed reactor equipped with quartz reactor. Specifically, the direct conversion of syngas to light hydrocarbons was carried out in a dual bed, the upper layer is the physical mixed CZA and A©$_3$(C) catalyst, after reduction at 280 °C, the reaction was performed at 260 °C. The lower layer is SAPO-34 zeolite, and the reaction was performed at 400 °C unless otherwise stated. The gas products were analyzed by an on-line gas chromatography Agilent 7890A, the C1~C5 hydrocarbon components were analyzed by using a capillary column HP-AL/S (30 m × 530 μm × 15 μm) and a hydrogen flame ionization detector (FID). The packed column Porapak-Q, capillary column HP-PLOT/Q, and HP-MOLESIEVE series and thermal conductivity detector (TCD) were used to analyze CO_2, CO, H_2 and N_2. The gas chromatograms are calibrated before being used to analyze samples. Chromatographic standard gases containing all hydrocarbon products were formulated and the chromatogram was calibrated using an external standard method.

For the STD reactions, the CO conversion and the selectivity of products were calculated as follows:

$$\text{CO conversion}(\%) = \frac{F_{-O,in} - F_{CO,out}}{F_{CO,in}} \times 100\%$$

$$\text{Sel}_{MeOH}(\%) = \frac{F_{MeOH,out}}{F_{C-,in} - F_{CO,out}} \times 100\%$$

$$\text{Sel}_{DME}(\%) = \frac{2 \times F_{DME,out}}{F_{CO,in} - F_{CO,out}} \times 100\%$$

$$Sel_{CO_2}(\%) = \frac{F_{CO_2,out}}{F_{CO,in} - F_{CO,out}} \times 100\%$$

Furthermore, for the syngas conversion to light hydrocarbons, the selectivity of hydrocarbons (C_nH_m) was calculated on a molar carbon basis in total hydrocarbons:

$$C_nH_m \text{ selectivity}(\%) = \frac{nFC_nH_{m,\,out}}{\sum_1^n nFC_nH_{m,\,out}} \times 100\%$$

where $F_{CO,in}$ and $F_{CO,out}$ are the molar flow rates of CO at the inlet and outlet, $F_{product,out}$ is the molar flow rate of product at the outlet.

For the reverse WGSR, the CO_2 conversion and the selectivity of products were calculated as follows:

$$CO_2 \text{ conversion}(\%) = \frac{F_{CO_2,in} - F_{CO_2,out}}{F_{CO_2,in}} \times 100\%$$

$$Sel_{CO}(\%) = \frac{F_{CO,out}}{F_{CO_2,in} - F_{CO_2,out}} \times 100\%$$

$$Sel_{CH_4}(\%) = \frac{F_{CH_4,out}}{F_{CO_2,in} - F_{CO_2,out}} \times 100\%$$

where $F_{CO_2,in}$ and $F_{CO_2,out}$ are the molar flow rate of CO_2 at the inlet and outlet, $F_{CO,out}$ and $F_{CH4,out}$ are the molar flow rates of CO and CH_4 at the outlet.

3.4. Thermodynamic Simulation

The thermodynamic calculations data for STM and STD were obtained by using the Equilibrium Compositions module in HSC Chemistry 6.0 software. The Gibbs free energy minimization method possesses the advantage that it is not needed to specify the reaction equation. The principle is to calculate the reaction equilibrium according to the minimum Gibbs free energy of the system when the reaction reaches equilibrium. It is only necessary to specify the state, temperature, pressure, and quantity of the initial reactants. The equilibrium conversion of CO or CO_2 is calculated as follows:

$$CO \text{ equilibrium conversion}(\%) = \frac{n_{CO,in} - n_{CO\,equilibrium}}{n_{CO,in}} \times 100\%$$

$$CO_2 \text{ equilibrium conversion}(\%) = \frac{n_{CO_2,in} - n_{CO_2\,equilibrium}}{n_{CO_2,in}} \times 100\%$$

where n_{in} in the formula refers to the molar amount of material of the raw material given initially, and $n_{equilibrium}$ refers to the molar amount of material of the material at equilibrium of the reaction.

4. Conclusions

In summary, a dual-bed strategy was adopted for the direct conversion of syngas to light paraffins by using CZA as methanol synthesis catalystAl_2O_3(C) as methanol dehydration catalyst to DME, and SP34-C zeolite as DME conversion catalyst. The CZA and Al_2O_3(C) catalysts were physically mixed and put in the upper bed for syngas conversion to DME, while SP34-C zeolite was put in the lower bed. A high mass ratio of CZA to Al_2O_3(C), high H_2/CO ratio, and low space velocity are conducive to the conversion of syngas to DME. For the simulated feed gas $CO/CO_2/DME/N_2/H_2$ with a molar ratio of 9/6/4/5 balanced with H_2 over SP34-C, a low space velocity is conducive to the formation of light paraffins, a high reaction temperature intensified the reverse WGSR, while slightly affect the selectivity of light paraffins. After optimizing the reaction condition, the selectivity of light paraffins reached 90.8%, and the selectivity of propane was as high as 76.7%. For the direct conversion of syngas to light paraffins in a dual-bed reactor, 88.9% of the light paraffins selectivity in hydrocarbons was obtained at a CO conversion of 33.4%.

Author Contributions: Conceptualization, F.M. and L.W.; methodology, F.M. and L.W.; software, L.W.; validation, L.W.; formal analysis, L.W.; investigation, L.W., B.L., and J.Z.; data curation, L.W. and B.L.; writing—original draft preparation, L.W.; writing—review and editing, F.M.; supervision, F.M. and Z.L.; funding acquisition, F.M. and Z.L. All authors have read and agreed to the published version of the manuscript.

Funding: This research was funded by the Natural Science Foundation of Shanxi Province (202103021224073), the Key Research and Development Project of Shanxi Province (201803D421011), and the National Natural Science Foundation of China (U1510203).

Conflicts of Interest: The authors declare no conflict of interest.

References

1. Yerga, R.M.N. Catalysts for production and conversion of syngas. *Catalysts* **2021**, *11*, 752. [CrossRef]
2. Zhai, P.; Li, Y.; Wang, M.; Liu, J.; Cao, Z.; Zhang, J.; Xu, Y.; Liu, X.; Li, Y.-W.; Zhu, Q.; et al. Development of direct conversion of syngas to unsaturated hydrocarbons based on Fischer-Tropsch route. *Chem* **2021**, *7*, 3027–3051. [CrossRef]
3. Pan, X.; Jiao, F.; Miao, D.; Bao, X. Oxide–zeolite-based composite catalyst concept that enables syngas chemistry beyond Fischer–Tropsch synthesis. *Chem. Rev.* **2021**, *121*, 6588–6609. [CrossRef] [PubMed]
4. Torres Galvis, H.M.; de Jong, K.P. Catalysts for production of lower olefins from synthesis gas: A review. *ACS Catal.* **2013**, *3*, 2130–2149. [CrossRef]
5. Lu, S.; Yang, H.; Zhou, Z.; Zhong, L.; Li, S.; Gao, P.; Sun, Y. Effect of In_2O_3 particle size on CO_2 hydrogenation to lower olefins over bifunctional catalysts. *Chin. J. Catal.* **2021**, *42*, 2038–2048. [CrossRef]
6. Meng, F.; Li, X.; Zhang, P.; Yang, L.; Liu, S.; Li, Z. A facile approach for fabricating highly active ZrCeZnOx in combination with SAPO-34 for the conversion of syngas into light olefins. *Appl. Surf. Sci.* **2021**, *542*, 148713. [CrossRef]
7. Meng, F.; Li, X.; Zhang, P.; Yang, L.; Yang, G.; Ma, P.; Li, Z. Highly active ternary oxide ZrCeZnOx combined with SAPO-34 zeolite for direct conversion of syngas into light olefins. *Catal. Today* **2020**, *368*, 118–125. [CrossRef]
8. Zhang, P.; Meng, F.; Yang, L.; Yang, G.; Liang, X.; Li, Z. Syngas to olefins over a CrMnGa/SAPO-34 bifunctional catalyst: Effect of Cr and Cr/Mn ratio. *Ind. Eng. Chem. Res.* **2021**, *60*, 13214–13222. [CrossRef]
9. Wang, Y. A new horizontal in C1 chemistry: Highly selective conversion of syngas to light olefins by a novel OX-ZEO process. *J. Energy Chem.* **2016**, *25*, 169–170. [CrossRef]
10. Cheng, K.; Zhou, W.; Kang, J.; He, S.; Shi, S.; Zhang, Q.; Pan, Y.; Wen, W.; Wang, Y. Bifunctional catalysts for one-Step conversion of syngas into aromatics with excellent selectivity and stability. *Chem* **2017**, *3*, 334–347. [CrossRef]
11. Zhou, W.; Shi, S.; Wang, Y.; Zhang, L.; Wang, Y.; Zhang, G.; Min, X.; Cheng, K.; Zhang, Q.; Kang, J.; et al. Selective conversion of syngas to aromatics over a Mo–ZrO_2/H-ZSM-5 bifunctional catalyst. *ChemCatChem* **2019**, *11*, 1681–1688. [CrossRef]
12. Feng, J.; Miao, D.; Ding, Y.; Jiao, F.; Pan, X.; Bao, X. Direct synthesis of isoparaffin-rich gasoline from syngas. *ACS Energy Lett.* **2022**, *7*, 1462–1468. [CrossRef]
13. Lin, Y.-G.; Hsu, Y.-K.; Chen, S.-Y.; Chen, L.-C.; Chen, K.-H. O_2 plasma-activated CuO-ZnO inverse opals as high-performance methanol microreformer. *J. Mater. Chem.* **2010**, *20*, 10611–10614. [CrossRef]
14. van den Berg, R.; Prieto, G.; Korpershoek, G.; van der Wal, L.I.; van Bunningen, A.J.; Lægsgaard-Jørgensen, S.; de Jongh, P.E.; de Jong, K.P. Structure sensitivity of Cu and CuZn catalysts relevant to industrial methanol synthesis. *Nat. Commun.* **2016**, *7*, 13057. [CrossRef]
15. Park, J.; Cho, J.; Lee, Y.; Park, M.-J.; Lee, W.B. Practical microkinetic modeling approach for methanol synthesis from syngas over a Cu-based catalyst. *Ind. Eng. Chem. Res.* **2019**, *58*, 8663–8673. [CrossRef]
16. Bhan, A.; Iglesia, E. A link between reactivity and local structure in acid catalysis on zeolites. *Acc. Chem. Res.* **2008**, *41*, 559–567. [CrossRef]
17. Aloise, A.; Marino, A.; Dalena, F.; Giorgianni, G.; Migliori, M.; Frusteri, L.; Cannilla, C.; Bonura, G.; Frusteri, F.; Giordano, G. Desilicated ZSM-5 zeolite: Catalytic performances assessment in methanol to DME dehydration. *Microporous Mesoporous Mater.* **2020**, *302*, 110198. [CrossRef]
18. Bae, J.W.; Kang, S.-H.; Lee, Y.-J.; Jun, K.-W. Synthesis of DME from syngas on the bifunctional Cu–ZnO–Al_2O_3/Zr-modified ferrierite: Effect of Zr content. *Appl. Catal. B* **2009**, *90*, 426–435. [CrossRef]
19. Liu, Z.; Ni, Y.; Fang, X.; Zhu, W.; Liu, Z. Highly converting syngas to lower olefins over a dual-bed catalyst. *J. Energy Chem.* **2021**, *58*, 573–576. [CrossRef]
20. Yang, M.; Fan, D.; Wei, Y.; Tian, P.; Liu, Z. Recent progress in methanol-to-olefins (MTO) catalysts. *Adv. Mater.* **2019**, *31*, 1902181. [CrossRef]
21. Ni, Y.; Liu, Z.; Tian, P.; Chen, Z.; Fu, Y.; Zhu, W.; Liu, Z. A dual-bed catalyst for producing ethylene and propylene from syngas. *J. Energy Chem.* **2022**, *66*, 190–194. [CrossRef]

22. Arora, S.S.; Nieskens, D.L.S.; Malek, A.; Bhan, A. Lifetime improvement in methanol-to-olefins catalysis over chabazite materials by high-pressure H_2 co-feeds. *Nat. Catal.* **2018**, *1*, 666–672. [CrossRef]
23. Zhao, X.; Li, J.; Tian, P.; Wang, L.; Li, X.; Lin, S.; Guo, X.; Liu, Z. Achieving a superlong lifetime in the zeolite-catalyzed MTO reaction under high pressure: Synergistic effect of hydrogen and water. *ACS Catal.* **2019**, *9*, 3017–3025. [CrossRef]
24. Zhang, X.; Hao, X.; Hao, J.; Wang, Q. Heat transfer and mechanical properties of wood-plastic composites filled with flake graphite. *Thermochim. Acta* **2018**, *664*, 26–31. [CrossRef]
25. Zhang, Y.; Ren, Z.; Wang, Y.; Deng, Y.; Li, J. Synthesis of small-sized SAPO-34 crystals with varying template combinations for the conversion of methanol to olefins. *Catalysts* **2018**, *8*, 570. [CrossRef]
26. Liu, H.; Chen, T.; Wang, G. Effect of preparation method on the structure and catalytic performance of CuZnO catalyst for low temperature syngas hydrogenation in liquid phase. *Catal. Lett.* **2018**, *148*, 1462–1471. [CrossRef]
27. Unutulmazsoy, Y.; Cancellieri, C.; Lin, L.; Jeurgens, L.P.H. Reduction of thermally grown single-phase CuO and Cu_2O thin films by in-situ time-resolved XRD. *Appl. Surf. Sci.* **2022**, *588*, 152896. [CrossRef]
28. Park, S.; Inagaki, S.; Kubota, Y. Selective formation of light olefins from dimethyl ether over MCM-68 modified with phosphate species. *Catal. Today* **2016**, *265*, 218–224. [CrossRef]
29. Cheng, K.; Gu, B.; Liu, X.; Kang, J.; Zhang, Q.; Wang, Y. Direct and highly selective conversion of synthesis gas into lower olefins: Design of a bifunctional catalyst combining methanol synthesis and carbon–carbon coupling. *Angew. Chem. Int. Ed.* **2016**, *55*, 4725–4728. [CrossRef]
30. Meng, F.; Liang, X.; Wang, L.; Yang, G.; Huang, X.; Li, Z. Rational design of SAPO-34 zeolite in bifunctional catalysts for syngas conversion into light olefins. *Ind. Eng. Chem. Res.* **2022**, *61*, 11397–11406. [CrossRef]
31. Thommes, M.; Kaneko, K.; Neimark, A.V.; Olivier, J.P.; Rodriguez-Reinoso, F.; Rouquerol, J.; Sing, K.S.W. Physisorption of gases, with special reference to the evaluation of surface area and pore size distribution (IUPAC Technical Report). *Pure Appl. Chem.* **2015**, *87*, 1051–1069. [CrossRef]

Article

Catalytic Performance for CO Methanation over Ni/MCM-41 Catalyst in a Slurry-Bed Reactor

Guoqiang Zhang [1,2], Jinyu Qin [2], Yuan Zhou [1], Huayan Zheng [1,2,*] and Fanhui Meng [2,*]

[1] Department of Food Science and Engineering, Moutai Institute, Renhuai 564502, China
[2] State Key Laboratory of Clean and Efficient Coal Utilization, Taiyuan University of Technology, Taiyuan 030024, China
* Correspondence: andyzheng1109@163.com (H.Z.); mengfanhui@tyut.edu.cn (F.M.)

Abstract: The Ni-based catalyst has been intensively studied for CO methanation. Here, MCM-41 is selected as support to prepare xNi/MCM-41 catalysts with various Ni contents and the catalytic performance for CO methanation in a slurry-bed reactor is investigated under different reaction conditions. The CO conversion gradually increases as the reaction temperature or pressure rises. As the Ni content increases, the specific surface area and pore volume of xNi/MCM-41 catalysts decrease, the crystallite sizes of metallic Ni increase, while the metal surface area and active Ni atom numbers firstly increase and then slightly decrease. The 20Ni/MCM-41 catalyst with the Ni content of 20 wt% exhibits the highest catalytic activity for CO methanation, and the initial CH_4 yield rate is well correlated to the active metallic Ni atom numbers. The characterization of the spent xNi/MCM-41 catalysts shows that the agglomeration of Ni metal is accountable for the catalyst deactivation.

Keywords: CO methanation; Ni/MCM-41 catalyst; deactivation; synthetic natural gas; slurry-bed reactor

---

Citation: Zhang, G.; Qin, J.; Zhou, Y.; Zheng, H.; Meng, F. Catalytic Performance for CO Methanation over Ni/MCM-41 Catalyst in a Slurry-Bed Reactor. *Catalysts* **2023**, *13*, 598. https://doi.org/10.3390/catal13030598

Academic Editor: Leonarda Liotta

Received: 20 February 2023
Revised: 7 March 2023
Accepted: 15 March 2023
Published: 16 March 2023

Copyright: © 2023 by the authors. Licensee MDPI, Basel, Switzerland. This article is an open access article distributed under the terms and conditions of the Creative Commons Attribution (CC BY) license (https://creativecommons.org/licenses/by/4.0/).

1. Introduction

Natural gas, primarily consisting of methane, is a highly efficient and clean energy source [1,2]. However, the natural gas reserves are poor in some regions of the world, especially in China, thus the production of synthetic natural gas (SNG) from syngas (CO + H_2) has attracted much attention in recent decades [3,4]. Generally, SNG is produced via gasification of coal or biomass, followed by subsequent gas cleaning, conditioning and methanation process. Among them, CO methanation is an essential step [5], and it is a highly exothermic and thermodynamically feasible reaction [6,7]. Due to the poor heat transfer of fixed-bed reactor and the high amount of CO in the syngas, one of the two major challenges in developing the methanation reactor is to remove the highly exothermic heat effectively and in time, and the other is to produce high-efficiency catalysts retaining high catalytic performance at low reaction temperatures.

Aimed at the exothermic characteristics of methanation, a slurry-bed reactor was introduced in the process of CO methanation. In a slurry-bed reactor, catalyst powder suspends in an inert liquid medium, and the system keeps at an isothermal condition because of the high heat transfer coefficients of the liquid medium, which is particularly suitable for the highly exothermic CO methanation reaction, indicating that it can effectively prevent the carbon deposition and catalyst sintering [8–10].

Since 1902, many catalysts, including Ni-, Fe-, Ru- and Co-based catalysts, have been developed to catalyze the methanation reaction [11,12]. Until now, the Ni-based catalyst is still the most commonly-used catalyst to produce SNG due to its good catalytic performance and relatively low price [13,14]. The Ni/Al_2O_3 catalyst was commonly used; however, it was designed particularly for the fixed-bed reactor, and it showed low catalytic activity in a slurry-bed reactor [15]. It has been reported that the catalyst with high Ni dispersion and moderate interactions between Ni species and support exhibited high catalytic activity for CO methanation. In response to these problems, one of the potential candidates for Ni-based catalysts is to select MCM-41 mesoporous molecular sieve as the support. The reason is that MCM-41 possesses high specific surface area, large pore volume, and homogeneous hexagonal pore array, which can highly disperse the active phase and enhance the metal surface area [16,17]. The Ni-based MCM-41 catalyst with high Ni content and dispersion could be prepared by the incipient wetness impregnation method [18]. It has been reported that the Ni-based MCM-41 catalyst, prepared via a wet impregnation method, exhibited good catalytic activity for naphthalene hydrogenation [19]. Zhang et al. [20] found that the nickel based MCM-41 catalyst showed good activity for syngas methanation in a fixed-bed reactor.

In this work, MCM-41-supported Ni catalysts were prepared via the impregnation method where the Ni content varied from 4 to 28 wt% and the catalytic performance for CO methanation in a slurry-bed reactor was investigated. The obtained catalysts were characterized by N_2 adsorption-desorption, XRD, H_2-TPR, H_2 chemisorption and TEM. The effects of reaction temperature and pressure were studied, and the reason for the catalyst deactivation after reaction was analyzed. The aim is to clarify the relationship between the structures and the catalytic methanation performance.

2. Results and Discussion

2.1. Catalytic Performance for CO Methanation in a Slurry-Bed Reactor

The Ni content greatly affects the catalytic methanation performance. Thus, the methanation performance over xNi/MCM-41 catalysts in a slurry-bed reactor was investigated, and the results are shown in Figure 1. Figure 1a shows that the 4Ni/MCM-41 catalyst exhibits an initial CO conversion of 71.8% which then decreased to 32.1% after a 45 h reaction. As the Ni content increased to 12 wt%, the initial CO conversion of 12Ni/MCM-41 catalyst reached 89.0%; moreover, the stability was also enhanced. Further increasing the Ni content to 20 wt% or 28 wt%, the initial CO conversion reached up to ~94.4%. For comparison, the 20Ni/SiO_2 catalyst prepared by using silica as support shows much lower CO conversion than the 20Ni/MCM-41 catalyst. The main gas products during the CO methanation reaction are CH_4, CO_2 and C_{2-4}. Figure 1b–d show the variations in selectivity of CH_4, CO_2, and C_{2-4} as reaction time goes on. It could be found in Figure 1b that the selectivity of CH_4 slightly increased during the reaction, and all the xNi/MCM-41 catalysts exhibit higher CH_4 selectivity than 20Ni/SiO_2. Figure 1c shows that the selectivity of CO_2 is in the range of 0.5% and 6.1%, and it slightly decreased during the reaction. Figure 1d shows that the C_{2-4} selectivity of xNi/MCM-41 catalyst was less than 1.8%, and it only slightly changed with the variation in Ni content, whereas the C_{2-4} selectivity of the 20Ni/SiO_2 catalyst was significantly higher than that of 20Ni/MCM-41.

For the CO methanation reaction, the active center Ni metal may form the [Ni(CO)$_4$] specie at low temperature, which deactivated the catalysts [21,22]. Thus, to minimize the generation of [Ni(CO)$_4$] specie during the methanation reaction, the reaction was generally conducted at the temperatures higher than 250 °C. Figure 2 displays the effect of reaction temperature on catalytic methanation performance over xNi/MCM-41 catalysts. Figure 2a shows that the CO conversion decreased as the temperature decreased, especially when the temperature was lower than 280 °C. The 4Ni/MCM-41 catalyst exhibits an initial CO conversion of 77.5% at 320 °C, and it shows a dramatic and continuous drop when the reaction temperature decreased from 320 °C to 260 °C. Both 20Ni/MCM-41 and 28Ni/MCM-41 catalysts show high catalytic activity and stability when the temperature reached 300 °C or above. However, the activity decreased quickly when the temperature was reduced to 260 °C. The CH_4 selectivity of the xNi/MCM-41 catalyst is relatively stable at each reaction temperature.

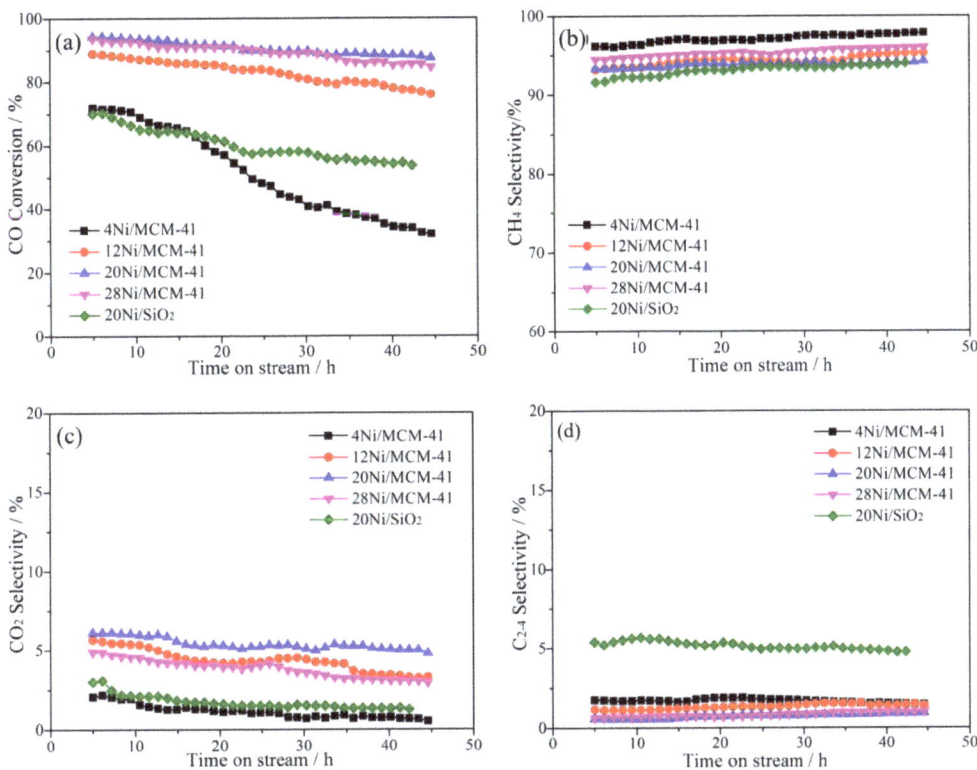

Figure 1. Effect of Ni content on catalytic performance for CO methanation over Ni-based catalysts. (**a**) CO conversion, (**b**) CH_4 selectivity, (**c**) CO_2 selectivity, (**d**) C_{2-4} selectivity. Reaction conditions: 300 °C, 1.0 MPa, and 3000 mL/(g·h).

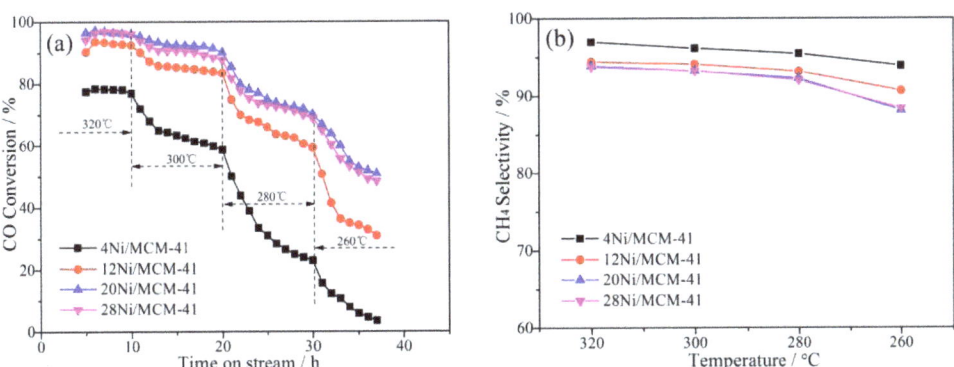

Figure 2. Effect of reaction temperature on catalytic performance for CO methanation over xNi/MCM-41 catalysts. (**a**) CO conversion, (**b**) CH_4 selectivity.

The average value of CH_4 selectivity is displayed in Figure 2b. As the reaction temperature decreased from 320 °C to 260 °C, the selectivity of CH_4 slightly decreased. Furthermore, it is interesting to find that the catalyst with low Ni content exhibits a high CH_4 selectivity, which is probably due to the low Ni content that avoids the side reaction.

The effect of reaction pressure on the catalytic methanation performance over the xNi/MCM-41 catalyst was studied by varying the pressure from 0.5 MPa to 3.0 MPa. The performance tests were carried out at 300 °C and 3000 mL/(g·h), and the results are displayed in Figure 3. Figure 3a shows that the CO conversion increased as the pressure increased from 0.5 MPa to 3.0 MPa. The 4Ni/MCM-41 catalyst exhibited the lowest CO conversion of 21.7% at 0.5 MPa; however, it increased to 94.9% as the pressure increased to 3.0 MPa. When the pressure was 1.0 MPa or higher, the CO conversion of xNi/MCM-41 catalyst slightly increased as the reaction pressure increased.

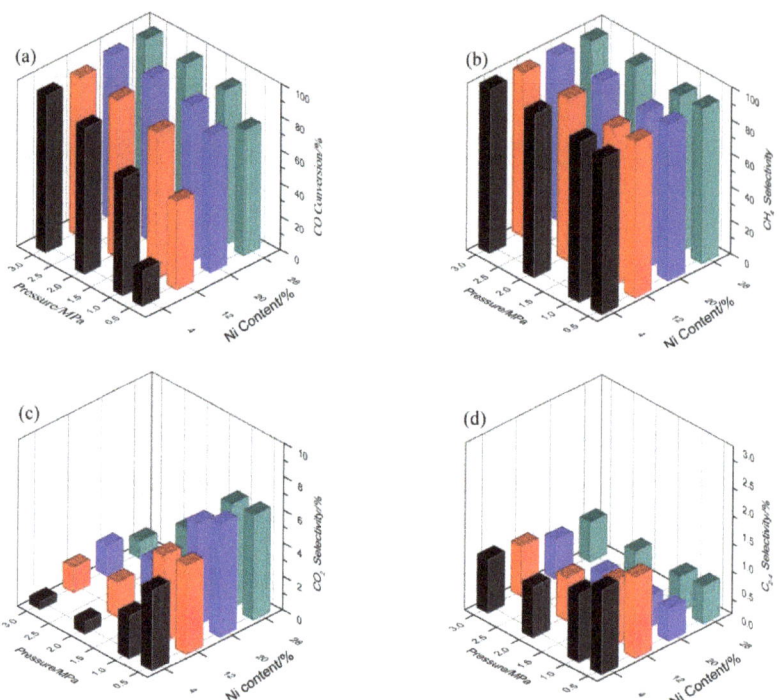

Figure 3. Effect of reaction pressure on catalytic performance for CO methanation over xNi/MCM-41 catalysts. (**a**) CO conversion, (**b**) CH_4 selectivity, (**c**) CO_2 selectivity, (**d**) C_{2-4} selectivity. Black: 4Ni/MCM-41, red: 12Ni/MCM-41, blue: 20Ni/MCM-41, dark cyan: 28Ni/MCM-41.

Figure 3b shows that the selectivity of CH_4 of all the catalysts exceeded 93.0%, and the selectivity gradually increased as the reaction pressure increased from 0.5 MPa to 3.0 MPa, while the CH_4 selectivity changed slightly as the Ni content increased. This is because the CO methanation is a gas molecular number reducing reaction, where the high pressure could facilitate the positive reaction, which improves the CO conversion and CH_4 selectivity [20]. Figure 3c shows that the selectivity of CO_2 decreased as the reaction pressure was enhanced, and when the pressure reached 2.0 MPa or higher, the selectivity of CO_2 decreased to 3.3% or less. The selectivity of C_{2-4} in Figure 3d was very low, and it changed very slightly as the pressure or Ni content varied.

2.2. Structure and Textural Properties Analysis

X-ray diffraction is one of the most important techniques for characterizing the crystallite size and catalyst structure. Figure 4a shows the wide-angle XRD patterns of MCM-41 support and the calcined Ni-based catalysts. The MCM-41 support exhibits a broad amorphous silica peak at 2θ angles of around 23°, whereas the MCM-41 supported catalysts show diffraction peaks at 2θ of 37.3°, 43.3°, 62.9°, 75.4° and 79.4°, which are assigned to the characteristic peaks of NiO(111), NiO(200), NiO(220), NiO(311) and NiO(222) (JPCDS, No. 47-1049), respectively. The 4Ni/MCM-41 catalyst exhibits the weakest peak intensity of NiO, indicating the presence of the smallest crystallite size of NiO. As the Ni content increased, the diffraction peaks of NiO become stronger. The crystallite size of NiO at 2θ of 37.3° was estimated using the Scherrer equation, and the results are shown in Table 1. As the Ni content increased, the crystallite sizes of NiO increased, and the 28Ni/MCM-41 catalyst shows the largest crystallite size of 20.9 nm. For comparison, the 20Ni/SiO$_2$ catalyst shows almost the same peak intensity and crystallite size as that of 20Ni/MCM-41.

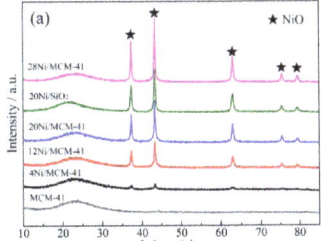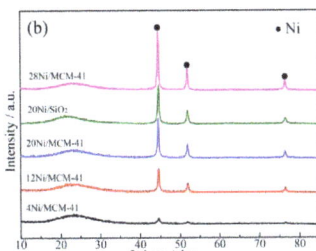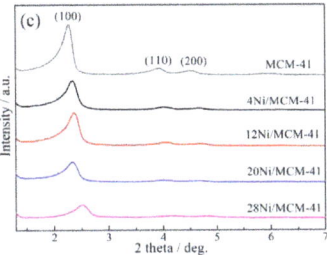

Figure 4. Wide-angle XRD patterns of the (**a**) calcined samples and (**b**) reduced samples, and (**c**) small-angle XRD patterns of calcined samples.

Table 1. Textural properties and crystallite sizes of various samples.

Catalyst	Specific Surface Area (m²/g) [a]	Pore Volume (cm³/g) [b]	Average Pore Size (nm) [c]	Crystallite Size of NiO/Ni (nm) [d]
MCM-41	1188	1.41	3.1	-
4Ni/MCM-41	979	1.40	3.3	10.7/9.0
12Ni/MCM-41	921	1.39	4.9	15.8/15.6
20Ni/MCM-41	838	1.38	5.2	17.4/17.2
28Ni/MCM-41	804	1.35	5.1	20.9/19.2
20Ni/SiO$_2$	153	1.38	24.1	17.2/16.8

[a] Calculated by BET (Brunauer-Emmett-Teller) equation. [b] BJH (Barret-Joyner-Hallender) desorption pore volume. [c] BJH (Barret-Joyner-Hallender) desorption average pore diameter. [d] calculated from NiO (200) and Ni (111) plane using the Scherrer equation.

Figure 4b shows the XRD patterns of reduced Ni-based catalysts. Each catalyst shows three diffraction peaks at 2θ of 44.5°, 51.9° and 76.4°, which are assigned to Ni(111), Ni(200) and Ni(220) (JPCDS, No. 04-0850), respectively. The peaks of metallic Ni became stronger as the Ni content rose. The crystallite sizes of metallic Ni at 2θ of 44.5° are summarized in Table 1, and it could be found that the crystallite sizes of metallic Ni increased as the Ni content increased.

The low-angle XRD patterns of MCM-41 support and the calcined xNi/MCM-41 catalysts are shown in Figure 4c. Three diffraction peaks were observed for MCM-41 support, indexed as the reflections of (100), (110) and (200) crystal face, characteristic of a highly ordered mesoporous structure with hexagonal pore array [23]. With the increase in Ni content, the (100) peak intensity decreased, and the position of the interplanar spacing of (100) reflection shifts toward higher angles, indicating the decrease of the lattice parameter. The (110) and (200) peaks almost disappeared, which indicates a consequential loss of a long-range order form concerning to the mesopores [24].

N$_2$ adsorption-desorption is a routine technique to probe the texture of porous solids. The isotherms and pore size distributions for MCM-41 support and the reduced xNi/MCM-41 catalysts are presented in Figure 5a,b, respectively. According to the IUPAC nomenclature, the N$_2$ adsorption-desorption isotherms in Figure 5a can be categorized as type IV isotherm, which is typical for the purely siliceous material MCM-41 with mesoporous structure, the hysteresis loops with parallel and almost horizontal branches can be classified as H4-type [25]. A sharp step in the range of relative pressure between 0.2 and 0.4 indicates that the catalyst possesses uniform mesopores [25,26]. As the Ni content rose, the hysteresis loops of relative pressures p/p$_0$ changed slightly except the high relative pressures between 0.9 and 1.0, which confirmed that these catalysts have narrow pore size distributions.

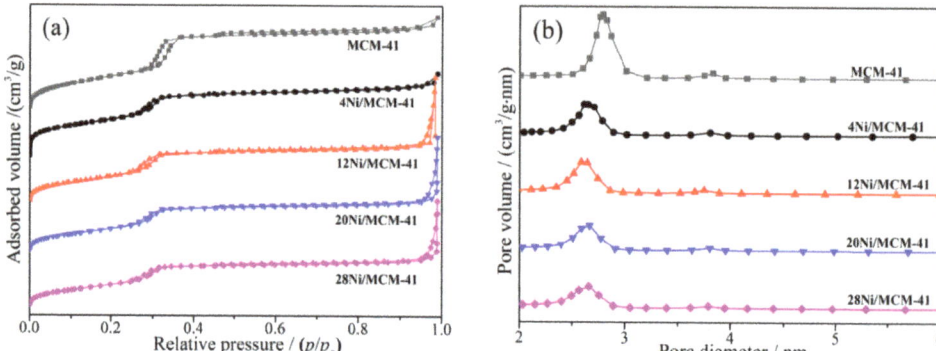

Figure 5. (a) Pore size distributions and (b) N$_2$ adsorption-desorption isotherms of MCM-41 support and reduced xNi/MCM-41 catalysts.

Figure 5b shows that all samples exhibit quite narrow pore size distribution mostly in the range of 2 to 4 nm. MCM-41 support shows a large and broad pore size distribution centered at 2.8 nm and a small one centered at 3.8 nm. After the introduction of Ni, the pore size distributions decreased and centered at 2.6 nm; further increasing the Ni content slightly changed the pore size distributions.

The textural properties of MCM-41 support and xNi/MCM-41 catalysts are summarized in Table 1. MCM-41 support possesses the highest specific surface area and pore volume of 1188 m^2/g and 1.41 cm^3/g, respectively, as well as the lowest average pore size of 3.1 nm. While for the xNi/MCM-41 catalysts, the specific surface area and pore volume decreased. The reason is probably due to that the pores were blocked by the Ni species. The average pore size firstly increased and then decreased as the Ni content rose, which was probably due to the accumulation of nickel species existed on the external surface and formed new holes, which led to a textural change in Ni/MCM-41.

2.3. Catalyst Reducibility and Surface Properties Analysis

The catalyst reducibility and metal-support interaction can be characterized by H$_2$-TPR technique. Figure 6 shows the TPR profiles of calcined catalysts xNi/MCM-41 and 20Ni/SiO$_2$. All catalysts exhibit three reduction peaks, indicating that there were three different interactions between the metal oxide and support [27]. The first peak centered at 310–350 °C could be attributed to the reduction of NiO species, which had no or very weak interaction with MCM-41 support [28]; the second peak around 360–400 °C ascribed to the reduction of bulk NiO [28,29], and the last peak at high temperature could be belonged to the reduction of very small NiO particles and/or NiO species strongly interacted with support [29]. The 4Ni/MCM-41 catalyst shows a very weak peak intensity, and the peak intensity significantly enhanced as the Ni content increased. Moreover, the reduction peaks shift to higher temperatures as the Ni content increased. The 20Ni/MCM-41 catalyst shows reduction peaks at high temperatures, indicating the presence of small NiO particles and/or a strong interaction between the Ni species and MCM-41 support. Generally, the strong interaction between the Ni species and support favors the enhancement of Ni dispersion and the formation of small active Ni particles, which benefits the catalytic activity. The referenced 20Ni/SiO$_2$ catalyst shows a higher reduction temperature compared with 20Ni/MCM-41. Interestingly, the 20Ni/SiO$_2$ catalyst did not present the third reduction peaks at high temperatures, which is probably due to the poor dispersion of NiO species.

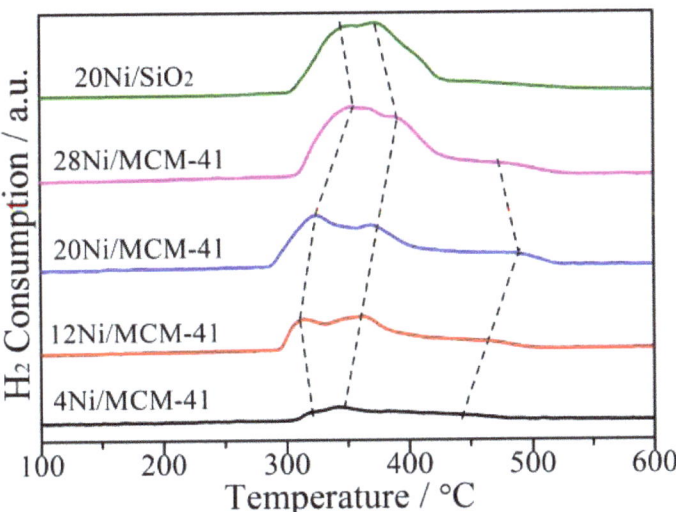

Figure 6. H_2-TPR profiles of calcined xNi/MCM-41 and 20Ni/SiO$_2$ catalysts.

An H_2 chemisorption measurement was conducted to determine the Ni dispersion, metal surface area, number of active Ni atoms and average particle size of Ni. All results were estimated by assuming that one hydrogen atom is adsorbed on one active nickel atom [30,31]. Table 2 shows the H_2 chemisorption results for the reduced xNi/MCM-41 and 20Ni/SiO$_2$ catalysts. Obviously, the Ni dispersion of the xNi/MCM-41 catalysts gradually decreased as the Ni content rose, while the average particle diameter of Ni increased. The results indicate that the active Ni were highly dispersed with small particle size at low Ni contents. The metal surface area and active metallic Ni atom numbers of the catalysts firstly increased and then slightly decreased, indicating that there was an optimum threshold value of Ni content on the MCM-41 support. The 20Ni/SiO$_2$ catalyst shows the Ni dispersion of 1.66% and the values of metal surface area and active metallic Ni atom numbers were all lower than those of 20Ni/MCM-41 catalyst. Among the catalysts tested, the 20Ni/MCM-41 catalyst with a Ni content of 20 wt% shows the highest metal surface area (2.97 m$^2 \cdot$g^{-1}$_{cat}$) and active metallic Ni atom numbers (4.57 × 10^{19} g^{-1}), which is probably responsible for the high catalytic methanation activity [32].

Table 2. H_2 chemisorption results of the reduced xNi/MCM-41 catalysts.

Catalyst	D_{Ni}(%) [a]	S_{cat}(m$^2 \cdot$g^{-1}$_{cat}$) [b]	N_{Ni}(×10$^{19} \cdot$g^{-1}) [c]	d_H(nm) [d]
4Ni/MCM-41	3.34	0.86	1.32	30.1
12Ni/MCM-41	3.18	2.27	3.49	31.8
20Ni/MCM-41	2.67	2.97	4.57	37.8
28Ni/MCM-41	1.80	2.62	4.04	56.3
20Ni/SiO$_2$	1.66	1.85	2.84	60.7

[a] D_{Ni}: amount of exposed Ni on the surface of the catalysts. [b] S_{cat}: active metal surface area per gram of catalyst. [c] N_{Ni}: numbers of active metallic Ni atom. [d] d_H: average Ni particle size.

TEM technology was employed to observe the catalyst morphology and Ni dispersion. The TEM images of MCM-41 support and representative calcined 4Ni/MCM-41 and 20Ni/MCM-41 catalysts are shown in Figure 7. The TEM image of MCM-41 in Figure 7a exhibits well-ordered mesoporous structure, which was consistent with the results of BET and XRD. Figure 7b shows that the NiO was highly dispersed on MCM-41 and part of NiO was located inside the well-ordered channels pores as clusters or very small nanoparticles. As the Ni content increased up to 20 wt%, the TEM image in Figure 7c exhibits small

NiO particles in the channels of MCM-41 support, which had strong interaction with the support. The result was confirmed by the high temperature reduction peak in H_2-TPR profile. However, parts of NiO particles accumulated and formed large particles on the external surface of the support, which might block the pores and decrease the specific surface areas [33].

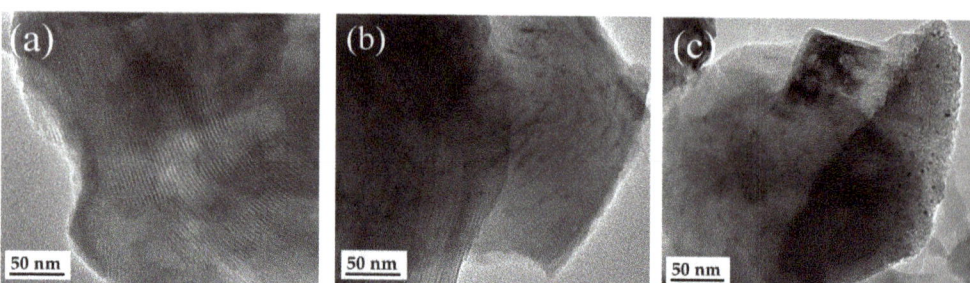

Figure 7. TEM images of MCM-41 support and calcined xNi/MCM-41 catalysts. (**a**) MCM-41 support, (**b**) 4Ni/MCM-41 catalyst, (**c**) 20Ni/MCM-41 catalyst.

2.4. Relationship between Catalytic Activity and Surface Properties

There are many factors affecting the catalytic methanation performance. However, it can be found that the active metallic Ni atom numbers of xNi/MCM-41 catalysts served well as a correlating parameter for the catalytic activity in a slurry-bed reactor. Figure 8 shows the relationship between the active metallic Ni atom numbers and the initial yield rate of CH_4 per mass of xNi/MCM-41 catalyst (r_{CH4}(mmol·h^{-1}·g^{-1})) at 300 °C, 1.0 MPa, and 3000 mL·g^{-1}·h^{-1}. As mentioned above, the initial catalytic activity of CO conversion in Figure 1a firstly increased and then slightly decreased as the Ni content increased, reaching the maximum value at the Ni content of 20 wt%, while the selectivity of CH_4 of each xNi/MCM-41 catalyst kept almost the same value. It should be noted that the initial CH_4 yield rate is well correlated to the active metallic Ni atom numbers (as shown in Table 2). In other words, xNi/MCM-41 catalysts with more amounts of active metallic Ni atom were favorable for improving the catalytic activity. The 20Ni/SiO_2 catalyst exhibited a lower yield rate of CH_4, which was probably due to the poor dispersion of active metallic Ni and the large Ni particles as shown in Table 2.

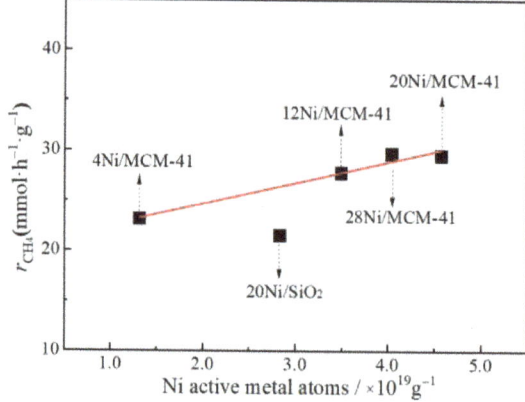

Figure 8. Relationship between the numbers of active Ni atom and initial CH_4 yield rate per unit mass of catalyst at 300 °C.

2.5. Characterization of Spent Catalysts

Figure 9a shows the XRD patterns of spent xNi/MCM-41 catalysts after a reaction time for 45 h. All catalysts exhibit the diffraction peaks at 2θ angles of 44.5°, 51.9°, and 76.4°, corresponding to characteristic peaks of metallic Ni (JPCDS, No. 04-0850), and the peak intensity increased as the Ni content rose. The crystallite sizes of metallic Ni at 2θ of 44.5° are summarized and compared in Table 3. It is found that the Ni crystallite sizes of spent xNi/MCM-41 catalysts slightly larger than those of the freshly reduced ones, especially for the freshly reduced 4Ni/MCM-41 catalyst, the increase in Ni crystallite size is probably responsible for the deactivation of xNi/MCM-41 catalysts. The results were confirmed by the TEM image of the spent 20Ni/MCM-41 catalyst in Figure 9b. It is displayed that the Ni particle of the spent 20Ni/MCM-41 catalyst was agglomerated and the particle size was larger than that of the fresh 20Ni/MCM-41 catalyst.

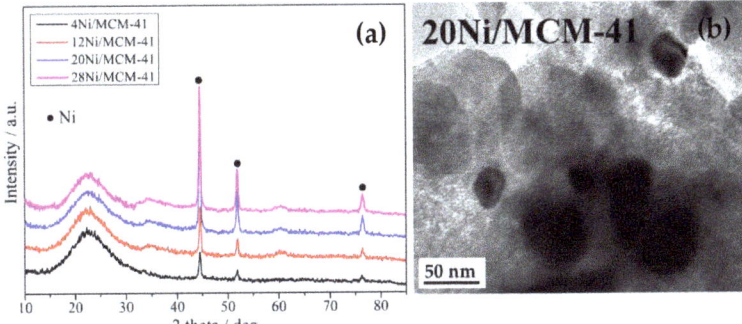

Figure 9. XRD patterns (**a**) and TEM image (**b**) of spent xNi/MCM-41 catalysts after 45 h reaction.

Table 3. Textural properties and Ni crystallite size of fresh and spent xNi/MCM-41 catalysts.

Catalyst	Specific Surface Area (m²/g) [a]	Pore Volume (cm³/g) [b]	Average Pore Size (nm) [c]	Ni Crystallite Size (nm) [d]
4Ni/MCM-41	783	0.78	3.25	9.0/17.7
12Ni/MCM-41	584	0.53	5.00	15.6/21.1
20Ni/MCM-41	488	0.53	4.07	17.2/23.0
28Ni/MCM-41	322	0.50	4.84	19.2/24.1

[a] Calculated by BET (Brunauer-Emmett-Teller) equation. [b] BJH (Barret-Joyner-Hallender) desorption pore volume. [c] BJH (Barret-Joyner-Hallender) desorption average pore diameter. [d] calculated from Ni(111) plane of fresh/spent catalyst using the Scherrer equation.

The textural properties of spent xNi/MCM-41 catalysts are summarized in Table 3. As the Ni content increased, the specific surface area significantly decreased from 783 m²/g for 4Ni/MCM-41 to 322 m²/g for 28Ni/MCM-41. Compared with the textural properties of fresh xNi/MCM-41 catalysts in Table 1, it could be found that the specific surface area and pore volume of spent xNi/MCM-41 catalysts all decreased. The catalyst with high Ni content shows a large decrease, the reason was probably due to the blockage of the micropores of MCM-41 by the agglomeration of Ni species.

3. Materials and Methods

3.1. Catalyst Preparation

The nickel catalysts supported on MCM-41 molecular sieve (supplied by Tianjin Nankai catalyst company, China, 100–200 mesh) were prepared by the incipient wetness impregnation method as follows: first, the MCM-41 was calcined in air at 550 °C for 4 h and used as support. Then, the required amount of Ni(NO$_3$)$_2$·6H$_2$O (purchased from Sinopharm mbcvx, and used without further treatment) was completely dissolved in a certain amount of distilled water, and followed by addition of 10.0 g of MCM-41, the resulting mixture was sealed with plastic film to avoid the quick evaporation of water and stirred continuously at room temperature for 24 h. Finally, the mixture was held at 80 °C for 5 h and dried at 120 °C for 12 h, and then calcined in air at 550 °C for 4 h at a heating rate of 2 °C·min^{-1} to form NiO/MCM-41; the obtained samples were designated as xNi/MCM-41 catalysts, where x represents the weight content of metallic Ni (4 wt%, 12 wt%, 20 wt%, and 28 wt%). For comparison, the 20Ni/SiO$_2$ catalyst with the Ni content of 20 wt% was prepared by the impregnation method as described above by using silica (Degussa, Aerosil 200) as support.

3.2. Catalyst Characterization

The N$_2$ adsorption-desorption measurement was performed by a Beishide 3H-2000PS specific surface and pore size analyzer, using N$_2$ as the adsorbing medium at −196 °C. Prior to the test, the sample was degassed at 250 °C for 3 h. The BET surface area was determined by the Brunauer-Emmett-Teller (BET) method, the average pore diameter was evaluated with the Barrett-Joyner-Halenda (BJH) method using the desorption isotherm branch.

X-ray diffraction (XRD) data were obtained on DX-2800A diffractometer (Dandong, China) with a scanning step of 4°/min using the Kα radiation of Cu (λ = 0.154056 nm) at 40 kV and 30 mA. The crystallite size of Ni metal or NiO was calculated using the Scherrer equation.

Temperature-programmed reduction of H$_2$ (H$_2$-TPR) was carried out on a Micromeritics Autochem II 2920 instrument (Micromeritics Instrument Corporation, Atlanta, USA). Prior to the measurement, 20 mg of the sample was pretreated in flowing He (50 mL/min) at 350 °C for 0.5 h, after cooling down to room temperature, the He was switched to 10%H$_2$/90%Ar (50 mL/min) and heated at 800 °C at a heating rate of 10 °C/min.

The H$_2$-chemisorption experiment was also conducted on a Micromeritics Autochem II 2920 instrument (Micromeritics Instrument Corporation, Atlanta, USA). Prior to the test, 200 mg of the sample was reduced by flowing 10%H$_2$/90%Ar at 550 °C for 2 h, then the sample was cooled down to 50 °C and the loop 10%H$_2$/90%Ar gas was pulsed over the sample and the TCD signals were recorded until the peak area remain constant. A 5 μL loop was used for this purpose and the loop was calibrated to determine its precise volume under local conditions. The metallic Ni surface area and Ni dispersion were calculated by assuming that one hydrogen atom occupies one surface metallic Ni atom [31]. The Ni dispersion, metal surface area, number of active nickel atoms and average particle size of Ni are calculated from Equations (1)–(4), respectively:

$$D_{Ni}(\%) = \left(\frac{V_{ad}}{V_m}\right) \times \left(\frac{SF \times M_{Ni}}{W_s \times F_{Ni}}\right) \times 100\% \quad (1)$$

$$S_{cat}(m^2/g_{cat}) = \frac{V_{ad}}{W_s \times V_m} \times SF \times N_A \times R_A \quad (2)$$

$$N_{Ni}(\text{Ni atoms/mol}) = \frac{V_{ad}}{W_s \times V_m} \times SF \times N_A \quad (3)$$

$$d_H(nm) = 6 \times 10^3/(S_{cat}/F \times Ni\rho_{Ni}) \quad (4)$$

where V_{ad} = volume of H$_2$ chemisorbed at STP (mL) to form a monolayer, V_m = molar volume of H$_2$ gas (22,414 mL/mol), SF = stoichiometric factor, i.e., Ni:H atomic ratio in the chemisorption, which is taken as 1; M_{Ni} = formula weight of Ni (58.69 g/mol), W_s = weight of the sample (g), F_{Ni} = weight percentage of Ni in the sample. S_{cat} = active metal atom; N_A = 6.023 × 10^{23}; R_A is the atomic cross-sectional area of Ni, which is 0.0649 nm^2; ρ_{Ni} = density of Ni metal (8.9 g/cm^3).

Transmission electron microscopy (TEM) was performed on JEOL JEM-2100F microscope (Tokyo, Japan) operated at 200 kV. Prior to the measurement, the solid was dispersed in ethanol, ultrasonicated and deposited on a sample holder.

3.3. Catalyst Performance Evaluation

The catalyst was firstly reduced at 550 °C for 6 h in a flow of 25% H_2 diluted with N_2, and then used for activity test. The CO methanation reaction was carried out in a 250 mL slurry-bed reactor (Dalian Tongda Reactor Factory, Dalian, China). In each experiment, 2.0 g of reduced catalyst and 120 mL of paraffin were introduced into the reactor and rotated at a speed of 750 r/min. N_2 was used to purge the air, and then the syngas of H_2 and CO was switched as feed gas. The reaction was performed at H_2/CO molar ratio of 3:1 and 3000 mL/g_{cat}·h in the temperature range of 260–320 °C with the interval of 20 °C. The gas products were firstly cooled down at 2 °C, the cooled liquid was then removed in a gas-liquid separator, the outlet gas was quantitatively analyzed by an online gas chromatography (Agilent 7890A), which was equipped with a flame ionization detector (FID) and a thermal conductivity detector (TCD) using He (99.999%) as the carrier gas. The HP-AL/S column was used to analyze CH_4 and C_{2-4}, while Porapak-Q column, HP-PLOT/Q column and HP-MOLESIEVE column were used to analyze CO, N_2 and CO_2.

The CO conversion, CH_4, CO_2, and C_{2-4} selectivity were calculated as follows:

$$X_{CO} = \frac{F(CO_{in}) - F(CO_{out})}{F(CO_{in})} \times 100\% \tag{5}$$

$$S_{CH4} = \frac{F(CH_{4,\,out})}{F(CO_{in}) - F(CO_{out})} \times 100\% \tag{6}$$

$$S_{CO2} = \frac{F(CO_{2,\,out})}{F(CO_{in}) - F(CO_{out})} \times 100\% \tag{7}$$

$$S_{Ci} = \frac{i \times F(C_{i,\,out})}{F(CO_{in}) - F(CO_{out})} \times 100\% \,(i = 2,\,3,\,4) \tag{8}$$

where X_{CO}, S_{CH4}, S_{CO2} and S_{Ci} represent the CO conversion, the CH_4, CO_2, and C_{2-4} selectivity, respectively. C_{2-4} represents the hydrocarbons contain 2 to 4 carbons. F represents the volume flow of CO, CH_4, and CO_2 (mL/min, STP), respectively.

4. Conclusions

xNi/MCM-41 catalysts with various Ni contents were prepared by the impregnation method, and applied for CO methanation in a slurry-bed reactor. As the Ni content increased, the specific surface area and pore volume of xNi/MCM-41 catalysts decreased, the crystallite size of metallic Ni increased, whereas the metal surface area and active Ni atom numbers firstly increased and then slightly decreased. The 20Ni/MCM-41 catalyst with the Ni content of 20 wt% exhibited the highest catalytic activity for CO methanation, and the initial CH_4 yield rate was well correlated to the active metallic Ni atom numbers. Moreover, the 20Ni/MCM-41 catalyst shows much higher activity than the referenced 20Ni/SiO_2 catalyst. The results of the optimization of reaction conditions show that the CO conversion gradually increases as the reaction temperature or pressure rises. The deactivation of xNi/MCM-41 catalysts was attributed to the agglomeration of Ni.

Author Contributions: Conceptualization, F.M.; methodology, G.Z. and J.Q.; software, J.Q. and F.M.; formal analysis, J.Q.; investigation, G.Z., J.Q. and Y.Z.; data curation, F.M.; writing—review and editing, F.M.; supervision, F.M. and H.Z.; funding acquisition, G.Z. H.Z. and F.M. All authors have read and agreed to the published version of the manuscript.

Funding: The authors are grateful to the financial support of National Natural Science Foundation of China (22262020), Natural Science Foundation of Guizhou Province (ZK [2023]448 and ZK [2023]447), Zunyi Technology and Big data Bureau, Moutai institute Joint Science and Technology Research and Development Project ([2021]328), and Natural Science Foundation of Shanxi Province (202103021224073).

Data Availability Statement: The data presented in this study are available on request from the corresponding author.

Conflicts of Interest: The authors declare no conflict of interest.

References

1. Zeng, Y.; Ma, H.; Zhang, H.; Ying, W.; Fang, D. Ni-Ce-Al composite oxide catalysts synthesized by solution combustion method: Enhanced catalytic activity for CO methanation. *Fuel* **2015**, *162*, 16–22. [CrossRef]
2. Li, J.; Li, P.; Li, J.; Tian, Z.; Yu, F. Highly-dispersed Ni-NiO nanoparticles anchored on an SiO_2 support for an enhanced CO methanation performance. *Catalysts* **2019**, *9*, 506. [CrossRef]
3. Meng, F.; Li, X.; Lv, X.; Li, Z. CO hydrogenation combined with water-gas-shift reaction for synthetic natural gas production: A thermodynamic and experimental study. *Int. J. Coal Sci. Technol.* **2018**, *5*, 439–451. [CrossRef]
4. Hatta, A.H.; Jalil, A.A.; Hassan, N.S.; Hamid, M.Y.S.; Rahman, A.F.A.; Teh, L.P.; Prasetyoko, D. A review on recent bimetallic catalyst development for synthetic natural gas production via CO methanation. *Int. J. Hydrog. Energy* **2022**, *47*, 30981–31002. [CrossRef]
5. Xu, B.; Meng, X.; Xin, Z.; Gao, W.; Yang, D.; Jin, D.; Zhao, R.; Dai, W. A novel CO methanation catalyst system based on acid-etched natural halloysites as supports. *Ind. Eng. Chem. Res.* **2022**, *61*, 13328–13340. [CrossRef]
6. Shinde, V.M.; Madras, G. CO methanation toward the production of synthetic natural gas over highly active Ni/TiO_2 catalyst. *AIChE J.* **2014**, *60*, 1027–1035. [CrossRef]
7. Liu, S.-S.; Jin, Y.-Y.; Han, Y.; Zhao, J.; Ren, J. Highly stable and coking resistant Ce promoted Ni/SiC catalyst towards high temperature CO methanation. *Fuel Process. Technol.* **2018**, *177*, 266–274. [CrossRef]
8. Zhang, J.; Bai, Y.; Zhang, Q.; Wang, X.; Zhang, T.; Tan, Y.; Han, Y. Low-temperature methanation of syngas in slurry phase over Zr-doped $Ni/\gamma\text{-}Al_2O_3$ catalysts prepared using different methods. *Fuel* **2014**, *132*, 211–218. [CrossRef]
9. Meng, F.; Li, Z.; Liu, J.; Cui, X.; Zheng, H. Effect of promoter Ce on the structure and catalytic performance of Ni/Al_2O_3 catalyst for CO methanation in slurry-bed reactor. *J. Nat. Gas Sci. Eng.* **2015**, *23*, 250–258. [CrossRef]
10. Ji, K.; Meng, F.; Xun, J.; Liu, P.; Zhang, K.; Li, Z.; Gao, J. Carbon deposition behavior of Ni catalyst prepared by combustion method in slurry methanation reaction. *Catalysts* **2019**, *9*, 570. [CrossRef]
11. Mills, G.A.; Steffgen, F.W. Catalytic methanation. *Catal. Rev.* **1974**, *8*, 159–210. [CrossRef]
12. Somorjai, G.A. The catalytic hydrogenation of carbon monoxide. The formation of C1 hydrocarbons. *Catal. Rev.* **1981**, *23*, 189–202. [CrossRef]
13. Guo, C.; Wu, Y.; Qin, H.; Zhang, J. CO methanation over ZrO_2/Al_2O_3 supported Ni catalysts: A comprehensive study. *Fuel Process. Technol.* **2014**, *124*, 61–69. [CrossRef]
14. Zhang, J.; Jia, X.; Liu, C.-J. Structural effect of Ni/TiO_2 on CO methanation: Improved activity and enhanced stability. *RSC Adv.* **2022**, *12*, 721–727. [CrossRef]
15. Meng, F.; Li, X.; Li, M.; Cui, X.; Li, Z. Catalytic performance of CO methanation over La-promoted Ni/Al_2O_3 catalyst in a slurry-bed reactor. *Chem. Eng. J.* **2017**, *313*, 1548–1555. [CrossRef]
16. Shen, S.; Chen, J.; Koodali, R.T.; Hu, Y.; Xiao, Q.; Zhou, J.; Wang, X.; Guo, L. Activation of MCM-41 mesoporous silica by transition-metal incorporation for photocatalytic hydrogen production. *Appl. Catal. B* **2014**, *150–151*, 138–146. [CrossRef]
17. Yang, M.; Lingjun, Z.; Xiaonan, Z.; Prasert, R.; Shurong, W. CO_2 methanation over nickel-based catalysts supported on MCM-41 with in situ doping of zirconium. *J. CO2 Util.* **2020**, *42*, 101304. [CrossRef]
18. Lensveld, D.J.; Gerbrand Mesu, J.; Jos van Dillen, A.; de Jong, K.P. Synthesis and characterisation of MCM-41 supported nickel oxide catalysts. *Microporous Mesoporous Mater.* **2001**, *44–45*, 401–407. [CrossRef]
19. Qiu, S.; Zhang, X.; Liu, Q.; Wang, T.; Zhang, Q.; Ma, L. A simple method to prepare highly active and dispersed Ni/MCM-41 catalysts by co-impregnation. *Catal. Commun.* **2013**, *42*, 73–78. [CrossRef]
20. Zhang, J.; Xin, Z.; Meng, X.; Tao, M. Synthesis, characterization and properties of anti-sintering nickel incorporated MCM-41 methanation catalysts. *Fuel* **2013**, *109*, 693–701. [CrossRef]
21. Zyryanova, M.M.; Snytnikov, P.V.; Gulyaeva, R.V.; Amosov, Y.I.; Boronin, A.I.; Sobyanin, V.A. Performance of Ni/CeO_2 catalysts for selective CO methanation in hydrogen-rich gas. *Chem. Eng. J.* **2014**, *238*, 189–197. [CrossRef]
22. Munnik, P.; Velthoen, M.E.Z.; de Jongh, P.E.; de Jong, K.P.; Gommes, C.J. Nanoparticle growth in supported nickel catalysts during methanation reaction—Larger is better. *Angew. Chem. Int. Ed.* **2014**, *53*, 9493–9497. [CrossRef] [PubMed]
23. Carraro, P.; Elías, V.; García Blanco, A.; Sapag, K.; Moreno, S.; Oliva, M.; Eimer, G. Synthesis and multi-technique characterization of nickel loaded MCM-41 as potential hydrogen-storage materials. *Microporous Mesoporous Mater.* **2014**, *191*, 103–111. [CrossRef]
24. Park, S.J.; Lee, S.Y. A study on hydrogen-storage behaviors of nickel-loaded mesoporous MCM-41. *J. Colloid. Interf. Sci.* **2010**, *346*, 194–198. [CrossRef] [PubMed]
25. Carraro, P.; Elías, V.; Blanco, A.A.G.; Sapag, K.; Eimer, G.; Oliva, M. Study of hydrogen adsorption properties on MCM-41 mesoporous materials modified with nickel. *Int. J. Hydrog. Energy* **2014**, *39*, 8749–8753. [CrossRef]
26. Lehmann, T.; Wolff, T.; Hamel, C.; Veit, P.; Garke, B.; Seidel-Morgenstern, A. Physico-chemical characterization of Ni/MCM-41 synthesized by a template ion exchange approach. *Microporous Mesoporous Mater.* **2012**, *151*, 113–125. [CrossRef]

27. Tao, M.; Meng, X.; Xin, Z.; Bian, Z.; Lv, Y.; Gu, J. Synthesis and characterization of well dispersed nickel-incorporated SBA-15 and its high activity in syngas methanation reaction. *Appl. Catal. A* **2016**, *516*, 127–134. [CrossRef]
28. Zhang, Q.; Wang, T.; Li, B.; Jiang, T.; Ma, L.; Zhang, X.; Liu, Q. Aqueous phase reforming of sorbitol to bio-gasoline over Ni/HZSM-5 catalysts. *Appl. Energy* **2012**, *97*, 509–513. [CrossRef]
29. van de Loosdrecht, J.; van der Kraan, A.M.; van Dillen, A.J.; Geus, J.W. Metal-support interaction: Titania-supported and silica-supported nickel catalysts. *J. Catal.* **1997**, *170*, 217–226. [CrossRef]
30. Meng, F.; Li, X.; Shaw, G.M.; Smith, P.J.; Morgan, D.J.; Perdjon, M.; Li, Z. Sacrificial carbon strategy toward enhancement of slurry methanation activity and stability over Ni-Zr/SiO$_2$ catalyst. *Ind. Eng. Chem. Res.* **2018**, *57*, 4798–4806. [CrossRef]
31. Meng, F.; Wang, L.; Li, X.; Perdjon, M.; Li, Z. Mesoporous nano Ni-Al$_2$O$_3$ catalyst for CO$_2$ methanation in a continuously stirred tank reactor. *Catal. Commun.* **2022**, *164*, 106437. [CrossRef]
32. Jia, C.; Gao, J.; Li, J.; Gu, F.; Xu, G.; Zhong, Z.; Su, F. Nickel catalysts supported on calcium Titanate for enhanced CO methanation. *Catal. Sci. Technol.* **2013**, *3*, 490–499. [CrossRef]
33. Wu, C.; Wang, L.; Williams, P.T.; Shi, J.; Huang, J. Hydrogen production from biomass gasification with Ni/MCM-41 catalysts: Influence of Ni content. *Appl. Catal. B* **2011**, *108–109*, 6–13. [CrossRef]

Disclaimer/Publisher's Note: The statements, opinions and data contained in all publications are solely those of the individual author(s) and contributor(s) and not of MDPI and/or the editor(s). MDPI and/or the editor(s) disclaim responsibility for any injury to people or property resulting from any ideas, methods, instructions or products referred to in the content.

Review

Research Progress of Carbon Deposition on Ni-Based Catalyst for CO$_2$-CH$_4$ Reforming

Yuan Ren [1,†], Ya-Ya Ma [1,2,†], Wen-Long Mo [1,*], Jing Guo [3,*], Qing Liu [4], Xing Fan [4] and Shu-Pei Zhang [5]

1. State Key Laboratory of Chemistry and Utilization of Carbon-Based Energy Resources and Key Laboratory of Coal Clean Conversion & Chemical Engineering Process (Xinjiang Uyghur Autonomous Region), School of Chemical Engineering and Technology, Xinjiang University, Urumqi 830046, China
2. Qingdao Institute of Bioenergy and Bioprocess Technology, Chinese Academy of Sciences, Qingdao 266101, China
3. School of Chemistry and Chemical Engineering, Ningxia Normal University, Guyuan 756000, China
4. College of Chemical and Biological Engineering, Shandong University of Science and Technology, Qingdao 266590, China
5. Xinjiang Yihua Chemical Industry Co., Ltd., Changji 831700, China
* Correspondence: mowenlong@xju.edu.cn (W.-L.M.); guojingsn@163.com (J.G.)
† These authors contributed equally to this work.

Abstract: As we all know, the massive emission of carbon dioxide has become a huge ecological and environmental problem. The extensive exploration, exploitation, transportation, storage, and use of natural gas resources will result in the emittance of a large amount of the greenhouse gas CH$_4$. Therefore, the treatment and utilization of the main greenhouse gases, CO$_2$ and CH$_4$, are extremely urgent. The CH$_4$ + CO$_2$ reaction is usually called the dry methane reforming reaction (CRM/DRM), which can realize the direct conversion and utilization of CH$_4$ and CO$_2$, and it is of great significance for carbon emission reduction and the resource utilization of CO$_2$-rich natural gas. In order to improve the activity, selectivity, and stability of the CO$_2$-CH$_4$ reforming catalyst, the highly active and relatively cheap metal Ni is usually used as the active component of the catalyst. In the CO$_2$-CH$_4$ reforming process, the widely studied Ni-based catalysts are prone to inactivation due to carbon deposition, which limits their large-scale industrial application. Due to the limitation of thermodynamic equilibrium, the CRM reaction needs to obtain high conversion and selectivity at a high temperature. Therefore, how to improve the anti-carbon deposition ability of the Ni-based catalyst, how to improve its stability, and how to eliminate carbon deposition are the main difficulties faced at present.

Keywords: CO$_2$-CH$_4$ reforming; Ni-based catalyst; carbon deposition

1. Introduction

With the increasing use of fossil resources, such as coal, oil, and natural gas, the global CO$_2$ emissions will continue to rise. In recent years, some countries have been using renewable energy; however, this trend has not been enough to prevent the climate change, polar ice sheet melting, and hurricane intensification caused by the increase in CO$_2$ emissions. The massive emission of CO$_2$ not only accelerates the deterioration of the greenhouse effect but also wastes valuable carbon resources. Therefore, CO$_2$ reduction and resource utilization have become the most noticeable research issues. According to the proportions of different greenhouse gases in the total greenhouse gas emissions calculated by CO$_2$ equivalent, CH$_4$ accounts for 16% and is therefore the second villain of the greenhouse effect. Although the emission of CH$_4$ is far less than that of CO$_2$, its potential to produce a greenhouse effect is about 20 times more than that of CO$_2$. Therefore, how to convert the above two greenhouse gases into useful chemicals or chemical raw materials has attracted the great attention of governments and scientists around the world [1–3].

As early as 1928, Fischer and Tropsch discovered the carbon dioxide reforming of methane reaction, that is, $CO_2 + CH_4 = 2CO + 2H_2$ (carbon dioxide reforming of methane, CRM, ΔH = 247 kJ/mol), also known as methane dry reforming. CRM can produce synthesis gas; it is a strongly endothermic reaction, and the synthesis gas obtained has a low H_2/CO ratio, which is more suitable for the subsequent Fischer–Tropsch synthesis reaction [4]. The ratio of H_2/CO in the CRM product is about 1, and it can also be used in chemical reactions such as carbonyl synthesis and hydrocarbon production and to produce clean liquid fuels and high-value chemicals [5,6]. Hence, this reaction can also utilize CO_2 and CH_4, which are two main greenhouse gases; therefore, it has significant industrial value and ecological and environmental significance. The research on this reaction has been further developed, particularly in the past 30 years [7].

CRM and steam reforming of methane (SRM) are catalytic reactions at a high temperature (about 800 °C). The ΔH of CRM = 247 kJ/mol, which is greater than that of SRM (206 kJ/mol), indicating that both CRM and SRM are strongly endothermic reactions, and the endothermic capacity of CRM is nearly 20% higher than that of SRM. Therefore, the reverse reaction of CRM can theoretically release up to 247 kJ/mol of energy. Therefore, the reaction can be used as a good chemical energy transmission system (CETS) to store energy. On the other hand, CRM can be realized through fossil fuels (such as coal, petroleum, etc.), light energy, or nuclear energy, and the above energy can be stored in the product (synthesis gas); then, the synthesis gas can be transported to the place where it is needed for a reverse reaction to release energy.

Over a long period of time, researchers have conducted many studies on the selection and optimization of CRM catalysts, and have achieved fruitful results, making the research on this reaction increasingly broad and deep. Without loss of generality, the relationship between the chemical reaction itself, the type of active component and carrier, the modification of additives, the carbon deposition, and the catalyst performance have been consistently discussed by many researchers. As far as the CRM process is concerned, the main side reaction is the reverse water gas shift reaction ($CO_2 + H_2 = CO + H_2O$, reverse water gas shift reaction, RWGS), which consumes the H_2 ($CH_4 = C + 2H_2$) generated by CH_4 cracking and generates a large amount of CO, which can cause carbon deposition on the catalyst through another side reaction—the CO disproportionation reaction ($2CO = C + CO_2$) [8,9]. Thus, the proportion of CO_2 can be increased from the perspective of chemical equilibrium to inhibit the formation of carbon deposition (that is, the carbon elimination reaction, $C + CO_2 = 2CO$). Thermodynamic calculation showed that the RWGS reaction can be inhibited or avoided at temperatures above 820 °C, but this needs a lot of energy [10]. Therefore, how to optimize the catalyst structure, match the process conditions, and selectively control the degree of carbon deposition, the reverse water gas shift reaction, carbon elimination, and other chemical reactions involving carbon is of great significance in improving the activity of the CO_2-CH_4 reforming catalyst, inhibiting catalyst deactivation, and extending the service life of the catalyst.

Pakhare et al. [11] introduced DRM literature on a catalyst based on Rh, Ru, Pt, and Pd metals. This includes the effect of these noble metals on the kinetics, mechanism, and deactivation of these catalysts. The inert support catalysts are more prone to deactivation due to carbon deposition than the acidic or basic supports.

At present, the widely used CO_2-CH_4 reforming catalyst is still dominated by the non-noble metal catalyst, especially the Ni-based catalyst. Its activity is equivalent to that of the noble metal, but it is very easily inactivated due to carbon deposition. Therefore, the development of the Ni-based catalyst with high carbon deposition resistance is the key to realizing the industrialization of CO_2-CH_4 reforming. A large number of studies have shown that carbon deposition in the hydrocarbon conversion process is mainly affected by such factors as the acid–base property of the carrier, the dispersion of the active component, and the interaction between the carrier and the active metal [12–16]. The acid sites on the catalyst surface are not conducive to the adsorption of CO_2, resulting in the carbon deposition rate on the catalyst surface being much higher than the carbon elimination. Conversely, the surface basic sites can inhibit the carbon deposition caused by CO disproportionation to a certain extent, thereby improving the stability of the catalyst. It was found that when the particle size of the active component was less than 10 nm, the catalyst could present a high anti-coking performance [14]. In addition, strong metal-support interaction is also beneficial in improving the anti-carbon deposition performance of the catalyst [15,16].

In this paper, the thermodynamics, kinetics, and reaction mechanism of the CO_2-CH_4 reforming reaction are reviewed. Because Ni-based catalysts exhibit high activity but have the problem of the easy deactivation of carbon deposition, this paper further summarizes the research situation regarding carbon deposition on Ni-based catalysts, including the types of carbon deposition, the amount of carbon deposition, and the elimination of carbon deposition. As to how to improve the anti-carbon deposition ability of the Ni-based catalyst and how to eliminate carbon deposition, this paper focuses on two aspects: one is the resistance of carbon deposition from the perspective of catalyst optimization; the other is the elimination of carbon deposition from the perspective of process condition matching. Finally, how to improve the carbon deposition resistance of Ni-based CRM catalyst is prospected.

2. Thermodynamics of CO_2-CH_4 Reforming

CO_2-CH_4 reforming mainly includes the following reactions:

$$CH_4 + CO_2 = 2H_2 + 2CO, \Delta H (298 \text{ K}) = 247 \text{ kJ/mol}, \tag{1}$$

$$H_2 + CO_2 = H_2O + CO, \Delta H (298 \text{ K}) = 41 \text{ kJ/mol}, \tag{2}$$

$$2CO = CO_2 + C, \Delta H (298 \text{ K}) = -172 \text{ kJ/mol}, \tag{3}$$

$$CH_4 = C + 2H_2, \Delta H (298 \text{ K}) = 75 \text{ kJ/mol}, \tag{4}$$

$$CO + H_2 = C + H_2O, \Delta H (298 \text{ K}) = -131 \text{ kJ/mol}, \tag{5}$$

Zhang et al. [17] obtained the thermodynamic equilibrium constants of the above reactions as a function of temperature through thermodynamic calculations, as shown in Figure 1.

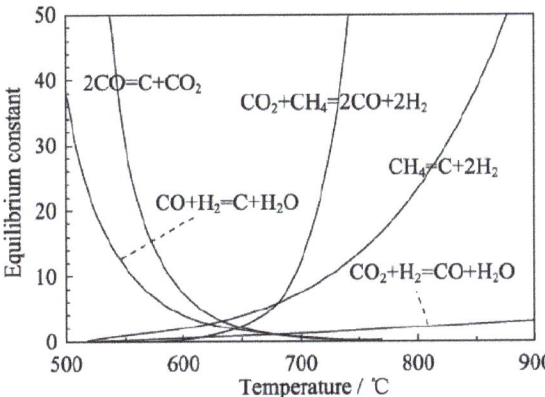

Figure 1. Equilibrium constants of reactions as a function of temperature during DRM [17].

As both methane and carbon dioxide are very stable, CRM is an extremely strong endothermic reaction. Meanwhile, the RWGS reaction is an important side reaction in the CRM process which reduces the H_2/CO ratio in the product [18,19]. In the CRM reaction process, high temperature (>1000 K) and low pressure (~1 atm) are usually required to obtain the efficient conversion of methane and carbon dioxide to syngas. At a higher pressure, it can promote the RWGS reaction and reduce the H_2/CO ratio. Therefore, the ratio of carbon dioxide and methane in the feed gas has a great influence on the H_2/CO ratio. Many experiments showed that the ideal H_2/CO ratio of 1 can be achieved when the CO_2/CH_4 ratio is about 1. With the increase in the CO_2/CH_4 ratio, the H_2/CO ratio decreases. This trend is more obvious at higher pressures (10 atm).

In addition to RWGS, there are two other side reactions, methane decomposition and CO disproportionation, which also occur in the CRM process [11]. These two reactions lead to the formation of carbon deposition, which leads to the deactivation of the catalyst. In order to study the formation of the carbon deposition, Lu et al. [19] calculated the limit temperatures of the two side reactions (where the Gibbs free energy change is zero). The CO_2-CH_4 reaction can be accompanied by methane cracking at above 640 °C, while the reverse water gas shift reaction starts at above 820 °C, without CO disproportionation (2CO = C + CO_2, the Boudouard reaction). In the range of 557–700 °C, carbon is mainly formed by methane cracking or the Boudouard reaction. Under the pressure of 0.01–0.1 atm, the feed ratio of CO_2/CH_4 = 1 can reach the equilibrium conversion rate. At a fixed temperature, the rate at a low pressure is always higher than that at a high pressure. Under the pressure of 0.01 atm, the rate reached 90% at 550 °C, and did not reach 90% until 700 °C at 0.1 atm. There is an upper temperature limit for carbon deposition, and the temperature increases with the increase in reaction pressure and the decrease in the CO_2/CH_4 ratio. Therefore, the formation of carbon deposition at a certain temperature can be inhibited by reducing the reaction pressure and increasing the proportion of carbon dioxide in the feed gas.

The main reactions of carbon deposition are methane decomposition and the carbon monoxide disproportionation reaction. Severe carbon deposition will lead to blockage or even deactivation of the catalyst bed. According to thermodynamic analysis, the carbon monoxide disproportionation reaction is a strongly exothermic reaction, which mainly occurs in a relatively low temperature range (<650 °C), and methane cracking reaction is a strongly endothermic reaction, which mainly occurs in a relatively high temperature range. Therefore, low temperature and high pressure are beneficial to carbon monoxide disproportionation, and high temperature and low pressure are beneficial to methane cracking, and the main reaction temperature is in the range of 557–700 °C. When the temperature is higher than 600 °C, the amount of carbon deposition will increase rapidly. However, with the increase in reaction temperature, the disproportionation reaction of carbon monoxide will be inhibited, and methane cracking will become the main reaction of the carbon deposition [11,20,21]. However, a high reaction temperature often leads to both the sintering of the active metal and carbon deposition on the catalyst. Sometimes, the fibrous carbon formed during the reaction process has a high mechanical strength, which will damage the catalyst and lead to rapid deactivation of the catalyst [18,19,22].

Adding O_2 to CRM system can remove the carbon deposition formed to promote the regeneration of the catalyst, and the heat released from the reaction can also accelerate the decomposition of the methane. For periodic operation, the addition of oxygen (CO_2/O_2 ratio of 7/3) during the regeneration process at 750 °C significantly improved the stability and activity of the catalyst. During the stability experiment, the catalytic performance of the Ni/SiO_2·MgO catalyst for CRM in the presence of O_2 increased with the increase in O_2 content and reaction temperature [23]. In addition, the introduction of another auxiliary means, such as light and plasma treatment, can break the energy barrier of the reaction and improve the conversion rate of the reactants. Some researchers used Au as the plasma promoter for the first time to improve the reforming performance of noble metal-based catalysts. The results showed that visible light irradiation could significantly improve the reforming activity of the Rh-Au/SBA-15 catalyst. The maximum CO_2 conversion rate under light conditions is 1.7 times higher than that under dark conditions. The test at 400 °C showed that the CO_2 conversion rate under light conditions was 2.4 times higher than that under dark conditions. Kinetic measurement showed that the activation energy was reduced by 30% under light conditions [24].

3. Kinetics of CO_2-CH_4 Reforming (Reaction Mechanism)

With the development of research technology at the micro-level, the perspective of the theoretical research has gradually shifted from thermodynamics to kinetics. The view on the adsorption and dissociation of reactants and products provides an ideal explanation for the kinetics of the CRM reaction.

Presently, the focus of the research on CO_2-CH_4 reforming has two aspects. On the one hand, it is necessary to find new catalysts and additives to improve the catalyst activity and carbon deposition resistance. On the other hand, it is necessary to study the reaction mechanism in detail by the kinetic method, with the aim of deeply understanding the reforming reaction and designing a new catalyst according to the mechanism. Kinetic study is one of the best methods to reveal the intrinsic activity of the used catalyst. Due to the high activity and low price of Ni in CO_2-CH_4 reforming, the study of Ni kinetics has attracted more and more attention. Most studies have focused on exploring the rate-determining step in the CO_2-CH_4 reforming process. Discussing the rate-determining step of the reaction can lead to further understanding of the catalytic reaction mechanism and the design and improvement of the catalyst and reaction conditions. It was found that the kinetic process of CO_2-CH_4 reforming mainly includes the adsorption and dissociation of the reactants, CH_4 and CO_2, as well as the formation and desorption of the products, H_2 and CO. In order to correctly understand the kinetics and reaction mechanism of the reaction, we should understand the adsorption, desorption, and reaction properties of the reactants and products on the catalyst.

CRM reaction is a complex process. The mechanism of CRM varies greatly according to the difficulty in forming reaction intermediates in different catalyst systems and reaction conditions [25]. Many studies have shown that the key step of CRM is the adsorption and dissociation of CH_4 on the catalyst surface.

The reaction mechanism of CO_2-CH_4 reforming is closely related to the type and composition of the catalyst. At present, there is no uniform conclusion on the reaction mechanism of CO_2-CH_4 dioxide reforming. Many researchers have explored the mechanisms of different catalysts.

Bodrov et al. [26] first proposed the CO_2-CH_4 reforming reaction principle on a Ni-based catalyst, including the following basic steps:

$$CH_4 + * \rightarrow CH_2^* + H_2, \tag{6}$$

$$CO_2 + * \longleftrightarrow CO + O^*, \tag{7}$$

$$O^* + H_2 \longleftrightarrow H_2O + *, \tag{8}$$

$$CH_2^* + H_2O \longleftrightarrow CO^* + 2H_2, \tag{9}$$

$$CO^* \longleftrightarrow CO + *, \tag{10}$$

In the above formula, "*" represents the active site, (6) represents an irreversible slow reaction, and the other steps are reversible reactions.

Later, Hansen et al. [27] improved the above mechanism and proposed that the dissociation of CH_4 on Ni/MgO can be divided into two steps:

$$CH_4 + * \rightarrow CH_3^* + H^*, \tag{11}$$

$$CH_3^* + (3-x)^* \rightarrow CH_x^* + (3-x)H^*, \tag{12}$$

Osaki et al. [28] analyzed the surface pulse reaction rate of a Ni/MgO catalyst and found that the adsorption and dissociation of CH_4 can directly generate gaseous H_2, namely

$$CH_4 + * \rightarrow CH_x^* + (4-x)/2H_2^*, \tag{13}$$

It was experimentally concluded that the dissociation of CH_4 was the rate-determining step of the CO_2-CH_4 reforming reaction.

For Ni and Pt-based catalysts, Bradford et al. [29] believed that CH_4 first dissociated into CH_x^* and H_2, while CO_2 dissociated under the action of H^* to form CO^* and OH^*; then, CH_x^* reacted with OH^* to form CH_xO^*, and finally, CH_xO^* decomposed into CO and H_2.

Wei et al. [30] studied the kinetics of the CO_2-CH_4 reaction on a Ni/MgO catalyst by using carbon dioxide and methane isotopes. They observed an obvious isotope effect of CH_4 dissociation and speculated that CH_4 dissociation was the rate-determining step of the reforming reaction. In the field of theoretical research, Wang et al. [31] used the DFT (density functional theory) method to study the principle of the CO_2-CH_4 reaction on the surface of perfect Ni(111). The calculation results showed that the principle of CO_2-CH_4 reforming is:

1. CO_2 decomposes to generate O and CO, while CH_4 gradually cleaves H on the surface to generate CH and H_2;
2. CH is oxidized to obtain CHO;
3. The main product CO is obtained by the dissociation reaction of CHO;
4. H_2 and CO are desorbed from Ni(111) to form free H_2 and CO.

Although the activation mechanisms of different catalytic systems are quite different, the existing research results show that the dehydrogenation cracking of CH_4 on the metal surface is a common and critical process. That is, CH_4 is decomposed into surface CH_x (x = 1–3) and H ($CH_4 \rightarrow CH_3 \rightarrow CH_2 \rightarrow CH$). On the other hand, the adsorption and activation mechanism of CO_2 is very important for the decarbonization process, because it not only generates the key product CO, but also provides a surface oxygen species for CH_4 reforming, which is the core intermediate for the elimination of carbon deposition. CO_2 activation consists of two steps: the first step is CO_2 chemical adsorption and the formation of an anionic $CO_2^{\delta-}$ precursor on the surface [32,33]; in the second step, the $CO_2^{\delta-}$ precursor is dissociated into surface adsorbed CO and O species. Therefore, CO_2 is the only source of the oxygen atom in the reaction gas and is the supplier of active oxygen species on the catalyst surface [34–38]. In addition, the activation path of CO_2 varies with the acidity and alkalinity of the support, which has a certain effect on the anti-carbon deposition performance of a Ni-based catalyst. Generally, at the interface between the active metal and the support, the acidic support can promote the dissociation of CH_4, but the stronger the acidity, the easier it is to produce carbon deposition [39].

It is known that increasing the adsorbed oxygen species over the catalyst surface is really effective in promoting the catalytic activity and restraining the side reaction (RWGS). Simultaneously, the adsorbed oxygen species are effective in suppressing/removing the deposited carbon, thereby alleviating catalyst deactivation [40].

The higher CO_2 activity enhanced the oxidation rate of the surface carbon generated from the side reactions, thereby resulting in a higher reforming rate and in the inhibition of the coke formation, especially the detrimental graphitic encapsulating carbon on an active nickel surface [41].

Based on the combined results of catalytic testing and characterization, Ni/$Ce_{0.9}Eu_{0.1}O_{1.95}$-HT can accelerate the rate of CO_2 activation and promote the conversion of CH_4 into CO instead of into coke deposition, leading to a relatively good performance for the DRM reaction [42].

Burghaus [43] clarified the correlation between CO_2 adsorption kinetics and the surface structure characteristics of various metals and oxides, including metals (Cu, Cr), metal oxides (ZnO, TiO_2, CaO), model catalysts (Cu/ZnO, Zn/Cu), and nano-catalysts. The binding energy of CO_2 and metal oxides is generally greater than that of metals. When CO_2 chemisorption occurs on CaO, the C atom combines with the O site of CaO, and the surface carbonate formed is very stable. Its decomposition and desorption temperature is as high as 1100 K.

Pan et al. [44] used DFT calculation and found that the adsorption and activation of a 3D transition metal dimer (M_2/γ-Al_2O_3, M = Sc, Ti, V, Cr, Mn, Fe, Co, Ni, Cu) supported by γ-Al_2O_3 were consistent with the experimental results reported in many of the studies.

CO_2 adsorbed on M_2/γ-Al_2O_3, a negatively charged species is formed, forming a metal dimer; γ-Al_2O_3 supports could provide electrons to the adsorbed CO_2 to activate it, and the most favorable adsorption position was at the interface between the metal dimer and the support; so, the highly dispersed metal particles showed good activity. In addition, the hydroxyl group on the surface of the carrier reduces the amount of charge transferred from the metal dimer to the CO_2 and weakened the chemical adsorption of CO_2.

The addition of La_2O_3 to the Ni/γ-Al_2O_3 catalyst could inhibit the carbon deposition in CO_2-CH_4 reforming. Some researchers have found that in the CO_2-CH_4 reaction, La_2O_3 interacts with CO_2 on the Ni/La_2O_3 catalyst to generate $La_2O_2CO_3$ [45], and $La_2O_2CO_3$ decomposes CO and provides oxygen species, and the oxygen species can react with carbon species accumulated after the dissociation of CH_4 on Ni grains to generate CO, thus achieving the effect of inhibiting carbon deposition.

4. Carbon Deposition and Elimination on Ni-Based Catalyst

The Ni-based catalyst is widely used in industrial processes because of its high activity, good stability, and low cost, but the biggest problem is that the catalysts are easily inactivated. There are three main ways that deactivation occurs: carbon deposition, sintering, and poisoning. Among them, the most important factor causing catalyst inactivation in the carbon-related reaction process was carbon deposition. Hence, the reaction mechanism of carbon deposition and the inhibition of carbon deposition need to be further studied. The several existing inhibition methods can be divided into two types: one involves the resistance of carbon deposition from the perspective of catalyst optimization, and the other involves the elimination of carbon deposition from the perspective of process condition matching.

4.1. Formation and Type of Carbon Deposition

The deactivation of the Ni-based catalyst is mainly due to carbon deposition on the catalyst surface. The raw materials for the CO_2-CH_4 reforming reaction are all carbon-containing gases. Methane cracking (CH_4 = C + $2H_2$) and carbon monoxide disproportionation (2CO = C + CO_2) will inevitably form carbon on the catalyst surface [11].

According to existing studies, the types of carbon deposition can be divided into amorphous carbon, polymerized carbon, carbon nanotubes, graphitized carbon, and filamentous carbon [46]. Amorphous carbon is composed of carbon atoms adsorbed on the metal active center; these atoms have high reactivity and can be removed by an oxidation reaction (C + O_2 = CO_2) at about 200 °C. Polymerized carbon composed of partially hydrogenated carbon-carbon chains has low reactivity, but it is still a kind of carbon species that can be oxidized and eliminated under mild conditions or eliminated under appropriate process conditions (such as excessive CO_2). Graphitized carbon is a ring structure composed of six carbon atoms and needs higher a reaction temperature to be oxidized and eliminated. It belongs to inert carbon deposition. Filamentous carbon and carbon nanotubes can block the pores of the catalyst and gradually reduce its activity until it is completely deactivated [7,47,48].

Mo et al. [49] studied the effect of reaction time on carbon deposition on a Ni-Al_2O_3 catalyst. Figure 2 shows the TPH spectra of samples after a reforming reaction at different times. According to the report [50], 200–350 °C is the first type of hydrogenation peak, which belongs to the amorphous α type of carbon species, which is the active intermediate of the decarbonization reaction (CO_2 + C = 2CO) and is also the desired type of carbon deposition for a carbon-related reaction. It is easy for this type of carbon to be converted into a slightly less active carbon at a high-temperature β type of carbon species; 350–500 °C is the second type of hydrogenation peak, which belongs to the β_1 type carbon species, which do not form a strong interaction with the carrier and are easy to deposit in the catalyst pore, or they enter the catalyst lattice to form carbon nanotubes or filamentous carbon; they have decarbonization reaction activity at higher temperatures but easily become inert when aggregated for a long time at high-temperature γ carbon species; 500–700 °C is the third type of hydrogenation peak, belonging to the γ type of carbon species, such as graphite carbon, whose activity is lower than α carbon and β carbon and is an important reason for the irreversible deactivation of the catalyst due to carbon deposition. The literature also showed that [51], after the CRM reaction occurred on the surface of the Ni-CaO-ZrO$_2$ catalyst for 1 h, the temperature-programmed hydrogenation reaction characterization (TPH) found that the coking hydrogenation peak at about 800 °C was attributable to β$_2$ types of carbon species, namely the fourth type of hydrogenation peak. It can be seen from Figure 2 that in the process of the CO$_2$-CH$_4$ reforming reaction on the Ni-Al$_2$O$_3$ catalyst, the surface hydrogenation peaks of the carbon species generated are all at 200–500 °C, that is, the amorphous carbon C$_α$ and filamentous carbon C$_{β1}$ type exists [52,53]. However, no obvious hydrogenation peak was found at about 600 and 800 °C. Therefore, it can be considered that in the range of 10 h, the CO$_2$-CH$_4$ reforming reaction almost did not form the C$_γ$ and C$_{β2}$ types of carbon species.

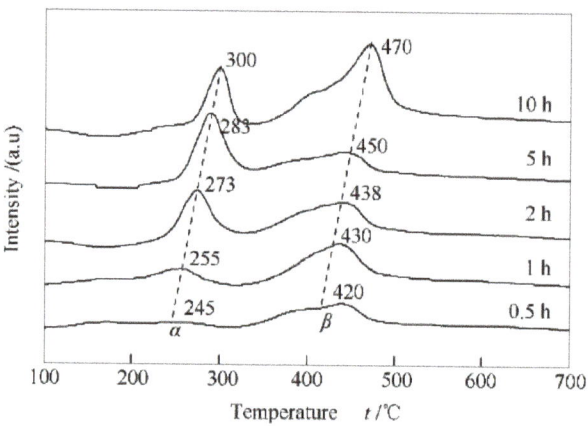

Figure 2. TPH profiles of various spent Ni-Al$_2$O$_3$ catalysts after carrying out the CO$_2$-CH$_4$ reforming reaction for different times [49].

According to the degree of difficulty of carbon elimination under the conditions of the reforming reaction, the above carbon deposition can be classified as active carbon deposition (amorphous carbon), transitional carbon deposition (polymeric carbon, carbon nanotubes, and filamentous carbon), and inactive carbon deposition (graphitized carbon).

Mo et al. [49] observed the carbon deposition on the catalyst surface at different reaction times and found that the fibrous carbon deposition began to accumulate after 2 h of reaction, and after 5 h, it was observed that there were many fibrous or rod-shaped morphologies with larger diameters. With the extension of reaction time, the amount of fibrous carbon deposition increased significantly, while the diameter of the carbon fibers decreased significantly, which is probably because of the occurrence of a carbon elimination reaction (C + CO$_2$ = 2CO) in CO$_2$-CH$_4$ reforming (Figure 3). The results also showed that each sample had two hydrogenation peaks, one low-temperature hydrogenation peak and one high-temperature hydrogenation peak, corresponding to two types of carbon deposition species [49]. With the increase in reaction temperature, the high-temperature hydrogenation peak of the carbon deposition moves to a high temperature. The reaction temperature increased from 650 °C to 850 °C, and the peak temperature of the high-temperature hydrogenation peak increased from 425 °C to 455 °C, which may be due to the fact that filamentous carbon is easily converted into graphite carbon at high temperatures, and the hydrogenation reaction needs to be carried out at a higher temperature [54].

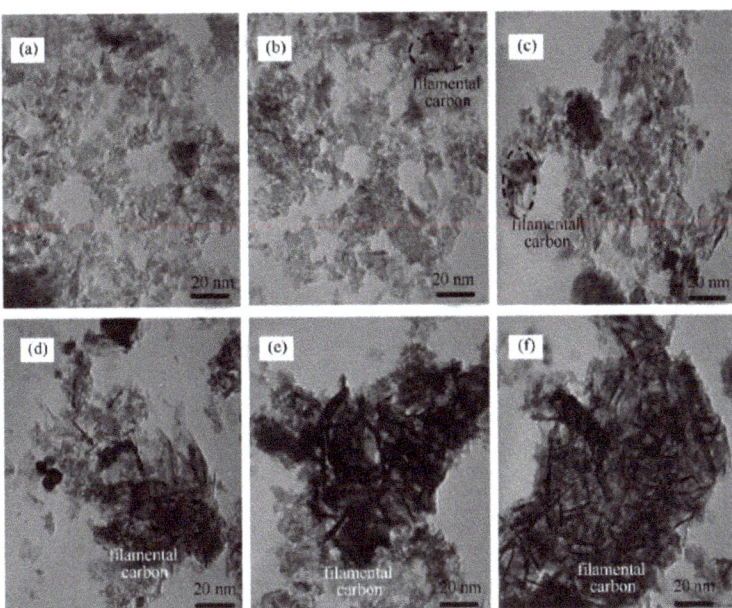

Figure 3. TEM photos of different Ni-Al$_2$O$_3$ catalysts after CO$_2$-CH$_4$ reforming reaction for different times [49]. (**a**–**f**) Fresh catalyst and catalyst after reaction for 0.5 h, 1 h, 2 h, 5 h, and 10 h, respectively.

Figure 4 showed the morphological characteristics of carbon on the catalyst surface after the reforming reaction. It can be seen from the figure that a large amount of filamentous carbon is formed on the catalyst surface and is even covering the catalyst surface in a large range. It can be speculated that if the carbon deposition continues, it will completely cover the catalyst surface and cause catalyst deactivation [46].

Figure 4. SEM of CO$_2$-CH$_4$ reforming reaction after 20 h [46].

4.2. Resistance and Elimination of Carbon Deposition

In CO$_2$-CH$_4$ reforming, the carbon deposition rate usually depends on its formation rate and elimination rate. When the elimination rate of carbon deposition is higher than the formation, the carbon deposition can be inhibited [55]. According to the cause of carbon deposition in the CO_2-CH_4 reaction, the carbon deposition can be suppressed by two aspects: catalyst modification and process conditions optimization.

In the process of CO_2-CH_4 reforming, carbon deposition mainly comes from methane cracking and carbon monoxide disproportionation. The dehydrogenation of CH_4 on the metal surface generates carbon species CH_x (x = 0–3, $CH_4 \rightarrow C + 2H_2$), which do not react with the surface oxygen species generated by the timely adsorption and dissociation of CO_2 to generate carbon species of CO (C + $CO_2 \rightarrow 2CO$), and the carbon species may accumulate on the metal surface, causing carbon deposition. As the carbon dioxide content in the reactant gas increases, the adsorption rate of methane and its subsequent dissociation rate (i.e., the cracking of methane) decrease. At the same time, the oxidation rate of carbon species on the catalyst surface increases. Therefore, the amount of carbon deposited on the active site is reduced, which can significantly improve the stability of the Ni-based catalyst. In an atmosphere with sufficient CO_2, during a long catalytic process, carbon will migrate and accumulate in the center of the Ni crystal, forming a hollow fiber type structure from bottom to top. The process of carbon deposition and carbon elimination on a Ni/CeO_2 catalyst is shown in Figure 5 [56].

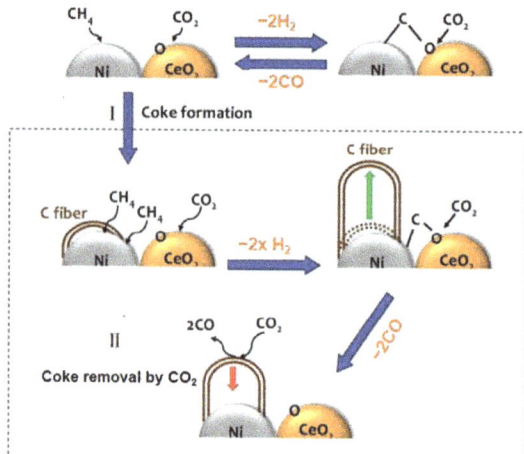

Figure 5. Carbon deposition and decarbonization on Ni/CeO_2 catalyst [56].

Kuijpers [57] found the size sensitivity of CH_4 activation in the study of Ni-based catalysts, that is, CH_4 preferentially dissociates on smaller Ni grains. Osaki et al. [28,58] reported the x value of CH_x species on different catalysts: Ni/MgO = 2.7, Ni/ZnO = 2.5, Ni/Al_2O_3 = 2.4, Ni/TiO_2 = 1.9, Ni/SiO_2 = 1.0, and Co/Al_2O_3 = 0.75. It can be seen that for the same metal, the x value of basic carrier is higher, that is, the degree of CH_4 dissociation increases with the increase in carrier acidity. On the other hand, the activation of the C-H bond requires electrons from the surface of the Ni; so, the electronic environment around the Ni is also extremely important [59]. For example, the strong metal-support interaction will significantly affect the activity of Ni to dissociate CH_4 [60,61].

Horiuchi et al. [7] believe that alkaline metal oxides can enhance the adsorption capacity of CO_2 and generate more active oxygen atoms (O_{ad}); O_{ad} can effectively prevent the adsorption of $CH_{x,ad}$ on the active center of Ni metal through the reaction of $CH_{x,ad}$ + $O_{ad} \rightarrow CO + H_2$, thus avoiding the surface carbon deposition caused by the cracking of $CH_{x,ad}$. Therefore, adding additives such as alkali or alkaline oxide can enhance the basicity of the carrier surface and the ability to absorb CO_2, change the electronic density of the metal active center, effectively improve the catalytic activity of the catalyst, and inhibit the carbon deposition on the catalyst surface [62].

4.2.1. Resistance of Carbon Deposition from the Perspective of Catalyst
Effect of Ni Grain Size on the Deposition of Carbon

Studies have shown that high dispersion of active metal on the surface of the support can reduce the agglomeration size and effectively inhibit carbon deposition [11,63,64]. It has also been shown that only when the size of the active component is larger than a certain critical size (for example, ≥9 nm) can lead the carbon simple substance to form nuclei [7]. The small size and high dispersion of the active component can effectively inhibit the nucleation and growth of carbon whiskers. Therefore, by selecting the appropriate support, additive, and preparation method, the metal dispersion and particle size can be adjusted to effectively inhibit the occurrence of carbon deposition and improve the anti-carbon deposition performance [65]. In addition, the addition of an appropriate additive can also improve the surface alkalinity of the catalyst, strengthen the adsorption of CO_2, promote the elimination of deposited carbon species, and enhance the anti-carbon deposition performance of the catalyst.

XU et al. [66] prepared $Ni/La_2O_3/\gamma-Al_2O_3$ and $Ni/La_2O_3/\alpha-Al_2O_3$ catalysts and found that when the size of the Ni particles was less than 15 nm, the carbon deposition was significantly reduced (Figure 6), which made the catalyst have higher catalytic activity and stability. CRNIVEC et al. [67] found that the catalyst surface with metal active particles less than 6 nm has excellent resistance to carbon deposition. LIU et al. [68] found that when nickel particles were less than 5 nm, they had an obvious inhibition effect on carbon deposition.

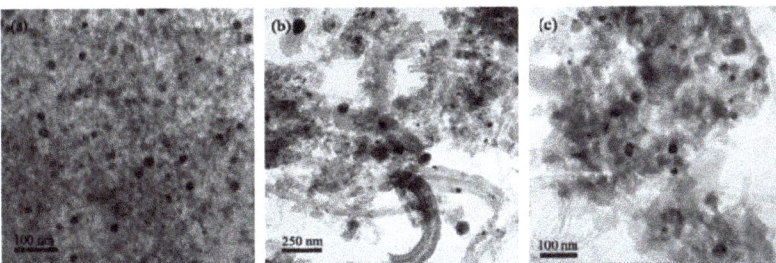

Figure 6. TEM images of catalyst after reaction [66]: (**a**) $Ni/La_2O_3/\gamma-Al_2O_3$; (**b**) $Ni/La_2O_3/Al_2O_3$; (**c**) $Ni/La_2O_3/\alpha-Al_2O_3$.

Li et al. [69] analyzed the surface carbon on a Ni/MgO catalyst based on density functional theory and found that the size of the active component Ni had a great influence on the anti-carbon deposition performance of the catalyst. The three different sizes of active metals, Ni4, Ni8, and Ni12 were loaded on the surface of MgO, showing significant differences in catalytic activity and stability. Small size Ni4 can reduce the activation energy of the CH_4 dissociation adsorption, the CH dissociation, and the Coxidation, thus improving the CO_2-CH_4 reforming performance.

In their study, Mo et al. [70,71] found that after the reduction in the Ni-based catalyst based on $NiAl_2O_4$ spinel, the size of the Ni crystal was small and could effectively prevent the high-temperature sintering of the active component. By increasing the calcination temperature, the proportion of $NiAl_2O_4$ spinel (the active component precursor) in the catalyst was increased, and the size of the active component was effectively reduced; the stability of the catalyst was improved, and the amount of carbon deposited on the catalyst was obviously reduced. It was also found that there were two Ni precursors, crystalline NiO (calcinated at lower temperature (≤600 °C) and spinel $NiAl_2O_4$ (calcinated at higher temperature (≥700 °C), indicating that the calcination temperature significantly affected the interaction force between the metal and the support. The higher the calcination temperature, the stronger the force, and the more the spinel phase Ni species that formed. This type of Ni species gave a small size of active component after reduction, which could obtain higher CO_2 and CH_4 conversion and H_2 selectivity [72].

Inhibition of Carbon Deposition from the Stability of Ni Component

In general, researchers add some additives to the Ni-based catalyst to improve the dispersion and stability of the nickel, promoting the reforming reaction and inhibiting the formation of carbon deposition, which is a research hotspot in this reaction. The additives commonly used are mainly divided into two categories. The first type of additive is alkaline oxides, including MgO, CaO, K_2O, etc. [73,74]. It was found that the addition of alkaline oxides to a Ni-based catalyst can promote dispersion of the Ni and inhibit carbon deposition. On the other hand, the alkaline medium can improve the adsorption performance of CO_2 (weak acid gas). Some researchers [74] believed that the addition of alkali metals can inhibit carbon deposition and improve the activity and stability of the catalyst. Mo et al. [75] studied the effect of CaO on the structure, reforming performance, and carbon deposition of the Ni-Al_2O_3 catalyst. The results showed that the activity of Ni-Ca-4 was higher, with the conversion rate of CH_4 and CO_2 of 52.0% and 96.7%, respectively. The amount of carbon deposited on the catalyst was lower, and the type of the carbon was attributed to an amorphous one, presenting a good anti-carbon deposition performance.

Another kind of additive is rare earth oxides [73], such as CeO_2 and La_2O_3, which can achieve both high activity and stability for CO_2-CH_4 reforming. By adding rare earth oxide, the crystal phase, pore structure, and mechanical strength of the catalyst could be significantly changed, thereby improving the activity, stability, and selectivity of the catalyst. For example, the addition of La_2O_3 to the Ni/γ-Al_2O_3 catalyst could inhibit the carbon deposition in CO_2-CH_4 reforming. Some researchers found that in the CO_2-CH_4 reaction, La_2O_3 interacts with CO_2 on the Ni/La_2O_3 catalyst to generate $La_2O_2CO_3$ [52]; $La_2O_2CO_3$ decomposes CO and provides oxygen species, and the oxygen species can react with the carbon species accumulated after the dissociation of CH_4 on Ni grains to generate CO, thus achieving the effect of inhibiting carbon deposition. Other additives, such as metal Cr [76] and mixed oxide CeO_2-ZrO_2 [77,78], can also improve the activity and stability of the catalyst.

Mo et al. [70] prepared a series of La_2O_3-NiO-Al_2O_3 catalysts with different La loading to improve the performance of the Ni-based catalyst for CO_2-CH_4 reforming. The results showed that the precursor of the active component mainly exists in the form of $NiAl_2O_4$ spinel. The "confinement effect" of La_2O_3 on Ni grains can inhibit the sintering of the active component, prevent carbon deposition, and improve the reforming performance. Mo et al. [79] also prepared a Ni-Al_2O_3 catalyst with Ca, Co, and Ce as additives by the combustion method. The results showed that the activity order of the catalysts was followed by Co-Ni-Al_2O_3>Ca-Ni-Al_2O_3>Ni-Al_2O_3>Ce-Ni-Al_2O_3. Carbon deposition analysis showed that Ca-Ni-Al_2O_3 presented poor carbon deposition resistance, and a certain amount of graphitic carbon was generated on the catalyst. The dry reforming performance of Ni catalysts supported by different supports is shown in Table 1

Table 1. Catalytic performance of Ni-based catalysts with different supports.

Active Component	Support	Mass Fraction of Active Component /%	Temperature /°C	Time /h	SV /(mL·g⁻¹·h⁻¹)	CH$_4$ Conversion Rate /%	CO$_2$ Conversion Rate /%	Carbon Deposition /%	Reference
LaNiO$_3$	SBA-15	10	700	60	36,000	78	73	4.47	[80]
LaNiO$_3$	MCM-41	10	700	60	36,000	75	71	4.83	[80]
LaNiO$_3$	SiO$_2$	10	700	60	36,000	68	64	5.67	[80]
Ni	MgO	20	750	2.3	168,000	46.13→34.3	51.4→37.6	2.648	[81]
Ni	Al$_2$O$_3$-T	10	700	5	24,000	80	90	—	[82]
Ni	Al$_2$O$_3$-S	10	700	5	24,000	68	79	—	[82]
Ni	Al$_2$O$_3$	10	700	5	24,000	72	75	—	[82]
Ni	γ-Al$_2$O$_3$-S	10	700	5	48,000	56.0→52.2	—	—	[83]
Ni	γ-Al$_2$O$_3$-P	10	700	5	48,000	52.2→39.3	—	—	[83]
Co-Ni	CeO$_2$	—	600	10	12,000	77	80	—	[84]
Ni	MgO-ZrO$_2$	10	700	60	16,000	84.7	86.5	20	[85]
Ni	ZrO$_2$-RC-100	5	700	7	42,000	67.4→46.5	68.4→58.2	66.3	[86]
Ni	ZrO$_2$-ELTN	5	700	7	42,000	42.3→31.9	52.3→43.9	25.2	[86]
Ni	ZrO$_2$-Z-3215	5	700	7	42,000	62.2→45.3	69.5→58.4	38.3	[86]
Ni	ZrO$_2$(MK)	5	700	7	42,000	51.2→36.3	56.7→46.4	46.9	[86]
Ni	ZrO$_2$-O$_2$	10	750	10	24,000	78→64	86→73	—	[87]
Ni	ZrO$_2$	5	750	36	24,000	83→78	—	—	[87]

Application of High-Activity Bimetallic Catalysts

The introduction of a second metal to obtain a bimetallic Ni-based catalyst is also considered to be an effective and practical strategy to improve the performance of the CRM catalyst. The synergistic effect between Ni and the second metal can significantly improve the activity and carbon deposition resistance of the Ni-based catalyst [17,88–91].

In order to discuss the synergistic effect and the basic principle for improving the performance of the used catalyst, the researchers prepared a series of bimetallic Ni-based CRM catalysts. The results showed that Ni-Pt [92], Ni-Co [93], and Ni-Cu [94] showed better activity and carbon deposition resistance. In general, bimetallic Ni-based CRM catalysts include Ni-noble metals (Pt, Ru, etc.) and Ni-transition metals (Co, Fe, and Cu) [95,96]. Ni-noble metal bimetallic catalysts have three advantages: the promotion of reduction, surface modification, and surface reconstruction. Noble metals usually contribute to the reduction in NiO crystal, thereby increasing the number of active sites [97–100]. In terms of surface modification, the surface properties of Ni can be changed by adding a trace noble metal. In addition, the surface reconstruction of the bimetallic particles can be caused by temperature or adsorbate [101,102]. GARCIÁ-DIÉGUEZG et al. [103] prepared a Ni-Pt bimetallic catalyst for a CRM reaction. Compared with the Ni catalyst, the Ni-Pt bimetallic catalyst formed a Ni-Pt alloy with higher activity and lower carbon deposition. Although only a small amount of precious metals was added to the Ni-based catalyst, the production cost of the catalyst still increased. Therefore, some researchers doped transition metals such as Co, Fe, and Cu into a Ni-based catalyst to construct a CRM bimetallic catalyst to reduce the industrial production cost [104]. Co, Fe, and Cu have a strong synergistic effect in the bimetallic system. Of course, the specific effects of the three metals are different [105–109]. Some researchers discussed the effect of Ni-Co, Ni-Fe, and Ni-Cu bimetallic catalysts. The introduction of the second active component, Co or Cu, into the Ni-based catalyst helped to improve the catalytic activity and carbon deposition resistance [110].

The Ni-Co bimetallic catalyst shows a stronger synergistic effect [111–113]. Additionally, the Ni/Co ratio, which can adjust the surface composition of Ni-Co clusters, plays a crucial role in the Ni-Co bimetallic system [114–117]. Generally, a small amount of Co can optimize the adjustment process, while excessive Co will cause the catalyst to be oxidized. The promotive effect of Co is mainly due to its strong affinity for oxygen species, enhancing the ability to eliminate carbon deposition on the catalyst [111,118,119]. The Ni-Co/Al_2O_3 bimetallic catalyst showed high thermal stability at 800 °C and effectively inhibited the side reaction of RWGS [120]. Turap et al. [84] prepared a Ni-Co/CeO_2 bimetallic catalyst for CRM reaction and found that the strong oxygen affinity of Co and the strong oxygen storage capacity of CeO_2 were helpful in eliminating carbon deposition. As the Co/Ni ratio was up to 0.8, the catalyst presented better activity and stability. Li et al. [121] studied the catalytic performance of a bimetallic Ni-Co/Al_2O_3 catalyst for CRM and found that the addition of metal Co can form a Ni-Co alloy, increasing the activation energy of CH_4 dissociation, thus inhibiting the CH_4 cracking activity. At the same time, the addition of Co could improve the oxygen affinity of the catalyst and remove carbon deposition. Liang et al. [122] used a one-pot method to synthesize an attapulgite-derived MFI (ADM) zeolite-coated Ni-Co alloy. The results showed that the Ni-Co alloy existed stably in the CRM process, which was conducive to the formation of electron-rich Ni metal and significantly improved the fracture ability of the C-H bond. At the same time, ADM not only firmly anchors metal sites through pore structure or layered system, but also provides rich CO_2 adsorption/activation centers, realizing high CRM reaction activity and improving the anti-carbon deposition performance.

Cu can partly replace Ni to improve catalyst activity and carbon deposition resistance [110]. Song et al. [123] constructed a Ni-Cu bimetallic catalyst. The catalyst with a 0.25–0.50 Cu/Ni ratio showed good activity, stability, and carbon deposition resistance, while the catalyst with higher and lower Cu/Ni ratios would be deactivated due to serious carbon deposition. The excellent performance of the optimized Ni-Cu/Mg(Al)O catalyst was related to the synergistic effect between Ni and Cu. On the one hand, the alloying of

Ni and Cu inhibited the deep dissociation of methane, and the carbon species obtained were more easily gasified (carbon eliminated). On the other hand, Cu provided active sites for the dissociation of CO_2, leading to the formation of active oxygen species. The alloying of Ni and Cu reduced the decomposition rate of CH_4, promoted the dissociation of CO_2, and effectively inhibited carbon deposition. Other studies showed that [124], during the CRM reaction, the addition of Cu had a significant effect on the activity and anti-carbon deposition performance of the Ni/CeO_2 catalyst, and the formation of a Ni-O-Ce solid solution generated more oxygen vacancies, improving catalytic activity.

Fe has always played a certain role in promoting the CRM reaction. Both Fe and Ni are iron elements with similar element properties, and the two metals can be alloyed in a certain proportion to make a catalyst with good catalytic performance [93,125]. The research from Kim et al. [126] showed that the catalysts supported solely with Ni or Fe presented the problems of fast deactivation and a low conversion rate, respectively, while the bimetallic Ni-Fe catalyst showed good activity and stability in the CRM reaction. By further analysis, it was found that the promotion of Fe in a Ni-Fe alloy was due to the cracking of CH_4 on the active metal Ni to produce H_2 and carbon. A part of Fe reacts with CO_2 to generate FeO, which falls off from the alloy. Additionally, the carbon can react with FeO and be oxidized to generate CO. Then, FeO is reduced to Fe, which is the original Ni-Fe alloy. This decarburization reaction cycle is conducive to the reducing of the surface carbon on the catalyst. The anti-carbon deposition performance of different bimetallic catalysts is shown in Table 2.

Table 2. Carbon deposition resistance of bimetallic catalysts in DRM.

Catalysts	SV /(mL·g^{-1}·h^{-1})	Feed Ratio	Temperature /K	CH_4 Conversion Rate /%	CO_2 Conversion Rate /%	Carbon Deposition /%
Ru-Ni/Al_2O_3 [127]	60,000		1023	94.00	97.00	0.32
Co-Ni/CeO_2 [84]	30,000	CH_4:CO_2 = 1:1	1073	80.10	82.20	10.00
NiFe/Al_2O_3 [128]	12,000	CH_4:CO_2 = 1:1	823	26.60	37.80	2.30
NiCu/Al_2O_3 [129]	18,000	CH_4:CO_2:He = 1:1:8	923	65.00	64.34	6.40
Ni-Co/Al_2O_3 [130]	54,000	CH_4:CO_2:N_2 = 2:2:1	1023	96.10	92.20	1.00
NiPt/Al_2O_3 [100]		CH_4:CO_2:Ar = 45:45:10	1023	86.00	87.00	7.00

Selection of Support

The support is a very important part of a catalyst. In a CO_2-CH_4 reforming reaction, the commonly used supports are Al_2O_3, MgO, CeO_2, TiO_2, SiO_2, etc. Although the support itself has no activity in the reaction, it can change the overall performance of the catalyst. The physical and chemical properties of the support, such as surface morphology, pore structure, interaction with active component, and the resulting differences due to the support, such as surface–interface structure, surface composition, grain size, and dispersion of the active component, can affect the existence form of the active component precursor and the catalyst activity, selectivity, stability, and carbon deposition resistance. Many studies have pointed out that strong interaction between the support and the active component is conducive to improving the dispersion and sintering resistance of the active metal, resulting in a high carbon resistance performance.

The stronger the interaction between the support and the metal, the less likely the catalyst will be reduced. If it can be reduced under certain conditions, then the smaller the metal particles, the better the dispersion. The excessive surface acidity of the support leads to catalyst deactivation through methane decomposition. Similarly, excessive surface basicity leads to catalyst deactivation through the Boudouard reaction as well as through the formation of metal oxides [39]. Hao et al. [131] reported that a close combination of Ni and carrier caused by a strong metal-support interaction promoted the transfer of transition species at the interface and the transfer of electrons, leading to the transformation of non-inert carbon species in the reaction process and avoiding the forming of inert carbon deposition. At the same time, strong metal-support interaction can effectively inhibit the sintering and growth of Ni particles under the reaction conditions and can have a certain stabilizing effect on Ni particles, thereby improving the performance of the catalyst. Liu et al. [60] found through their research that Ni/CeO$_2$ was very active in the CRM reaction and that strong metal-support interaction enhanced the dissociation reaction activity of Ni to CH$_4$ and inhibited the formation of carbon deposition. Ruckenstein et al. [132] prepared a Ni/TiO$_2$ catalyst and found that there was a strong interaction between Ni and TiO$_2$, which led to the reduction in the free energy of the system. TiO$_x$ can promote the elimination of carbon to a certain extent, but TiO$_x$ molecules migrate on the surface during the reduction process, covering the active sites of Ni. A large amount of filamentous carbon was formed on Ni supported on CeO$_2$ and on CeO$_2$ doped with iso-valent Zr, while a negligible amount was formed on Ni supported on CeO$_2$ doped with aliovalent Sm or La (Figure 7). The ceria dopants can change the interaction of Ni with the support [133].

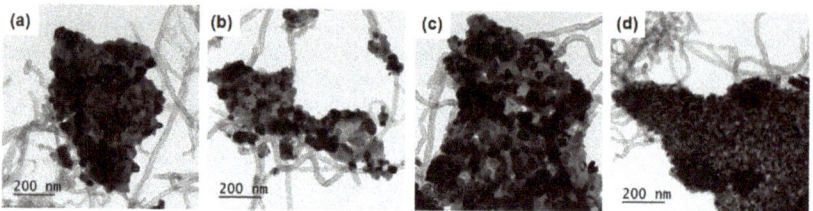

Figure 7. Bright field TEM images (a–d) [133].

The pore structure of the support has a great influence on the performance of the catalyst and has a limited domain effect on the active component. It has been found that micropores (<2nm) are not conducive to the dispersion of metal particles; mesopores (2–50 nm) can make the catalyst have a large specific surface area; macropores (>50nm) can promote the diffusion of reactant and product molecule, make gas molecules fully contact the catalyst, and increase the number of exposed Ni active sites [134]. Due to the limitation of the mesoporous structure of the support, Ni particles exist in the pores on the catalyst as much as possible, with high dispersion, which is conducive to strong metal-support interaction, thus reducing the formation of carbon deposition [135]. The catalyst with multistage pore structure has higher carbon capacity and a lower carbon deposition deactivation rate due to the addition of different levels of pores [134,136]. Du et al. [137] reported a CRM catalyst of HT-NiMgAl with a multistage pore structure. The multistage pore structure of this catalyst effectively increased the specific surface area of the catalyst, improving the dispersion of Ni particles; it could effectively inhibit carbon deposition due to its role in limiting the region of the active component.

After the reduction in the catalyst, the Ni atoms are easily sintered at high temperature, which leads to the reduction in the dispersion of Ni atoms on the surface and the increase in the concentration difference between the bulk Ni atoms and the surface Ni atoms so that the Ni atoms dissolved in the support will migrate to the surface under the promotion of the concentration gradient, supplementing the dispersion of the surface Ni atoms. Studies showed that the Co/MgO catalyst provides a strong Lewis alkaline environment due to the formation of solid solution CoMgO$_x$, which effectively stabilizes the Co nanoparticles on the surface of the support. Due to the alternating polar nanolayer structure of O^{2-} and Mg^{2+} and the existence of a large number of O^{2-} Lewis alkaline sites on the surface of MgO(111), the anti-sintering ability and anti-carbon deposition performance of the catalyst have been improved [138,139]. In addition, the support has a Lewis base, which can increase the alkalinity of the catalyst, promote the adsorption and dissociation of CO$_2$, and eliminate carbon deposition in the reaction. At the same time, the alkalinity of the support can inhibit the growth of Ni metal particles at high temperatures, thus improving the activity, stability, and carbon deposition resistance of the catalyst [140]. Jafarbegloo et al. [141] prepared a NiO-MgO catalyst and found that the strong Lewis base of MgO absorbed a large amount of carbon dioxide, improved the conversion rate of carbon dioxide, and eliminated carbon deposition on the catalyst surface.

Li et al. [142] prepared an iron-rich biomass-derived carbon for the CO_2-CH_4 reforming and found that it had higher activity than non-iron-rich carbon. Before 800 °C, the order of the iron-rich carbon promoting the reforming reaction was followed by Fe-C_2 (10% Fe content) > Fe-C_3 (20% Fe content) > Fe-C_1 (5% Fe content). After 800 °C, Fe-C_2 can still achieve the maximum CH_4 conversion rate. In addition, the catalytic activity of Fe-C_2 to CH_4 at 800 °C was better than that of other catalysts at higher temperature. By further measuring the carbon catalyst used, it was found that the weights of iron-rich carbon and non-iron-rich carbon increased by 0.2% and 0.9%, respectively. Therefore, it can be proved that the carbon deposition on the carbon catalyst is less, which effectively eliminates the carbon deposition. After the test, the iron-rich carbon had less carbon deposition, mainly in the form of filamentous carbon, which was more easily removed by carbon removal reaction.

Most biomass carbons are alkaline, and their ash contains a large amount of alkali metals (K, Na) and alkaline earth metals (Ca, Mg), which can promote the formation of alkaline sites, facilitate the adsorption and dissociation of CO_2, and inhibit the formation of carbon deposition. It has been reported that alkali/alkaline earth metals are one of the main reasons for biomass carbon to promote CO_2-CH_4 reforming [142]. Zhang et al. [143] studied the role of alkali/alkaline earth metals in tar reforming. The results showed that alkali/alkaline earth metals promoted the interaction between the active metal Ni and the carrier and inhibited the sintering of Ni. Alkali/alkaline earth metals cause more oxygen to be adsorbed on the surface of the catalyst, which has strong oxidizability. It can react with reaction intermediates or C, avoid the deposition of C on the catalyst, and inhibit carbon deposition [142–144]. San et al. [145] studied the role of alkali metal K and speculated that K can promote C gasification reaction and cover some active sites to inhibit CH_4 decomposition and reduce carbon deposition, but the coverage of active sites will also have a certain negative impact on the reforming reaction. Wu et al. [146] used CaO as the carrier to theoretically calculate that the presence of CaO adsorbed more CO_2 to the participate in the CO_2-CH_4 reforming and that CO_2 dissociated at the interface between Ni and CaO; in addition, the oxygen species produced by dissociation and carbon deposition on the surface of the catalyst generated CO, which extended the service life of the catalyst.

Application of Confined Catalyst

A confined catalyst can effectively confine the active center on the catalyst in different ways, which mainly include lattice limit, pore limit, core-shell limit, surface space limit, and multiple limits.

Lattice confinement can effectively anchor precious metal or non-precious metal on the regularly arranged spatial skeleton and can improve the dispersion of active centers. Ruitenbeek et al. [147] used a catalyst composed of a single iron atom in the lattice confined region; it had high activity and selectivity in the reaction and almost no carbon deposition. The surface confined catalyst had a high specific surface area, highly ordered pore structure, and narrow pore size distribution. Wang et al. [148] used dendritic mesoporous SiO_2 (DMS) as a carrier to prepare an alkali metal oxide modified low-temperature carbon deposition-resistant Ni-based catalyst and applied it to a reforming reaction, which showed excellent low-temperature carbon deposition resistance.

Kong et al. [149] prepared microporous molecular sieve S-1 with rich pores and high specific surface area, and effectively embedded the active component Ni in the pores and applied it to CO_2-CH_4 reforming. The results showed that the catalyst had excellent activity and stability at 650 °C and 0.5 MPa for 100 h. The thermogravimetric test of the catalyst after reaction did not find weight loss and indicated that the S-1-encapsulated Ni-based catalyst had excellent carbon deposition resistance. The main reason was that the catalyst channel effectively restricted the aggregation of Ni particles, which made the Ni disperse uniformly and reduced the size of the Ni.

Core-shell catalysts mainly include two types: one is the close contact type; the other is the eggshell type (the active component is separated from the shell). Zhang et al. [150] wrapped the perovskite $LaNiO_3$ nano-cube in the mesoporous silica shell to form a new core-shell structure catalyst, which was used in the CO_2-CH_4 reforming and showed excellent carbon deposition resistance. Compared with the eggshell catalyst, the core-shell catalyst had a contact interface between the core and shell, which resulted in enhanced interaction, inhibition of the movement of the active center, and reduction in the particle size. Liu et al. [151] designed a high-performance In-Ni@SiO_2 close-contact nanocore-shell catalyst. The In-Ni@SiO_2 catalyst had higher activity compared to the Ni@SiO_2 catalyst. CO_2 and CH_4 reacted at 800 °C for 430 h and still maintained 90% conversion. After reaction, compared with the other supported catalyst, they had less carbon deposition, better stability, and anti-carbon deposition performance.

Multiple restriction can limit the active center, reduce its exposure, and improve carbon deposition resistance on the catalyst. Wang et al. [152] prepared a Ni@La_2O_3/SiO_2 catalyst, and the results showed that an amorphous La_2O_3 layer was coated on the SiO_2, while small Ni nanoparticles were encapsulated in the La_2O_3 layer. As Ni nanoparticles were encapsulated in the La_2O_3 amorphous layer, it could effectively inhibit the formation of carbon deposition in CO_2-CH_4 reforming.

4.2.2. Eliminate Carbon Deposition from Process Condition Matching

The conversion rate of carbon dioxide and methane varies with the ratio of reaction gas, space velocity, reactor size, and catalyst dosage.

Selection of Operating Conditions (Temperature, Pressure, etc.)

Nematollahi et al. [153] conducted the same thermodynamic simulation under different pressures and found that the conversion rate of CH_4 and CO_2 and the amount ratio of the H_2/CO substances decreased significantly with the increase in operating pressure. This is due to the fact that the CRM reforming is a reaction with an increase in volume. The lower the pressure, the better the reaction. The high-pressure environment inhibits the conversion of the reactants. Some researchers conducted thermodynamic simulation on the influence of temperature, CH_4/CO_2 ratio, reaction pressure, and other oxidants on the formation of carbon deposition and proposed that high conversion and less carbon deposition could be obtained by operating at a high temperature, low pressure, and high CH_4/CO_2 ratio above 850 °C [17,154,155]. Bao et al. [156] prepared a NiCeMgAl double-porous (mesoporous–mesoporous) catalyst. When the space velocity was lower than 96,000 h^{-1}, the larger mesopores provides a fast transport channel for the reactants and product molecules. At a higher space velocity (such as 120,000 h^{-1}), the conversion rate was reduced because the reactants could not fully diffuse to the active center in the catalyst. The thermogravimetric analysis results showed that with the (Ni_{15}CeMgAl) the total weight loss of the dual porous catalyst after reaction was 16.8%, of which amorphous carbon accounted for 2.5%, carbon nanotubes accounted for 9%, graphite-like carbon accounted for 1%, and the others comprised the desorption of adsorbed small molecules. It was further found that carbon nanotubes could act as a carrier to continue the reaction and prolong the service life of the catalyst.

Adjustment and Matching of Reaction Gases

Carbon species deposited on the catalyst surface can usually be eliminated by the oxidation of CO_2 through the carbon elimination reaction (CO_2 + C→2CO). Therefore, the total amount of carbon deposition on the catalyst depends on the balance between methane cracking, carbon monoxide disproportionation, and the decarburization reaction [63], which can be considered from the two aspects of the inhibition of the carbon deposition reaction and the promotion of the decarburization reaction to improve the anti-carbon deposition performance of the catalyst. Of course, increasing the proportion of CO_2 can inhibit the formation of carbon deposition but increasing the proportion of CO_2 will promote the occurrence of side reactions and lead to increased separation costs in the later period. Therefore, the determination of the CO_2/CH_4 ratio should be combined with various factors. Mo et al. [49] found that with the increase in the CO_2/CH_4 ratio, the amount of carbon deposition on the catalyst surface gradually decreased, and the area and intensity of the high-temperature hydrogenation peak gradually weakened, indicating that low activity β carbon was significantly reduced due to the increase in the proportion of CO_2. The results also showed that the addition of CO_2 played an important role in preventing the transformation from active carbon to inactive carbon.

Adding steam or oxygen to the reaction for mixed reforming can also reduce carbon deposition on the catalyst. Li et al. [157] prepared a Ni/CeO_2-ZrO_2-Al_2O_3 catalyst, carried out a CO_2-CH_4 reforming reaction with and without steam, and measured the amount of carbon deposition. The results showed that the addition of steam to the reaction gas could significantly reduce the carbon deposition, improving the stability of the catalytic reaction. O'Connor et al. [158] found that the Ni/Al_2O_3 catalyst had high activity at 550–800 °C under the conditions of CO_2 reforming and the partial oxidation of methane. With the increase in the O_2 addition, almost no surface carbon deposition was found, but the activity of the catalyst decreased with time.

LI et al. [159] conducted a study employing the action of microwave-irradiated biological semi-coke; the experimental study of CO_2/steam-combined CH_4 reforming was carried out. The characteristics of the combined reforming reaction were examined, and the effects of the combined reforming reaction on the quality of the syngas, the loss of biochar, the surface characteristics, and the functional groups were discussed. The results showed that the combined reforming reaction could promote the conversion of the reaction gas, causing the the average value of the volume ratio of H_2/CO in the syngas to be within 90, the min reaction time to rise to 0.923, and the H_2/CO gas volume ratio to be closer to 1.

5. Conclusions and Prospect

CRM reforming not only promotes the utilization of CH_4 and CO_2 but also plays an important role in mitigating the greenhouse effect and reducing carbon emissions. It is an effective means of achieving carbon peaking and carbon neutralization and has good industrial value and application prospects. The key to the stable operation of the reaction is the construction of the catalyst, and the easy sintering of the active component and carbon deposition on the catalyst in the reaction is the core problem that needs to be solved urgently. As the preferred catalyst for this reaction, the Ni-based catalyst also faces the above problems. This paper briefly introduces the thermodynamics, kinetics, and reaction mechanism of the CRM reaction and focuses on the research progress of carbon deposition and carbon elimination on the used catalysts. The following prospects are put forward in terms of inhibiting carbon deposition in order to improve the activity and stability of the CRM catalyst:

1. More advanced characterization methods should be used to explore the reaction mechanism and carbon deposition mechanism of the CRM reaction on a Ni-based catalyst, and the reaction mechanism and anti-carbon deposition mechanism of the Ni-based catalyst should be further clarified.
2. By introducing different types of additives to regulate the number of alkaline sites on the surface of the catalyst, the adsorption performance of CO_2 may be enhanced, and more adsorbed oxygen may be generated; the gasification process of the carbon deposition may also be promoted.
3. By DFT or other calculations, the formation and elimination mechanism of carbon deposition can be discussed in depth, and the catalyst design scheme can correspondingly be optimized to inhibit carbon deposition.

Funding: This work was supported by the natural science foundation of Xinjiang Uyghur Autonomous Region (2022D01C23), the Ningxia natural science foundation (2022AAC03307), the high quality development special project for science and technology supporting industry from Changji (2022Z04), and the special project for the central government to guide local scientific and technological development (ZYYD2022C16).

Data Availability Statement: No new data were created or analyzed in this study. Data sharing is not applicable to this article.

Conflicts of Interest: The authors declare no conflict of interest.

References

1. Liu, Q.; Wang, S.J.; Zhao, G.M.; Yang, H.Y.; Yuan, M.; An, X.X.; Zhou, H.F.; Qiao, Y.Y.; Tian, Y.Y. CO_2 methanation over ordered mesoporous NiRu-doped CaO-Al_2O_3 nanocomposites with enhanced catalytic performance. *Int. J. Hydrogen Energy* **2018**, *43*, 239–250. [CrossRef]
2. Zain, M.M.; Mohamed, A.R. An overview on conversion technologies to produce value added products from CH_4 and CO_2 as major biogas constituents. *Renew. Sustain. Energy Rev.* **2018**, *98*, 56–63. [CrossRef]
3. Lougou, B.G.; Shuai, Y.; Chaffa, G.; Xing, H.; Tan, H.P.; Du, H.B. Analysis of CO_2 utilization into synthesis gas based on solar thermochemical CH_4-reforming. *J. Energy Chem.* **2019**, *28*, 61–72. [CrossRef]
4. Bradford, M.C.J.; Vannice, M.A. Catalytic reforming of methane with carbon dioxide over nickel catalysts II. Reaction kinetics. *Appl. Catal. A Gen.* **1996**, *142*, 97–122. [CrossRef]
5. Oezkara-Aydmoglu, S. Thermodynamic equilibrium analysis of combined carbon dioxide reforming with steam reforming of methane to synthesis gas. *Int. J. Hydrogen Energy* **2010**, *35*, 12821–12828. [CrossRef]
6. Kang, J.; He, S.; Zhou, W.; Shen, Z.; Li, Y.Y.; Chen, M.S.; Zhang, Q.H.; Wang, Y. Single-pass transformation of syngas into ethanol with high selectivity by triple tandem catalysis. *Nat. Commun.* **2020**, *11*, 827. [CrossRef]
7. Kawi, S.; Kathiraser, Y.; Ni, J.; Oemar, U.; Li, Z.W.; Saw, E.T. Progress in Synthesis of Highly Active and Stable Nickel-Based Catalysts for Carbon Dioxide Reforming of Methane. *ChemSusChem* **2015**, *8*, 3556–3575. [CrossRef]
8. Oyama, S.; Hacarlioglu, P.; Gu, Y.; Lee, D. Dry reforming of methane has no future for hydrogen production: Comparison with steam reforming at high pressure in standard and membrane reactors. *Int. J. Hydrogen Energy* **2012**, *37*, 10444–10450. [CrossRef]
9. Qin, Z.; Chen, J.; Xie, X.; Luo, X.; Su, T.M.; Ji, H.B. CO_2 reforming of CH_4 to syngas over nickel-based catalysts. *Environ. Chem. Lett.* **2020**, *18*, 997–1017. [CrossRef]
10. Abdulrasheed, A.; Jalil, A.A.; Gambo, Y.; Ibrahim, M.; Hambali, H.U.; Hamid, M.Y.S. A review on catalyst development for dry reforming of methane to syngas: Recent advances. *Renew. Sustain. Energy Rev.* **2019**, *108*, 175–193. [CrossRef]
11. Pakhare, D.; Spivey, J. A review of dry (CO_2) reforming of methane over noble metal catalysts. *Chem. Soc. Rev.* **2014**, *43*, 7813–7837. [CrossRef]
12. Li, W.Y.; Feng, J.; Xie, K.C.; Sun, Q. Study on carbon deposition performance of nickel catalyst for CH_4-CO_2 reforming reaction. *J. Fuel Chem. Technol.* **1997**, *25*, 460–464.
13. Li, X.C.; Li, S.G.; Yang, Y.F.; Wu, M.; He, F. Study on Coke Formation and Stability of Nickel-Based Catalysts in CO_2 Reforming of CH_4. *Catal. Lett.* **2007**, *118*, 59–63.
14. Ji, L.; Tang, S.; Zeng, H.C.; Lin, J.; Tan, K.L. CO_2 reforming of methane to synthesis gas over sol-gel-made Co/γ-Al_2O_3 catalysts from organometallic precursors. *Appl. Catal. A Gen.* **2001**, *207*, 247–255. [CrossRef]
15. Xu, Z.; Li, Y.; Zhang, J.Y.; Chang, L.; Zhou, R.Q.; Duan, Z.T. Bound-state Ni species-a superior form in Ni-based catalyst for CO_2-CH_4 reforming. *Appl. Catal. A Gen.* **2001**, *210*, 45–53. [CrossRef]
16. Yang, Y.L.; Xu, H.Y.; Li, W.Z. Pyrolysis and deposition performance of CH_4, C_2H_6 and C_2H_4 on Ni-based catalysts. *Acta Phys. Chim. Sin.* **2001**, *17*, 773–775. [CrossRef]
17. Zhang, J.; Wang, H.; Dalai, A.K. Development of stable bimetallic catalysts for carbon dioxide reforming of methane. *J. Catal.* **2007**, *249*, 300–310. [CrossRef]
18. Fraenkel, D.; Levitan, R.; Levy, M. A solar thermochemical pipe based on the CO_2-CH_4 (1:1) system. *Int. J. Hydrogen Energy* **1986**, *11*, 267–277. [CrossRef]
19. Wang, S.B.; Lu, G.Q.; Millar, G.J. Carbon Dioxide Reforming of Methane to Produce Synthesis Gas over Metal-Supported Catalysts: State of the Art. *Energy Fuels* **1996**, *10*, 896–904. [CrossRef]
20. Abdullah, B.; Ghani, N.A.A.; Vo, D.-V.N. Recent advances in dry reforming of methane over Ni-based catalysts. *J. Clean. Prod.* **2017**, *162*, 170–185. [CrossRef]
21. Al-Fatesh, A.; Singh, S.K.; Kanade, G.S.; Atia, H.; Fakeeha, A.H.; Ibrahim, A.A.; El-Toni, A.M.; Labhasetwar, N.K. Rh promoted and ZrO_2/Al_2O_3 supported Ni/Co based catalysts: High activity for CO_2 reforming, steam-CO_2 reforming and oxy-CO_2 reforming of CH_4. *Int. J. Hydrogen Energy* **2018**, *43*, 12069–12080. [CrossRef]
22. Oemar, U.; Kathiraser, Y.; Mo, L.; Ho, X.K.; Kawi, S. CO_2 reforming of methane over highly active La-promoted Ni supported on SBA-15 catalysts: Mechanism and kinetic modelling. *Catal. Sci. Technol.* **2016**, *6*, 1173–1186. [CrossRef]

23. Assabumrungrat, S.; Charoenseri, S.; Laosiripojana, N.; Kiatkittipong, W.; Praserthdam, P. Effect of oxygen addition on catalytic performance of Ni/SiO$_2$·MgO toward carbon dioxide reforming of methane under periodic operation. *Int. J. Hydrogen Energy* **2009**, *34*, 6211–6220. [CrossRef]
24. Wang, Z.J.; Song, H.; Liu, H.; Ye, J. Coupling of Solar Energy and Thermal Energy for Carbon Dioxide Reduction: Status and Prospects. *Angew. Chem. Int. Ed.* **2020**, *59*, 8016–8035. [CrossRef]
25. Paksoy, A.I.; Caglayan, B.S.; Aksoylu, A.E. An in situ FTIR-DRIFTS study on CDRM over Co-Ce/ZrO$_2$: Active surfaces and mechanistic features. *Int. J. Hydrogen Energy* **2020**, *45*, 12822–12834. [CrossRef]
26. Bodrov, N.N.; Apelbaum, L.O.; Temkin, M.I. Kinetics of the reaction of methane with steam on the surface of nickel. *Kinet. Catal.* **1964**, *5*, 614–621. [CrossRef]
27. Rostrupnielsen, J.R.; Hansen, J.H.B. CO$_2$-Reforming of Methane over Transition Metals. *J. Catal.* **1993**, *144*, 38–49. [CrossRef]
28. Osaki, T.; Masuda, H.; Mori, T. Intermediate hydrocarbon species for the CO$_2$-CH$_4$ reaction on supported Ni catalysts. *Catal. Lett.* **1994**, *29*, 33–37. [CrossRef]
29. Bradford, M.C.J.; Vannice, M.A. CO$_2$ reforming of CH$_4$ over supported Pt catalysts. *J. Catal.* **1998**, *173*, 157–171. [CrossRef]
30. Wei, J.M.; Iglesia, E. Isotopic and kinetic assessment of the mechanism of reactions of CH$_4$ with CO$_2$ or H$_2$O to form synthesis gas and carbon on nickel catalysts. *J. Catal.* **2004**, *224*, 370–383. [CrossRef]
31. Wang, S.G.; Liao, X.Y.; Jia, H.; Cao, D.B.; Li, Y.W.; Wang, J.; Jiao, H. Kinetic aspect of CO$_2$ reforming of CH$_4$ on Ni(111): A density functional theory calculation. *Surf. Sci.* **2007**, *601*, 1271–1284. [CrossRef]
32. Wang, S.G.; Liao, X.Y.; Cao, D.B.; Li, Y.W.; Wang, J.G.; Jiao, H.J. Formation of Carbon Species on Ni(111): Structure and Stability. *J. Phys. Chem. C* **2007**, *111*, 10894–10903. [CrossRef]
33. Freund, H.J.; Messmer, R.P. On the bonding and reactivity of CO$_2$ on metal surfaces. *Surf. Sci. Lett.* **1986**, *172*, 1–30. [CrossRef]
34. Muhammad, A.; Tahir, M.; Al-Shahrani, S.S.; Ali, A.M.; Rather, S.U. Template free synthesis of graphitic carbon nitride nanotubes mediated by lanthanum (La/g-CNT) for selective photocatalytic CO$_2$ reduction via dry reforming of methane (DRM) to fuels. *Appl. Surf. Sci.* **2020**, *504*, 144177. [CrossRef]
35. Xie, W.; Liang, D.; Li, L.; Qu, S.J.; Tao, W. Surface chemical ploperties and pore structure of the activated coke and their effects on the denitrification activity of selective catalytic reduction. *Int. J. Coal Sci. Technol.* **2019**, *6*, 595–602. [CrossRef]
36. Khavarian, M.; Chai, S.P.; Mohamed, A.R. Direct use of as-synthesized multi-walled carbon nanotubes for carbon dioxide reforming of methane for prodacing synthesis gas. *Chem. Eng. J.* **2014**, *257*, 200–208. [CrossRef]
37. He, L.; Hu, S.; Yin, X.; Jun, X.; Han, H.D.; Li, H.J.; Ren, Q.Q.; Su, S.; Wang, Y.; Xiang, J. Promoting effects of Fe-Ni alloy on co-production of H$_2$ and carbon nanotubes during steam reforming of biomass tar over Ni-Fe/α-Al$_2$O$_3$. *Fuel* **2020**, *276*, 118116. [CrossRef]
38. Ma, Q.X.; Wang, D.; Wu, M.B.; Zhao, T.S.; Yoneyama, Y.; Tsubaki, N. Effect of catalyticsite position: Nickel nanocatalyst selectively loaded inside or outside carbon nanotubes for methane dry reforming. *Fuel* **2013**, *108*, 430–438. [CrossRef]
39. Das, S.; Sengupta, M.; Patel, J.; Bordoloi, A. A study of the synergy between support surface properties and catalyst deactivation for CO$_2$ reforming over supported Ni nanoparticles. *Appl. Catal. A Gen.* **2017**, *545*, 113–126. [CrossRef]
40. Zhang, M.; Zhang, J.F.; Wu, Y.Q.; Pan, J.X.; Zhang, Q.D.; Tan, Y.S.; Han, Y.Z. Insight into the effects of the oxygen species over Ni/ZrO$_2$ catalyst surface on methane reforming with carbon dioxide. *Appl. Catal. B Environ.* **2019**, *244*, 427–437. [CrossRef]
41. Jin, B.T.; Li, S.G.; Liang, X.H. Enhanced activity and stability of MgO-promoted Ni/Al$_2$O$_3$ catalyst for dry reforming of methane: Role of MgO. *Fuel* **2021**, *284*, 119082. [CrossRef]
42. Wang, Y.N.; Zhang, R.J.; Yan, B.H. Ni/Ce$_{0.9}$Eu$_{0.1}$O$_{1.95}$ with enhanced coke resistance for dry reforming of methane. *J. Catal.* **2022**, *407*, 77–89. [CrossRef]
43. Burghaus, U. Surface science perspective of carbon dioxide chemistry-Adsorption kinetics and dynamics of CO$_2$ on selected model surfaces. *Catal. Today* **2009**, *148*, 212–220. [CrossRef]
44. Pan, Y.X.; Liu, C.J.; Wiltowski, T.S.; Ge, Q.F. CO$_2$ adsorption and activation over γ-Al$_2$O$_3$-supported transition metal dimers: A density functional study. *Catal. Today* **2009**, *147*, 68–76. [CrossRef]
45. Tsipouriari, V.A.; Verykios, X.E. Carbon and Oxygen Reaction Pathways of CO$_2$ Reforming of Methane over Ni/La$_2$O$_3$ and Ni/Al$_2$O$_3$ Catalysts Studied by Isotopic Tracing Techniques. *J. Catal.* **1999**, *187*, 85–94. [CrossRef]
46. Daza, C.E.; Gallego, J.; Mondragon, F.; Moreno, S.; Molina, R. High stability of Ce-promoted Ni/Mg-Al catalysts derived from hydrotalcites in dry reforming of methane. *Fuel Guildf.* **2010**, *89*, 592–603. [CrossRef]
47. Liu, C.J.; Ye, J.Y.; Jiang, J.J.; Pan, Y.X. Progresses in the preparation of coke resistant Ni-based catalyst for steam and CO$_2$ reforming of methane. *ChemSusChem* **2011**, *3*, 529–541.
48. Ferreira-Aparicio, P.; Fernandez-Garcia, M.; Guerrero-Ruiz, A.; Rodriguez-Ramos, I. Evaluation of the role of the metal-support interfacial centers in the dry reforming of methane on alumina-supported rhodium catalysts. *J. Catal.* **2000**, *190*, 296–308. [CrossRef]
49. Mo, W.L.; Ma, F.Y.; Liu, J.M.; Zhong, M.; Nulahong, A.S. Study on CO$_2$-CH$_4$ reforming reaction carbon deposition on Ni-Al$_2$O$_3$ catalyst based on programmed hydrogenation characterization. *J. Fuel Chem. Technol.* **2019**, *47*, 549–557.
50. Mo, W.L.; Ma, F.Y.; Liu, Y.E.; Liu, J.M.; Aisha, N. Preparation of porous Al$_2$O$_3$ by template method and its application in Ni-based catalyst for CH$_4$/CO$_2$ reforming to produce syngas. *Int. J. Hydrogen Energy* **2015**, *40*, 16147–16158. [CrossRef]
51. Wang, C.Z.; Sun, N.N.; Wei, W.; Zhang, Y.X. Carbon intermediates during CO$_2$ reforming of methane over Ni-CaO-ZrO$_2$ catalysts: Atemperature-programmed surface reaction study. *Int. J. Hydrogen Energy* **2016**, *41*, 19014–19024. [CrossRef]

52. Bodrov, I.M.; Apelbaum, L.O. Reaction kinetics of methane and carbon dioxide on a nickel surface. *Kinet. Catal.* **1967**, *8*, 379.
53. Li, D.L.; Xu, S.P.; Song, K.; Chen, C.Q.; Zhan, Y.Y.; Jiang, L.L. Hydrotalcite-derived Co/Mg(Al)O as a stable and coke-resistant catalystfor low-temperature carbon dioxide reforming of methane. *Appl. Catal. A Gen.* **2018**, *552*, 21–29. [CrossRef]
54. Wang, R.; Xu, H.Y.; Liu, X.B.; Ge, Q.J.; Li, W.Z. Role of redox couples of $Rh^0/Rh^{\delta+}$ and Ce^{4+}/Ce^{3+} in CO_2-CH_4 reforming over Rh-CeO_2/Al_2O_3 catalyst. *Appl. Catal. A Gen.* **2006**, *305*, 204–210. [CrossRef]
55. Ruckenstein, E.; Wang, H.Y. Carbon deposition and catalytic deactivation during CO_2 reforming of CH_4 over Co/γ-Al_2O_3 catalysts. *J. Catal.* **2002**, *205*, 289–293. [CrossRef]
56. Liang, T.Y.; Lin, C.Y.; Chou, F.C.; Wang, M.Q.; Tsai, D.H. Gas-phase synthesis of Ni-CeO_x hybrid nanopaticles and their synergistic catalysis for simultaneous reforming of methane and carbon dioxide to syngas. *J. Phys. Chem. C* **2018**, *122*, 11789–11798. [CrossRef]
57. Kuijpers, E.G.M.; Breedijk, A.K.; Van Der Wal, W.J.J.; Geus, J.W. Chemisorption of Methane on Ni/SiO_2 Catalysts and Reactivity of the Chemisorption Products toward Hydrogen. *J. Catal.* **1983**, *81*, 429–439. [CrossRef]
58. Osaki, T.; Masuda, H.; Horiuchi, T.; Mori, T. Highly hydrogen-deficient hydrocarbon species for the CO_2-reforming of CH_4 on Co/Al_2O_3 catalyst. *Catal. Lett.* **1995**, *34*, 59–63. [CrossRef]
59. Wang, Y.; Yao, L.; Wang, Y.N.; Wang, S.H.; Zhao, Q.; Mao, D.H.; Hu, C.W. Low-Temperature Catalytic CO_2 Dry Reforming of Methane on Ni-Si/ZrO_2 Catalyst. *ACS Catal.* **2018**, *8*, 6495–6506. [CrossRef]
60. Liu, Z.Y.; Lustemberg, P.; Gutierrez, R.A.; Carey, J.J.; Palomino, R.M.; Vorokhta, M.; Grinter, D.C.; Ramirez, P.J.; Matolin, V.; Nolan, M.; et al. In situ Investigation of Methane Dry Reforming on M-CeO_2(111) {M=Co, Ni, Cu} Surfaces: Metal-Support Interactions and the activation of C-H bonds at Low Temperature. *Angew. Chem. Int. Ed.* **2017**, *56*, 13041–13046. [CrossRef]
61. Liu, Z.Y.; Grinter, D.C.; Lustemberg, P.G.; Nguyen-Phan, T.-D.; Zhou, Y.H.; Luo, S.; Waluyo, I.; Crumlin, E.J.; Stacchiola, D.J.; Zhou, J.; et al. Dry Reforming of Methane on a Highly-Active Ni-CeO Catalyst: Effects of Metal-Support Interactions on C-H Bond Breaking. *Angew. Chem. Int. Ed.* **2016**, *55*, 7455–7459. [CrossRef]
62. Horiuchi, T.; Sakuma, K.; Fukui, T.; Kubo, Y.; Osaki, T.; Mori, T. Suppression of carbon deposition in the CO_2-reforming of CH_4 by adding basic metal oxides to a Ni/Al_2O_3 catalyst. *Appl. Catal. A* **1996**, *144*, 111–120. [CrossRef]
63. Gallego, G.S.; Batiot-Dupeyrat, C.; Barraault, J.; Florez, E.; Mondragon, F. Dry reforming of methane over $LaNi_{1-y}B_yO_{3\pm\delta}$ (B=Mg, Co) perovskites used as catalyst precursor. *Appl. Catal. A Gen.* **2008**, *334*, 251–258. [CrossRef]
64. Garcia-Dieguez, M.; Pieta, I.S.; Herrera, M.C.; Larrubia, M.A.; Malpartida, I.; Alemany, L.J. Transient study of the dry reforming of methane over Pt supported on different γ-Al_2O_3. *Catal. Today* **2010**, *149*, 380–387. [CrossRef]
65. Lucreda, F.; Assafj, M.; Assafe, M. Methane Conversion Reactions on Ni Catalysts Promoted with Rh: Influence of Support. *Appl. Catal. A Gen.* **2011**, *400*, 156–165. [CrossRef]
66. Xu, J.K.; Zhou, W.; Wang, J.H.; Li, Z.J.; Ma, J.X. Characterization and analysis of carbon deposited during the dry reforming of methane over Ni/La_2O_3/Al_2O_3 catalysts. *Chin. J. Catal.* **2009**, *30*, 1076–1084. [CrossRef]
67. Crnivec IG, O.; Djinovic, P.; Erjavec, B.; Pintar, A. Effect of synthesis parameters on morphology and activity of bimetallic catalysts in CO_2-CH_4 reforming. *Chem. Eng. J.* **2012**, *207*, 299–307. [CrossRef]
68. Liu, Z.C.; Zhou, J.; Cao, K.; Yang, W.M.; Gao, H.X. Highly dispersed nickel loaded on mesoporous silica: One-spot synthesis strategy and high performance as catalysts for methane reforming with carbon dioxide. *Appl. Catal. B Environ.* **2012**, *125*, 324–330. [CrossRef]
69. Guo, Y.P.; Feng, J.; Li, W.Y. Effect of the Ni size on CO_2-CH_4 reforming over Ni/MgO catalyst: A DFT study. *Chin. J. Chem. Eng.* **2017**, *25*, 1442–1448. [CrossRef]
70. Mo, W.L.; Ma, F.Y.; Ma, Y.Y.; Fan, X. The optimization of Ni-Al_2O_3 catalyst with the addition of La_2O_3 for CO_2-CH_4 reforming to produce syngas. *Int. J. Hydrogen Energy* **2019**, *44*, 24510–24524. [CrossRef]
71. Mo, W.L.; Ma, F.Y.; Liu, Y.E.; Liu, J.M.; Zhong, M.; Nulahong, A.S. Effect of preparation method on the catalytic performance of Ni-Al_2O_3 catalyst in CO_2-CH_4 reforming reaction. *J. Fuel Chem. Technol.* **2015**, *43*, 1083–1091.
72. Mo, W.L.; Ma, F.Y.; Liu, Y.E.; Liu, J.M.; Zhong, M.; Nulahong, A.S. Effect of roasting temperature on the performance of NiO/γ-Al_2O_3 catalyst for CO_2-CH_4 reforming syngas. *J. Inorg. Mater.* **2016**, *31*, 234–240.
73. Ding, R.G.; Yan, A.F. Structure characterization of the Co and Ni catalysts for carbon dioxide reforming of methane. *Catal. Today* **2001**, *68*, 135–143. [CrossRef]
74. Osaki, T.; Mori, T. Role of Potassium in Carbon-Free CO_2 Reforming of Methane on K-Promoted Ni/Al_2O_3 Catalysts. *J. Catal.* **2001**, *204*, 89–97. [CrossRef]
75. Wang, H.H.; Mo, W.L.; He, X.Q.; Fan, X.; Ma, F.Y.; Liu, S.; Tax, D. Effect of Ca Promoter on the Structure, Performance, and Carbon Deposition of Ni-Al_2O_3 Catalyst for CO_2-CH_4 Reforming. *ACS Omega* **2020**, *5*, 28955–28964. [CrossRef]
76. Wang, J.B.; Kuo, L.E.; Huang, T.J. Study of carbon dioxide reforming of methane over bimetallic Ni-Cr/yttria-doped ceria catalysts. *Appl. Catal. A Gen.* **2003**, *249*, 93–105. [CrossRef]
77. Potdar, H.S.; Roh, H.S.; Jun, K.W. Carbon Dioxide Reforming of Methane Over Co-precipitated Ni-CeO_2, Ni-ZrO_2, Ni-Ce-ZrO_2 Catalysts. *Catal. Today* **2004**, *93–95*, 39–44.
78. Roh, H.S.; Jun, K.W.; Baek, S.C.; Park, S.E. A Highly Active and Stable Catalyst for Carbon Dioxide Reforming of Methane: Ni/Ce-ZrO_2/θ-Al_2O_3. *Catal. Lett.* **2002**, *81*, 147–151.
79. Huang, X.J.; Mo, W.L.; He, X.Q.; Fan, X.; Ma, F.Y.; Tax, D. Effects of Promoters on the Structure, Performance, and Carbon Deposition of Ni-Al_2O_3 Catalysts for CO_2-CH_4 Reforming. *ACS Omega* **2021**, *6*, 16381–16390. [CrossRef]

80. Wang, N.; Yu, X.P.; Wang, Y.; Chu, W.; Liu, M. A comparison study on methane dry reforming with carbon dioxide over LaNiO$_3$ perovskite catalysts supported on mesoporous SBA-15, MCM-41 and silica carrier. *Catal. Today* **2013**, *212*, 98–107. [CrossRef]
81. Rashid, M.U.; Daud, W.M.A.W. Microemulsion based synthesis of Ni/MgO catalyst for dry reforming of methane. *RSC Adv.* **2016**, *6*, 38277–38289. [CrossRef]
82. Adans, Y.F.; Ballarini, A.D.; Martins, A.R.; Coelho, R.E.; Carvalho, L.S. Performance of nickel supported on gamma-Alumina obtained by aluminum recycling for methane dry reforming. *Catal. Lett.* **2017**, *147*, 2057–2066. [CrossRef]
83. Sun, J.W.; Wang, S.; Guo, Y.; Li, M.Z.; Zou, H.K.; Wang, Z.J. Carbon dioxide reforming of methane over nanostructured Ni/Al$_2$O$_3$ catalysts. *Catal. Commun.* **2018**, *104*, 53–56. [CrossRef]
84. Turap, Y.S.; Wang, I.W.; Fu, T.T.; Wu, Y.M.; Wang, Y.D.; Wang, W. Co-Ni alloy supported on CeO$_2$ as a bimetallic catalyst for dry reforming of methane. *Int. J. Hydrogen Energy* **2020**, *45*, 6538–6548. [CrossRef]
85. Huang, X.H.; Ji, Y.J.; Wei, T.; Jia, L.C.; Yan, D.; Li, J. High performance and stable mesoporous MgO-ZrO supported Ni catalysts for dry reforming of methane-ScienceDirect. *Curr. Res. Green Sustain. Chem.* **2021**, *4*, 100183. [CrossRef]
86. Ibrahim, A.A.; Fakeeha, A.H.; Abasaeed, A.E.; Al-Fatesh, A.S. Dry reforming of methane using Ni catalyst supported on ZrO$_2$, The effect of different sources of Zirconia. *Catalysts* **2021**, *11*, 827. [CrossRef]
87. Zhang, M.; Zhang, J.F.; Zhou, Z.L.; Zhang, Q.D.; Tan, Y.S.; Han, Y.Z. Effects of calcination atmosphere on the performance of the co-precipitated Ni/ZrO$_2$ catalyst in dry reforming of methane. *Can. J. Chem. Eng.* **2021**, *100*, 172–183.
88. Liu, D.P.; Quek, X.Y.; Cheo WN, E.; Lau, R.; Borgna, A.; Yang, Y. MCM-41 supported nickel-based bimetallic catalysts with superior stability during carbon dioxide reforming of methane: Effect of strong metal-support interaction. *J. Catal.* **2009**, *266*, 380–390. [CrossRef]
89. Bian, Z.F.; Das, S.; Wai, M.H.; Hongmanorom, P.; Kawi, S. A Review on Bimetallic Nickel-Based Catalysts for CO$_2$ Reforming of Methane. *ChemPhysChem* **2017**, *18*, 3117–3134. [CrossRef]
90. Mahboob, S.; Haghighi, M.; Rahmani, F. Sonochemically preparation and characterization of bimetallic Ni-Co/Al$_2$O$_3$-ZrO$_2$ nanocatalyst: Effects of ultrasound irradiation time and power on catalytic properties and activity in dry reforming of CH$_4$. *Ultrason. Sonochemistry* **2017**, *38*, 38–49. [CrossRef]
91. Dam, A.H.; Wang, H.M.; Niri, R.D.; Yu, X.F.; Walmsley, J.C.; Holmen, A.; Yang, J.; Chen, D. Methane Activation on Bimetallic Catalysts: Properties and Functions of Surface Ni-Ag Alloy. *ChemCatChem* **2019**, *11*, 3401–3412. [CrossRef]
92. Pawelec, B.; Damyanova, S.; Arishtirova, K.; Fierro, J.L.G.; Petrov, L. Structural and surface features of PtNi catalysts for reforming of methane with CO$_2$. *Appl. Catal. A Gen.* **2007**, *323*, 188–201. [CrossRef]
93. San-Jose-Alonso, D.; Juan-Juan, J.; Illan-Gomez, M.J.; Roman-Martinez, M.C. Ni, Co and bimetallic Ni-Co catalysts for the dry reforming of methane. *Appl. Catal. A Gen.* **2009**, *371*, 54–59. [CrossRef]
94. Wu, T.; Zhang, Q.; Cai, W.Y.; Zhang, P.; Song, X.F.; Sun, Z.; Cao, L. Phyllosilicate evolved hierarchical Ni-and Cu-Ni/SiO$_2$ nanocomposites for methane dry reforming catalysis. *Appl. Catal. A Gen.* **2015**, *503*, 94–102. [CrossRef]
95. Liu, H.L.; Nosheen, F.; Wang, X. Noble metal alloy complex nanostructures: Controllable synthesis and their electrochemical property. *Chem. Soc. Rev.* **2015**, *44*, 3056–3078. [CrossRef]
96. De, S.; Zhang, J.; Luque, R.; Yan, N. Ni-based bimetallic heterogeneous catalysts for energy and environmental applications. *Energy Environ. Sci.* **2016**, *9*, 3314–3347. [CrossRef]
97. Yu, X.P.; Zhang, F.B.; Wang, N.; Hao, S.X.; Chu, W. Plasma-Treated Bimetallic Ni-Pt Catalysts Derived from Hydrotalcites for the Carbon Dioxide Reforming of Methane. *Catal. Lett.* **2014**, *144*, 293–300. [CrossRef]
98. Hou, T.F.; Lei, Y.S.; Zhang, S.Y.; Zhang, J.H.; Cai, W.J. Ethanol dry reforming for syngas production over Ir/CeO$_2$ catalyst. *J. Rare Earths* **2015**, *33*, 42–45. [CrossRef]
99. Ma, Q.X.; Sun, J.; Gao, X.H.; Zhang, J.L.; Zhao, T.S.; Yoneyama, Y.; Tsubaki, N. Ordered mesoporous alumina-supported bimetallic Pd-Ni catalysts for methane dry reforming reaction. *Catal. Sci. Technol.* **2016**, *6*, 6542–6550. [CrossRef]
100. Oemar, U.; Hidajat, K.; Kawi, S. High catalytic stability of Pd-Ni/Y$_2$O$_3$ formed by interfacial Cl for oxy-CO$_2$ reforming of CH$_4$. *Catal. Today* **2017**, *281*, 276–294. [CrossRef]
101. Menning, C.A.; Chen, J.G. Thermodynamics and kinetics of oxygen-induced segregation of 3d metals in Pt-3d-Pt(111) and Pt-3d-Pt(100) bimetallic structures. *J. Chem. Phys.* **2008**, *128*, 164703. [CrossRef] [PubMed]
102. Menning, C.A.; Chen, J.G. General trend for adsorbate-induced segregation of subsurface metal atoms in bimetallic surfaces. *J. Chem. Phys.* **2009**, *130*, 174709. [CrossRef] [PubMed]
103. Garcia-Dieguez, M.; Pieta, I.S.; Herrera, M.C.; Larrubia, M.A.; Alemany, L.J. Nanostructured Pt- and Ni-based catalysts for CO$_2$-reforming of methane. *J. Catal.* **2010**, *270*, 136–145. [CrossRef]
104. Fan, M.S.; Abdullah, A.Z.; Bhatia, S. Utilization of greenhouse gases through carbon dioxide reforming of methane over Ni-Co/MgO-ZrO$_2$: Preparation, characterization and activity studies. *Appl. Catal. B Environ.* **2010**, *100*, 365–377. [CrossRef]
105. Qin, Z.Z.; Su, T.M.; Ji, H.B.; Jiang, Y.X.; Liu, R.W.; Chen, J.H. Experimental and theoretical study of the intrinsic kinetics for dimethyl ether synthesis from CO$_2$ over Cu-Fe-Zr/HZSM-5. *Aiche J.* **2015**, *61*, 1613–1627. [CrossRef]
106. Su, T.M.; Qin, Z.Z.; Ji, H.B.; Jiang, Y.X.; Huang, G. Recent advances in the photocatalytic reduction of carbon dioxide. *Environ. Chem. Lett.* **2016**, *14*, 99–112. [CrossRef]
107. Abukhadra, M.R.; Dardir, F.M.; Shaban, M.; Ahmed, E.A.; Soliman, M.F. Spongy Ni/Fe carbonate-fluorapatite catalyst for efficient conversion of cooking oil waste into biodiesel. *Environ. Chem. Lett.* **2017**, *16*, 665–670. [CrossRef]

108. Chen, L.; Huang, X.Y.; Tang, M.; Zhou, D.; Wu, F. Rapid dephosphorylation of glyphosate by Cu-catalyzed sulfite oxidation involving sulfate and hydroxyl radicals. *Environ. Chem. Lett.* **2018**, *16*, 1507–1511. [CrossRef]
109. Kumar, N.S.; Reddy, B.V.; Babu, M.S. Rapid synthesis of mono/bimetallic (Zn/Co/Zn-Co) zeolitic imidazolate frameworks at room temperature and evolution of their CO_2 uptake capacity. *Environ. Chem. Lett.* **2019**, *17*, 447–454.
110. Ray, K.; Sandupatla, A.S.; Deo, G. Activity and stability descriptors of Ni-based alloy catalysts for dry reforming of methane: A density functional theory study. *Int. J. Quantum Chem.* **2021**, *121*, e26580. [CrossRef]
111. Gonzalez-Delcruz, V.M.; Perenigue, R.; Ternero, F.; Holgado, J.P.; Caballero, A. In Situ XAS Study of Synergic Effects on Ni-Co/ZrO_2 Methane Reforming Catalysts. *J. Phys. Chem. C* **2012**, *116*, 2919–2926. [CrossRef]
112. Yu, M.; Zhu, K.; Liu, Z.; Xiao, H.; Deng, W.; Zhou, X. Carbon dioxide reforming of methane over promoted $Ni_xMg_{1-x}O(111)$ platelet catalyst derived from solvothermal synthesis. *Appl. Catal. B Environ.* **2014**, *148*, 177–190. [CrossRef]
113. Fan, X.; Liu, Z.; Zhu, Y.A.; Tong, G.S.; Zhang, J.D.; Engelbrekt, C.; Ulstrup, J.; Zhu, K.; Zhou, X.G. Tuning the composition of metastable $Co_xNi_yMg_{100-x-y}(OH)(OCH_3)$ nanoplates for optimizing robust methane dry reforming catalyst. *J. Catal.* **2015**, *330*, 106–119. [CrossRef]
114. Ay, H.; Uner, D. Dry reforming of methane over CeO_2 supported Ni, Co and Ni-Co catalysts. *Appl. Catal. B Environ.* **2015**, *179*, 128–138. [CrossRef]
115. Tsoukalou, A.; Imtiaz, Q.; Kim, S.M.; Abdala, P.M.; Yoon, S.; Muller, C.R. Dry-reforming of methane over bimetallic Ni-M/La_2O_3 (M=Co, Fe): The effect of the rate of $La_2O_2CO_3$ formation and phase stability on the catalytic activity and stability. *J. Catal.* **2016**, *343*, 208–214. [CrossRef]
116. Gao, X.Y.; Tan, Z.W.; Hidajat, K.; Kawi, S. Highly reactive Ni-Co/SiO_2 bimetallic catalyst via complexation with oleylamine/oleic acid organic pair for dry reforming of methane. *Catal. Today* **2017**, *281*, 250–258. [CrossRef]
117. Xu, L.L.; Wang, F.G.; Chen, M.; Fan, X.L.; Yang, H.M.; Nie, D.Y.; Qi, L. Alkaline-promoted Co-Ni bimetal ordered mesoporous catalysts with enhanced coke-resistant performance toward CO_2 reforming of CH_4. *J. CO2 Util.* **2017**, *18*, 1–14. [CrossRef]
118. Takanabe, K.; Nagaoka, K.; Nariai, K.; Aika, K. Influence of reduction temperature on the catalytic behavior of Co/TiO_2 catalysts for CO_2-CH_4 reforming and its relation with titania bulk crystal structure. *J. Catal.* **2005**, *230*, 75–85. [CrossRef]
119. Takanabe, K.; Nagaoka, K.; Nariai, K.; Aiak, K. Titania-supported cobalt and nickel bimetallic catalysts for carbon dioxide reforming of methane. *J. Catal.* **2005**, *232*, 268–275. [CrossRef]
120. Wu, Z.X.; Yang, B.; Miao, S.; Liu, W.; Xie, J.L.; Lee, S.; Pellin, M.J.; Xiao, D.Q.; Su, D.S.; Ma, D. Lattice strained Ni-Co alloy as high-performance catalyst for catalytic dry-reforming of methane. *ACS Catal.* **2019**, *9*, 2693–2700. [CrossRef]
121. Li, B.; Yuan, X.Q.; Li, L.Y.; Wang, X.J.; Li, B.T. Stabilizing Ni-Co alloy on bimodal mesoporous alumina to enhance carbon resistance for dry reforming of methane. *Ind. Eng. Chem. Res.* **2021**, *60*, 16874–16886. [CrossRef]
122. Liang, D.F.; Wang, Y.S.; Chen, M.Q.; Xie, X.L.; Li, C.; Wang, J.; Yuan, L. Dry reforming of methane for syngas production over attapulgite-derived MFI zeolite encapsulated bimetallic Ni-Co catalysts. *Appl. Catal. B Environ.* **2023**, *322*, 122088. [CrossRef]
123. Song, K.; Lu, M.M.; Xu, S.P.; Chen, C.Q.; Zhan, Y.Y.; Li, D.L.; Au, C.; Jiang, L.L.; Tomishige, K. Effect of alloy composition on catalytic performance and coke-resistance property of Ni-Cu/Mg(Al)O catalysts for dry reforming of methane. *Appl. Catal. B Environ.* **2018**, *239*, 324–333. [CrossRef]
124. Sagar, T.V.; Padmakar, D.; Lingaiah, N.; Prasad, P.S.S. Influence of Solid Solution Formation on the Activity of CeO_2 Supported Ni-Cu Mixed Oxide Catalysts in Dry Reforming of Methane. *Catal. Lett.* **2019**, *149*, 2597–2606. [CrossRef]
125. Wang, L.; Li, D.L.; Koike, M.; Watanabe, H.; Xu, Y.; Nakagawa, Y.; Tomishige, K. Catalytic performance and characterization of Ni-Co catalysts for the steam reforming of biomass tar to synthesis gas. *Fuel* **2013**, *112*, 654–661. [CrossRef]
126. Kim, S.M.; Abdala, P.M.; Margossian, T.; Hosseini, D.; Foppa, L.; Armutlulu, A.; Beek, W.V.; Comas-Vives, A.; Coperet, C.; Muller, C. Cooperativity and dynamics increase the performance of NiFe dry reforming catalysts. *J. Am. Chem. Soc.* **2017**, *139*, 1937–1949. [CrossRef]
127. Andraos, S.; Abbas-Ghaleb, R.; Chlala, D.; Vita, A.; Italiano, C.; Laganà, M.; Pino, L.; Nakhl, M.; Specchia, S. Production of hydrogen by methane dry reforming over ruthenium-nickel based catalysts deposited on Al_2O_3, $MgAl_2O_4$, and YSZ. *Int. J. Hydrogen Energy* **2019**, *44*, 25706–25716. [CrossRef]
128. Song, Z.; Wang, Q.; Guo, C.; Li, S.; Yan, W.; Jiao, W.; Qiu, L.; Yan, X.; Li, R. Improved effect of Fe on the stable NiFe/Al_2O_3 catalyst in low-temperature dry reforming of methane. *Ind. Eng. Chem. Res.* **2020**, *59*, 17250–17258. [CrossRef]
129. Chatla, A.; Ghouri, M.M.; El Hassan, O.W.; Mohamed, N.; Prakash, A.V.; Elbashir, N.O. An experimental and first principles DFT investigation on the effect of Cu addition to Ni/Al_2O_3 catalyst for the dry reforming of methane. *Appl. Catal. A* **2020**, *602*, 117699. [CrossRef]
130. Aghaali, M.H.; Firoozi, S. Enhancing the catalytic performance of Co substituted $NiAl_2O_4$ spinel by ultrasonic spray pyrolysis method for steam and dry reforming of methane. *Int. J. Hydrogen Energy* **2021**, *46*, 357–373. [CrossRef]
131. Hao, S.H.; Ma, F.Y.; Mo, W.L.; Li, M.F.; Zhu, W.J.; Zhang, J. Effect of additive La on the properties of CH_4/CO_2 reforming catalyst NiO/γ-Al_2O_3. *Nat. Gas Chem. Ind.* **2015**, *40*, 44–49.
132. Ruckenstein, E.; Hu, Y.H. Role of support in CO_2 reforming of CH_4 to syngas over Ni catalysts. *J. Catal.* **1996**, *162*, 230–238. [CrossRef]
133. Luisetto, I.; Tuti, S.; Romano, C.; Boaro, M.; Bartolomeo, E.D. Dry reforming of methane over Ni supported on doped CeO_2: New insight on the role of dopants for CO_2 activation. *J. CO2 Util.* **2019**, *30*, 63–78. [CrossRef]

134. Amin, M.H.; Sudarsanam, P.; Field, M.R.; Patel, J.; Bhargava, S.K. Effect of a Swelling Agent on the Performance of Ni/Porous Silica Catalyst for CH_4-CO_2 Reforming. *ACS Langmuir* **2017**, *33*, 10632–10644. [CrossRef]
135. Zhang, Q.L.; Zhang, T.F.; Shi, Y.Z.; Zhao, B.; Wang, M.Z.; Liu, Q.X.; Wang, J.; Long, K.X.; Duany, K.; Ning, P. A sintering and carbon-resistant Ni-SBA-15 catalyst prepared by solid-state grinding method for dry reforming of methane. *J. CO2 Util.* **2017**, *17*, 10–19. [CrossRef]
136. Tanggarnjanavalukul, C.; Donphai, W.; Witoon, T.; Chareonpanich, M.; Limtrakul, J. Deactivation of nickel catalysts in methane cracking reaction: Effect of bimodal mesomacropore structure of silica support. *Chem. Eng. J.* **2019**, *262*, 364–371. [CrossRef]
137. Du, X.J.; Zhang, D.S.; Shi, L.Y.; Gao, R.H.; Zhang, J.P. Coke- and sintering-resistant monolithic catalysts derived from in situ supported hydrotalcite-like films on Al wires for dry reforming of methane. *Nanoscale* **2013**, *5*, 2659–2663. [CrossRef]
138. Zhang, L.M.; Li, L.; Li, J.L.; Zhang, Y.H.; Hu, J.C. Carbon Dioxide Reforming of Methane over Nickel Catalyst Supported on MgO(111) Nanosheets. *Top. Catal.* **2014**, *57*, 619–626. [CrossRef]
139. Grigorkina, G.S.; Ramonova, A.G.; Kibizov, D.D.; Kozyrev, E.N.; Zaalishvili, V.B.; Fukutani, K.; Magkoev, T.T. Probing specific oxides as potential supports for metal/oxide model catalysts: MgO(111) polar film. *Solid State Commun.* **2017**, *257*, 16–19. [CrossRef]
140. Liu, S.G.; Guan, L.X.; Li, J.P.; Zhao, N.; Wei, W.; Sun, Y.H. CO_2 reforming of CH_4 over stabilized mesoporous Ni-CaO-ZrO_2 composites. *Fuel* **2008**, *87*, 2477–2481. [CrossRef]
141. Jafarbegloo, M.; Tarlani, A.; Mesbah, A.W.; Sahebdelfar, S. One-pot synthesis of NiO-MgO nanocatalysts for CO_2 reforming of methane: The influence of active metal content on catalytic performance. *J. Nat. Gas Sci. Eng.* **2015**, *27*, 1165–1173. [CrossRef]
142. Li, L.Z.; Yan, K.S.; Chen, J.; Feng, T.; Wang, F.M.; Wang, J.W.; Song, Z.L.; Ma, C.Y. Fe-rich biomass derived char for microwave-assisted methane reforming with carbon dioxide. *Sci. Total Environ.* **2019**, *657*, 1357–1367. [CrossRef] [PubMed]
143. Zhang, Z.H.; Ou, Z.L.; Qin, C.L.; Ran, J.Y.; Wu, C.F. Roles of alkali/alkaline earth metals in steam reforming of biomass tar for hydrogen production over perovskite supported Ni catalysts. *Fuel* **2019**, *257*, 116032. [CrossRef]
144. Li, L.; Yang, Z.; Chen, J.; Qin, X.; Jiang, X.; Wang, F.; Song, Z.; Ma, C. Performance of bio-char and energy analysis on CH_4 combined reforming by CO_2 and H_2O into syngas production with assistance of microwave. *Fuel* **2018**, *215*, 655–664. [CrossRef]
145. Jose-Alonso, D.S.; Illan-Gomez, M.J.; Roman-Martinez, M.C. K and Sr promoted Co alumina supported catalysts for the CO_2 reforming of methane. *Catal. Today* **2011**, *176*, 187–190. [CrossRef]
146. Wu, P.; Tao, Y.W.; Ling, H.J.; Chen, Z.B.; Ding, J.; Zeng, X.; Liao, X.Z.; Stampfl, C.; Huang, J. Cooperation of Ni and CaO at interface for CO_2 reforming of CH_4: A combined theoretical and experimental study. *ACS Catal.* **2019**, *9*, 10060–10069. [CrossRef]
147. Ruitenbeek, M.; Weckhuysen, B.M. A Radical twist to the versatile behavior of iron in selective methane activation. *Angew. Chem. International Ed.* **2014**, *53*, 11137–11139. [CrossRef]
148. Wang, J.; Mao, Y.R.; Zhang, L.Z.; Li, Y.L.; Liu, W.M.; Ma, Q.X.; Wu, D.S.; Peng, H.G. Remarkable basic-metal oxides promoted confinement catalysts for CO_2 reforming. *Fuel* **2022**, *315*, 123167. [CrossRef]
149. Kong, W.; Fu, Y.; Sun, Y.; Shi, L.; Li, S.G.; Vovk, E.; Zhou, X.H.; Si, R.; Pan, B.R.; Yuan, C.K.; et al. Nickel nanoparticles with interfacial confinement mimic noble metal catalyst in methane dry reforming. *Appl. Catal. B Environ.* **2021**, *285*, 119837. [CrossRef]
150. Zhang, L.; Lian, J.; Li, L.; Peng, C.; Liu, W.M.; Xu, X.L.; Fang, X.Z.; Wang, Z.; Wang, X.; Peng, H.E. $LaNiO_3$ nanocube embedded in mesoporous silica for dry reforming of methane with enhanced coking resistance. *Microporous Mesoporous Mater.* **2018**, *266*, 189–197. [CrossRef]
151. Liu, W.M.; Li, L.; Lin, S.X.; Luo, Y.W.; Bao, Z.H.; Mao, Y.R.; Li, K.Z.; Wu, D.S.; Peng, H.G. Confined Ni-In intermetallic alloy nanocatalyst with excellent coking resistance for methane dry reforming. *J. Energy Chem.* **2022**, *65*, 34–47. [CrossRef]
152. Wang, L.H.; Hu, R.; Liu, H.; Wei, Q.H.; Gong, D.D.; Mo, L.Y.; Tao, H.C.; Zhuang, Z.H. Encapsulated Ni@La_2O_3/SiO_2 catalyst with a one-pot method for the dry reforming of methane. *Catalysts* **2020**, *10*, 38. [CrossRef]
153. Nematollahi, B.; Rezaei, M.; Lay, E.N.; Khajenoori, M. Thermodynamic analysis of combined reforming process using Gibbs energy minimization method: In view of solid carbon formation. *J. Nat. Gas Chem.* **2012**, *21*, 694–702. [CrossRef]
154. Jang, W.J.; Jeong, D.W.; Shim, J.O.; Kim, H.M.; Roh, H.S.; Son, I.H.; Lee, S.J. Combined steam and carbon dioxide reforming of methane and side reactions: Thermodynamic equilibrium analysis and experimental application. *Appl. Energy* **2016**, *173*, 80–91. [CrossRef]
155. Nikoo, M.K.; Amin, N.A.S. Thermodynamic analysis of carbon dioxide reforming of methane in view of solid carbon formation. *Fuel Process. Technol.* **2011**, *92*, 678–691. [CrossRef]
156. Bao, Z.H.; Lu, Y.W.; Han, J.; Li, Y.B.; Yu, F. Highly active and stable Ni-based bimodal pore catalyst for dry reforming of methane. *Appl. Catal. A Gen.* **2015**, *491*, 116–126. [CrossRef]
157. Li, C.L.; Fu, Y.L.; Meng, M.; Bian, G.Z.; Xie, Y.N.; Hu, T.D.; Zhang, J. EXAFS study on the effect of water vapor addition on the structure of Ni components in the CH_4-CO_2 reforming catalyst Ni/CeO_2-ZrO_2-Al_2O_3. *Nucl. Tech.* **2002**, *(10)*, 879–882.
158. O'Connor, A.M.; Ross, J.R.H. The effect of O_2 addition on the carbon dioxide reforming of methane over Pt/ZrO_2 catalysts. *Catal. Today* **1998**, *46*, 203–210. [CrossRef]
159. Li, L.Z.; Jiang, X.W.; Wang, H.G.; Song, Z.L.; Ma, C.Y. Combined reforming of CH_4 with CO_2/steam assisted by microwave. *Combust. Sci. Technol.* **2017**, *23*, 293–298.

Disclaimer/Publisher's Note: The statements, opinions and data contained in all publications are solely those of the individual author(s) and contributor(s) and not of MDPI and/or the editor(s). MDPI and/or the editor(s) disclaim responsibility for any injury to people or property resulting from any ideas, methods, instructions or products referred to in the content.

Article

Characterization and Syngas Production at Low Temperature via Dry Reforming of Methane over Ni-M (M = Fe, Cr) Catalysts Tailored from LDH Structure

Manel Hallassi [1,2,3], Rafik Benrabaa [2,4], Nawal Fodil Cherif [5], Djahida Lerari [5], Redouane Chebout [5], Khaldoun Bachari [5], Annick Rubbens [3], Pascal Roussel [3], Rose-Noëlle Vannier [3], Martine Trentesaux [3] and Axel Löfberg [3,*]

1. Département de Génie des Procédés, Faculté de Technologie, Université 20 Août-Skikda, BP 26, Route Al-Hadaiek, El Hadaik 21000, Skikda, Algeria
2. Laboratoire de Physico-Chimie des Matériaux, Faculté des Sciences et de la Technologie, Université Chadli Bendjedid-El Tarf, BP 73, El Tarf 36000, Algiers, Algeria
3. Univ. Lille, CNRS, Centrale Lille, Univ. Artois, UMR 8181-UCCS-Unité de Catalyse et Chimie du Solide, F-59000 Lille, France
4. Laboratoire de Matériaux Catalytiques et Catalyse en Chimie Organique, Faculté de Chimie, Université des Sciences et de la Technologie Houari Boumediene, BP 32, El-Alia, Bab Ezzouar 16111, Alger, Algeria
5. Centre de Recherche Scientifique et Technique en Analyses Physico-Chimiques, BP 384, Siège ex-Pasna Zone Industrielle, Bou-Ismail 42004, Tipaza, Algeria
* Correspondence: axel.lofberg@univ-lille.fr; Tel.: +33-03-20-43-45-27

Abstract: Bimetallic layered double oxide (LDO) NiM (M = Cr, Fe) catalysts with nominal compositions of Ni/M = 2 or 3 were tailored from layered double hydroxides (LDH) using a coprecipitation method to investigate the effects of the trivalent metal (Cr or Fe) and the amount of Ni species on the structural, textural, reducibility, and catalytic properties for CH_4/CO_2 reforming. The solids before (LDH) and after (LDO) thermal treatment at 500 °C were characterized using TGA-TD-SM, HT-XRD, XRD, Raman, and IR-ATR spectroscopies; N_2 physical adsorption; XPS; and H_2-TPR. According to the XRD and Raman analysis, a hydrotalcite structure was present at room temperature and stable up to 250 °C. The interlayer space decreased when the temperature increased, with a lattice parameter and interlayer space of 3.018 Å and 7.017 Å, respectively. The solids fully decomposed into oxide after calcination at 500 °C. NiO and spinel phases (NiM_2O_4, M = Cr or Fe) were observed in the NiM (M = Cr, Fe) catalysts, and Cr_2O_3 was detected in the case of NiCr. The NiFe catalysts show low activity and selectivity for DRM in the temperature range explored. In contrast, the chromium compound demonstrated interesting CH_4 and CO_2 conversions and generally excellent H_2 selectivity at low reaction temperatures. CH_4 and CO_2 conversions of 18–20% with H_2/CO of approx. 0.7 could be reached at temperatures as low as 500 °C, but transient behavior and deactivation were observed at higher temperatures or long reaction times. The excellent activity observed during this transient sequence was attributed to the stabilization of the metallic Ni particles formed during the reduction of the NiO phase due to the presence of $NiCr_2O_4$, opening the path for the use of these materials in periodic or looping processes for methane reforming at low temperature.

Keywords: H_2 production; Ni-(Fe/Cr); layered double hydroxide; CO_2 reforming

---

Citation: Hallassi, M.; Benrabaa, R.; Cherif, N.F.; Lerari, D.; Chebout, R.; Bachari, K.; Rubbens, A.; Roussel, P.; Vannier, R.-N.; Trentesaux, M.; et al. Characterization and Syngas Production at Low Temperature via Dry Reforming of Methane over Ni-M (M = Fe, Cr) Catalysts Tailored from LDH Structure. *Catalysts* **2022**, *12*, 1507. https://doi.org/10.3390/catal12121507

Academic Editors: Fanhui Meng and Wenlong Mo

Received: 28 October 2022
Accepted: 18 November 2022
Published: 24 November 2022

Publisher's Note: MDPI stays neutral with regard to jurisdictional claims in published maps and institutional affiliations.

Copyright: © 2022 by the authors. Licensee MDPI, Basel, Switzerland. This article is an open access article distributed under the terms and conditions of the Creative Commons Attribution (CC BY) license (https://creativecommons.org/licenses/by/4.0/).

1. Introduction

The catalytic reactions for the transformation of natural gas into synthesis gas (CO + H_2) are currently highly strategic industrial targets for the production of alternative liquid fuels. One interesting way to valorize methane is through the dry reforming of methane (DRM), which is undertaken in the presence of carbon dioxide (CO_2) [1,2]. This process is mainly endothermic [3], and CO_2 is used as the oxidizing agent, as shown in the following equation:

$$CH_4 + CO_2 \rightarrow 2\,CO + 2H_2\ ;\ \Delta H_{298\,K} = +247\ kJ/mol, \tag{1}$$

DRM is of particular interest because it converts two greenhouse pollutant gases, CH_4 and CO_2, into synthesis gas or hydrogen, which can subsequently be converted into valuable chemicals [4]. The methane reforming reaction is commonly carried out in the presence of supported noble metal (Rh, Pt, and Pd) or nickel metal [5] catalysts. Noble metals show better resistance to coke formation, but nickel is known to be a less expensive metal, with great reactivity in reforming processes [3–5]. However, one of the major issues associated with CO_2 reforming is the rapid carbon deposition on the catalyst, which mainly results from the carbon monoxide dissociation (reaction (2)) and/or the methane decomposition (reaction (3)) [6]. This coke deposition brings a progressive catalyst deactivation.

$$2\,CO \rightarrow CO_2 + C, \tag{2}$$

$$CH_4 \rightarrow C + 2\,H_2, \tag{3}$$

To limit sintering and reduce coke formation, the stabilization of the particles at a nanoscale level is necessary. Among the various solutions available to increase metal particle dispersion on the catalyst surface, one consists in incorporating the active phase as a well-defined structure, such as a spinel, a perovskite, or a pyrochlore [7–10]. However, new classes of porous solids that can be used to obtain solid catalysts with high specific surface areas for DRM are being studied and investigated. Many studies have been published, or are still ongoing, on the design of efficient, more stable, and eco-friendly catalysts [11]. Among the various types of materials, layered double hydroxides (LDH) containing transition metals as active components seem to be good candidates for the dry reforming of methane [12,13]. Indeed, the interest in LDHs relates to their two-dimensional character, appropriate alkalinity and ability to form, through calcination, oxides with homogeneous mesoporous textures and proper specific surface areas. LDH materials are therefore desirable precursors for catalysts [14,15].

Thus, on one hand, this work was dedicated to the preparation of an efficient Ni-transition metal bimetallic catalyst for DRM. On the other hand, the reactivity in the CO_2 reforming reaction of two bimetallic catalysts based on nickel–iron (Ni-Fe) and nickel–chromium (Ni-Cr) compositions was investigated. The bimetallic catalysts were built up at 500 °C from LDH precursors. These latter were obtained through a coprecipitation method with Ni/Fe and Ni/Cr molar ratios equal to 2 and 3. Finally, the Ni-Fe and Ni-Cr mixed oxides were tested in the CO_2 reforming of methane at low reaction temperatures (400–650 °C). The influences of the molar ratio and the cationic composition in the preparation of the LDH precursors on the physicochemical properties of the target catalysts and on their performance in DRM were evaluated. A set of characterizations before (LDH) and after (LDO) thermal treatment at 500 °C using TGA-TD-SM, HT-XRD, XRD, Raman, and IR-ATR spectroscopies; N_2 physical adsorption; XPS; and H_2-TPR were also performed to attain structure–reactivity relationships and enhance hydrogen production.

2. Results

2.1. Structural Characterization (XRD, Raman, and FTIR) of LDH Precursors

The XRD patterns recorded at room temperature for LDH precursors (Figure 1) show the presence of Bragg reflections located at 2θ values of 11.27°, 22.79°, 33.8°, 38.6° and 60.6°, related to the (003), (006), (012), (015), and (110) crystallographic planes of LDH phase with rhombohedral symmetry (R3), in accordance with previous works [15]. The d-spacing values calculated from the position of (003) diffraction lines of Ni_2Cr, Ni_3Cr, Ni_2Fe, and Ni_3Fe-LDH are about 7.82 Å, 7.77 Å, 7.94 Å, and 7.94 Å, respectively. These values were ascribed to $CO_3{}^{2-}$ anions and water molecules intercalation in the LDH interlayer space, and hence the synthesis was performed an ambient atmosphere. Additionally, the relative weak intensities and wider peaks of (00l) diffractions suggest a low crystallinity of the as-prepared LDHs [16]. The lattice parameters "a" and "c" calculated for the four precursors are gathered in Table 1. The values are in good agreement with the literature; that is, the "c" and "a" values decrease with lower Ni/Cr and Ni/Fe molar ratios.

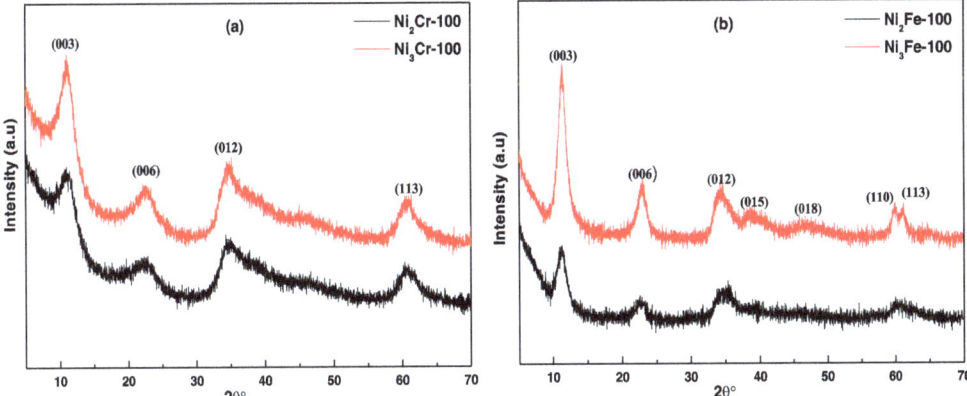

Figure 1. XRD patterns of Ni_RCr (a) and Ni_RFe (b) LDH precursors.

Table 1. Structural parameters of Ni_RFe and Ni_RCr LDH precursors obtained from XRD patterns.

LDH	d_{003} (Å)	d_{110} (Å)	a (Å) [1]	c (Å) [1]
Ni_2Fe	7.82	1.53	3.06	23.46
Ni_3Fe	7.77	1.54	3.08	23.31
Ni_2Cr	7.94	1.52	3.04	23.82
Ni_3Cr	7.94	1.53	3.06	23.83

[1] Lattices parameters a and c are equal to $2 \times d_{(110)}$ and $3 \times d_{(003)}$, respectively.

Figure 2 shows the Raman spectra of the precursors Ni_RFe LDH and Ni_RCr LDH (R = 2, 3) in the relevant spectral range to study hydrotalcites (300–1200 cm^{-1}), the water molecule is better detected by infrared absorption than by Raman scattering. The line observed at 1049 cm^{-1} accompanied by its shoulder at 1069 cm^{-1} is attributable to the elongation vibration of the carbon–oxygen bond of the CO_3^{2-} group in agreement with the work of Frost et al. [17].

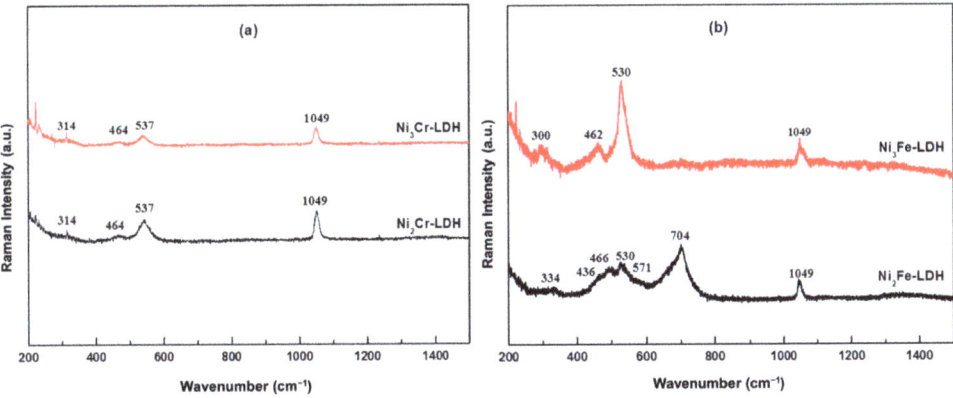

Figure 2. Raman spectra of Ni_RCr (a) and Ni_RFe (b) LDH precursors.

Comparing with the value of 1080 cm^{-1} generally obtained for pure carbonates [18], the band shift towards the low frequencies reveals interactions with the carbonate ion in the layers, with the shoulder indicating a slightly different environment for oxygen of this ion.

The Raman spectra of the Ni$_2$Fe and Ni$_3$Fe LDH precursors have different spectral features (Figure 2b). For the first, the observation of bands at 704, 571, 436, and 334 cm^{-1} wave numbers, indicates a strong dominance of the NiFe$_2$O$_4$ reverse spinel [19]. This strong presence is not found for Ni$_3$Fe LDH for which it is difficult to observe the most intense band of the NiFe$_2$O$_4$ spectrum at about 700 cm^{-1}, representative of the symmetrical elongation mode of the tetrahedral entity [FeO$_4$] composing the inverse spinel with the octahedral [FeO$_6$] and whose vibration modes are located in the spectral range of 300–600 cm^{-1} [20]. Nevertheless, for this spectral domain, three bands are recorded at 530, 462, and 300 cm^{-1}. The most intense one at 530 cm^{-1} for Ni$_3$Fe LDH, also observable for Ni$_2$Fe LDH, and with the broad band at 462 cm^{-1} constitute a doublet attributable to the Ni(OH)$_2$ species, according to the reference [21], which is in agreement with the presence of hydroxyl ions in the layers. The weak band at 300 cm^{-1} could be due to Fe$_2$O$_3$ [22].

The spectral features of the Ni$_2$Cr and Ni$_3$Cr LDH precursors are similar (Figure 2a). The bands with weaker intensity are comparable to the 1049 cm^{-1} intensity line assigned to the carbonate ion in the layers. The most intense and characteristic line of the tetrahedron [CrO$_4$], located in the spectral range 800–900 cm^{-1}, are not detected on these precursors. The band observed at 537 cm^{-1}, one of the most intense in the spectrum, is attributable to chromium oxide Cr$_2$O$_3$, according to the work of J. Singh et al. [23]. The band width can be due to an overlap with the characteristic line of Ni(OH)$_2$.

Figure 3 shows the IR-ATR absorption spectra of these different precursors for which the spectral feature is quite similar, except in the 400–1000 cm^{-1} domain. The wave numbers recorded at 3400 cm^{-1} for symmetrical elongation of the bond νs (O-H) and at 1630 cm^{-1} for angular deformation δs (H$_2$O) clearly indicate the formation of hydrogen bonds in the hydrotalcite layers and in particular with carbonate ions whose unsymmetrical elongation frequency ν_d (C=O) shifts towards the low wavenumbers 1351 cm^{-1} [24]. This is in agreement with the results obtained by Raman scattering for the symmetrical elongation frequency νs (C=O) observed at 1049 cm^{-1}.

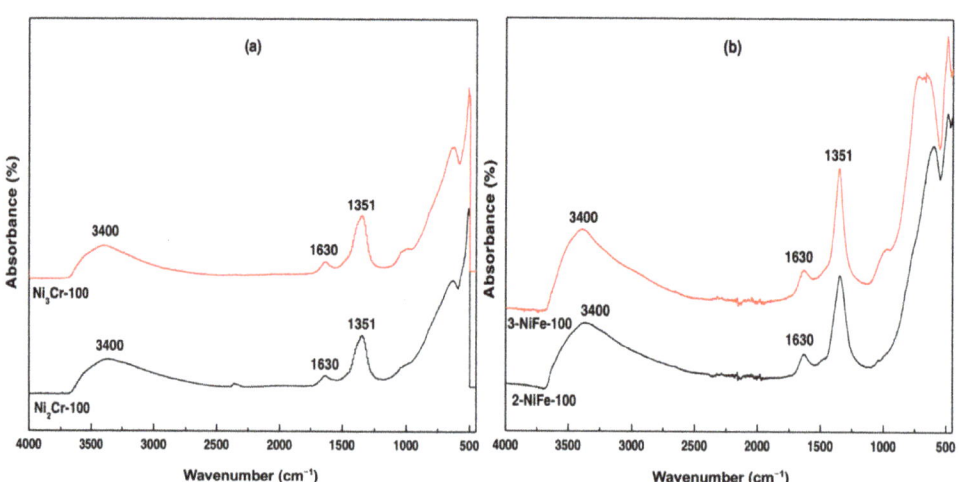

Figure 3. IR-ATR spectra of Ni$_R$Cr (a) and Ni$_R$Fe (b) LDH precursors.

2.2. Thermal Decomposition (HT-XRD and TG-DTA) of LDH Precursors

The recorded TG-DTA thermograms when Ni$_R$Fe and Ni$_R$Cr LDH were subjected to thermal decomposition are shown in Figure 4.

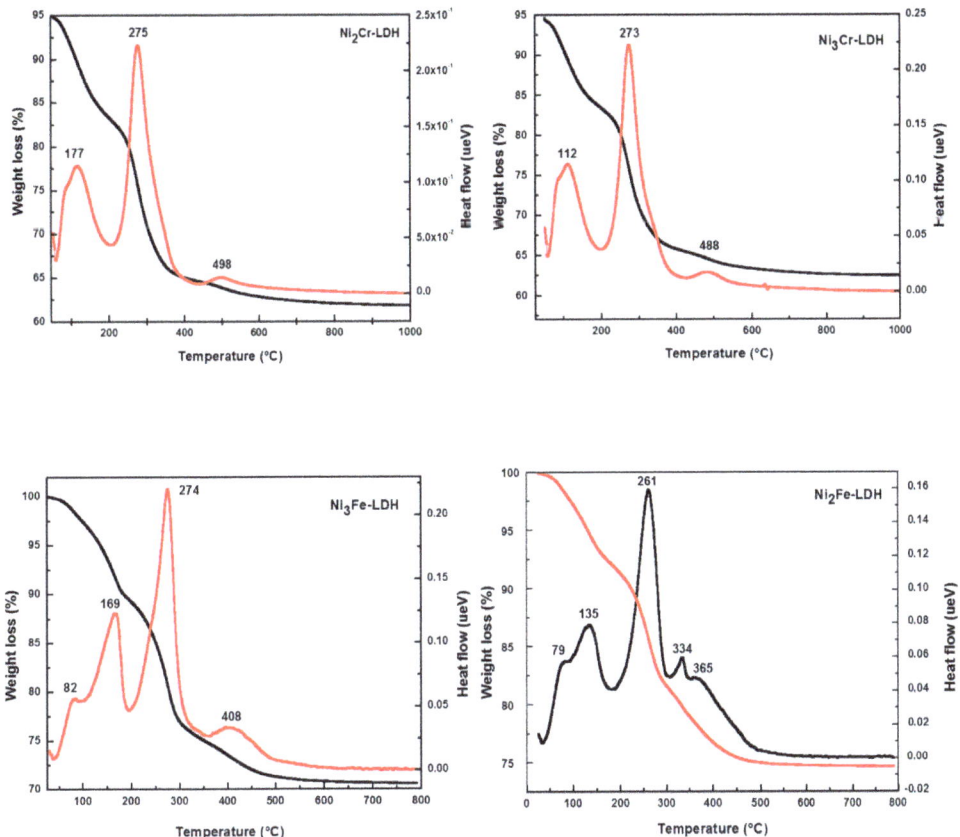

Figure 4. TG-DTA curves of Ni$_R$Cr and Ni$_R$Fe LDH precursors.

The thermal behavior analysis indicates mainly three steps of weight loss up to ≈600 °C for Ni$_R$Cr and Ni$_R$Fe LDH. There is no weight loss or heat flow observed above 600 °C. This implies that there is no phase change above 500 °C. The total weight loss is higher in the Ni$_R$Cr LDH case (≈32%) compared to Ni$_R$Fe LDH (≈24 and ≈27% for Ni$_2$Fe and Ni$_3$Fe, respectively).

The species responsible for the weight loss are water, carbonates, and nitrates. Through online mass spectroscopy (MS), m/z = 18, 30, 44, and 46, corresponding to H_2O, NO, CO_2 and NO_2, respectively, were followed and are reported in Figures S1 and S2 and summarized in Table S1. During the three steps, the release of H_2O was observed in the first and second stages, while NO, NO_2 and CO_2 occurred in the second and third steps. During the first step from ambient to 200 °C, MS analysis showed that only the removal of water physically adsorbed on the external surface of the crystallites and interlayer spaces with a weight loss of ~12% for Ni$_R$Cr and ~8–10% for Ni$_R$Fe. In the two other stages (200→320 °C and 320→600 °C), the release of H_2O is also observed and is accompanied by the departure of carbonates and nitrates from the LDH structure. Both steps correspond to the concurrent dehydroxylation of the brucite-like layers and the decomposition of the intercalated anions [25]. As the temperature further increases, the weights of the samples remain constant, with no obvious endothermic/exothermic peak, indicating that the structure of the materials reaches relative stability. The theoretical weight loss for the transformation of Ni(OH)$_2$, Cr(OH)$_3$ and Fe(OH)$_3$ hydroxides to Ni-Cr-O and to Ni-Fe-O oxides is Δm/m = 23–25%. This value was approximately reached in the Ni$_R$Fe samples, but a slight difference was observed for Ni$_R$Cr LDH formulations (Table S1).

The transformation of Ni-Fe-LDH precursors was studied by HT-XRD up to 800 °C in air (Figure 5). The hydrotalcite structure is observed at room temperature and up to 250 °C. At 275 °C, it fully collapsed into oxide (Layer Double Oxide-LDO) due to the dehydroxylation of the layer and removal of NO$_x$ and CO$_2$ from the interlayer, as observed at a similar temperature in TGA analysis (Figure 4). Until 800 °C, only the lines of the NiO (PDF: 01-080-5508) phase are observed and become sharper and more symmetric with increasing temperature. No other crystalline structure was detected, showing that Fe(III) is dispersed in the NiO rock salt phase as a solid solution.

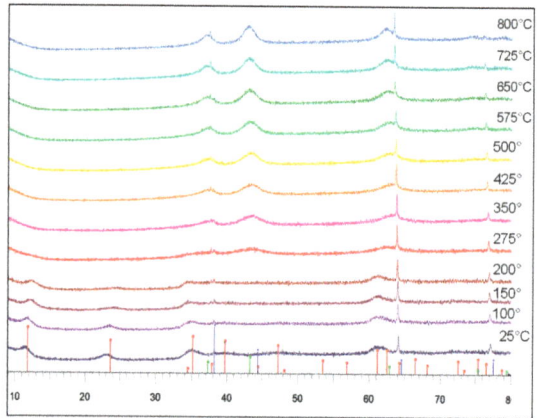

Figure 5. HT-XRD patterns of Ni$_3$Fe LDH precursor decomposition in air.

Lattice parameters a and c and interlayer space (d$_{003}$) were calculated as a function of temperature and are listed in Table 2. During HT-XRD measurements, from 25 to 250 °C, the values of the lattice parameter (a) are similar. However, the values of lattice parameter (c) and interlayer space (d$_{003}$) decrease markedly when the temperature increases as a result of a dehydration phenomenon and a strengthening of the interaction between interlayer anions and hydroxide layers during heating treatment. Benito et al. [26] and Kovanda et al. [27] reported the same observations concerning the change in lattice parameter (c) and interlayer space (d$_{003}$). Further heating of the powder induces a decrease in the intensity of LDH structure lines, which disappear completely at 275 °C (Figure 5). This result is in agreement with those obtained from TGA analysis. The temperature of decomposition is higher in TGA due to a different heating rate which yields different kinetics.

Table 2. Evolution of lattice parameters of Ni$_3$Fe-LDH as a function of temperature.

Temperature (°C)	d$_{003}$ (Å)	a (Å) [1]	c (Å) [1]
25	7.814	3.010	23.442
50	7.694	3.015	23.082
75	7.665	3.027	22.995
100	7.576	3.027	22.728
125	7.467	3.023	22.401
150	7.399	3.027	22.197
175	7.265	3.012	21.795
200	7.221	3.015	21.663
225	7.102	3.015	21.306
250	7.017	3.018	21.051

[1] Lattices parameters a and c are equal to $2 \times d_{(110)}$ and $3 \times d_{(003)}$, respectively.

2.3. Characterization (XRD, Raman, BET, XPS, and H_2-TPR) of Mixed Oxide Catalysts

X-Ray Diffraction and Laser Raman spectroscopy analyses were used to ascertain the structural properties of catalysts obtained after calcination at 500 °C. The LDH structure is fully destroyed due to the elimination of most interlayer anions (NO_x and CO_x) and water. As highlighted by TGA and HT-XRD analysis, LDH decomposition leads to mixed metal oxide structures.

As can be seen in Figure 6a, XRD patterns of Ni_2Cr-500 and Ni_3Cr-500 show similar diffractograms. Ni/Cr ratio used in the preparation has little effect on the structures of the resulting materials. The diffractograms (Figure 6a) confirm the presence of NiO and $NiCr_2O_4$ structures, where peak positions of $2\theta \approx 37.4, 43.3, 62.9,$ and $75.5°$ correspond to the (111), (200), (220), and (311) family of planes of NiO structure (PDF: 03-065-2901), while the characteristic diffraction peaks at 18.4°, 30.3°, 35.8°, and 57.4° belongs to the (111), (220), (311), and (511) planes of $NiCr_2O_4$ in accordance with PDF 85-0935. In contrast, for Ni_2Fe-500 and Ni_3Fe-500 samples, different diffractograms were obtained (Figure 6b), suggesting that the amount of Ni species used has a significant effect on the crystalline structure of Ni_RFe-500 catalysts. For the low amount of Ni-species (R = 2), a mixture of phases was detected containing NiO (PDF: 03-065-2901) ($2\theta \approx 37.4$ (111), 43.3 (200), 62.9 (220), and 75.5° (311)) and $NiFe_2O_4$ (PDF: 00-054-0964) spinel structure by the peaks located at $\approx$18.4 (111), 30.3 (220), 35.8 (311,) and 57.4° (511). However, the sample Ni_3Fe-500 matches only the NiO oxide phase (PDF: 03-065-2901) through the reflections located at $2\theta \sim 37.5, 43.6$ and 75.5°. The possibility of $NiFe_2O_4$ spinel oxide formation in Ni_3Fe-500 formulation cannot be excluded as it could be present in a very low amount or well-dispersed form which would make it difficult to be detected by the XRD.

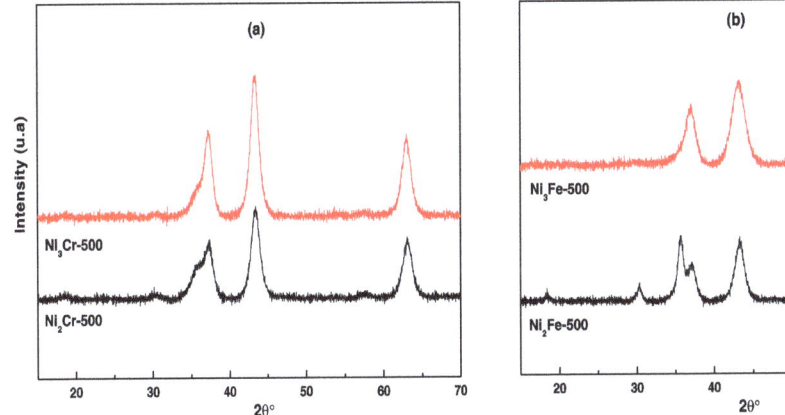

Figure 6. XRD patterns of Ni_RCr-500 (**a**) and Ni_RFe-500 (**b**) catalysts.

No characteristic peak corresponding to Cr_2O_3 and Fe_2O_3 phases could be detected for Ni_RCr-500 and Ni_RFe-500, respectively, which is probably due to their low crystallinity. In all cases, we observed NiO oxide as the dominant phase. No obvious diffraction peaks of the spinel phase were observed, which, considering the calcination temperature and the relatively low Fe and Cr loadings, may be related to the formation of amorphous or well-dispersed phases, not detected by X-ray diffraction.

The crystallite size (CS) for all samples has been calculated using XRD data and are reported in Table 3. Both samples, Ni_2Cr-500 and Ni_3Cr-500, which show the same crystalline structure (Figure 6a), exhibit similar crystallite size values (75–77 Å). In contrast, Ni_2Fe-500 and Ni_3Fe-500 samples show different crystallite sizes (69 Å for Ni_2Fe-500 against 53 Å for Ni_3Fe-500) suggesting an effect of the amount of Ni-species incorporated in the LDH structure.

Table 3. Textural properties of Ni-based catalysts.

Catalysts	Cs [1] (Å)	S_{BET} (m² g⁻¹)	Pore Volume (cm³ g⁻¹)	Mean Pore Diameter (Å)	
				B.E.T.	B.J.H.
Ni₂Fe-500	69	144	0.24	76	62
Ni₃Fe-500	53	160	0.39	91	76
Ni₂Cr-500	75	73	0.18	92	79
Ni₃Cr-500	77	74	0.23	124	107

[1] Crystallites size of NiO phase, BET surface area, pore volume, and pore diameter. The pore diameter parameter was obtained from BET and BJH methods.

Figure 7 shows the Raman spectra of the catalysts Ni_RFe-500 and Ni_RCr-500 (R = 2, 3) after calcination at 500 °C. In comparison with Figure 2b, one can note the disappearance of the bands located at 300, 462, and 530 cm⁻¹, accompanied by the reinforcement of the band intensity at 578 cm⁻¹. This indicates the transformation of $Ni(OH)_2$ into NiO [21]. For the Ni₂Fe-500 catalyst, a high proportion of $NiFe_2O_4$ in the mixture can be noted.

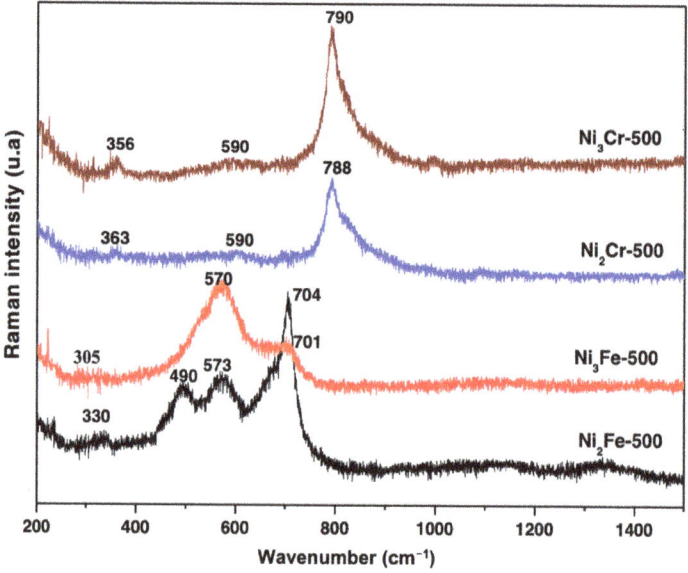

Figure 7. Raman spectra of Ni_RCr-500 and Ni_RFe-500.

The spectral feature (wide band) associated with the wavenumber values (707, 570 and ~305 cm⁻¹) for Ni₃Fe-500 suggests the coexistence of $NiFe_2O_4$ and $FeFe_2O_4$ spinels [28,29] with NiO nickel oxide.

For the Ni_RCr-500 samples, the spectra (Figure 7) show an intense and asymmetrical band, whose maximum is recorded at 791 cm⁻¹. It characterizes the symmetrical elongation movement of the tetrahedron [CrO_4]. The anti-symmetrical elongation vibrations of this same entity are represented by the different components forming the asymmetry of this band. The other modes of angular deformation of the tetrahedron, of lower intensity, are embedded in the wide and low band centered on 590 cm⁻¹. These results are in good agreement with the work of D'Ippolito et al. [30].

The textural properties of the catalysts after calcination at 500 °C were determined from the nitrogen adsorption–desorption isotherms at 77 K. The specific surface area values measured by BET are reported in Table 3. The N_2 adsorption–desorption isotherms of catalysts (Figure 8) are type IV according to IUPAC classification with H3-type hysteresis loop, indicating mesoporous materials. Furthermore, the hysteresis shape suggests slit-type pores with a void created by particle aggregation and attributed to open pores at both ends [31,32].

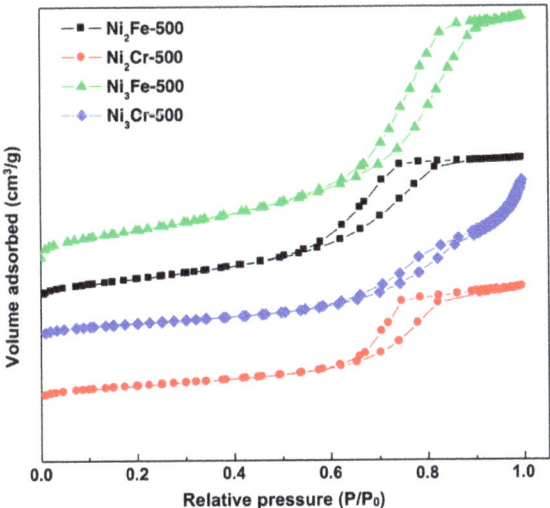

Figure 8. N_2 adsorption–desorption isotherms of Ni_RCr-500 and Ni_RFe-500 catalysts.

As can be seen in Table 3, the textural parameters of the solids follow the same trend; i.e., the values of BET surface area, pore volume, and pore diameter show a progressive increase with increasing Ni to trivalent metal ratios. The specific surface areas of Ni_RFe are approximately two times greater than that of Ni_RCr. The highest specific surface area value of Ni_3Fe catalyst (160 m^2/g) is in accordance with the smallest respective crystallite size (53 Å, Table 3).

Chemical state and surface compositions of the catalysts were examined by XPS analysis. Figures S3–S6 (see SI) represent the photoemission spectra of 2p levels of nickel (Ni2p3/2 line), 2p of iron (Fe2p3/2 line), 2p of chromium (line Cr2p3/2), and 1s of oxygen (O1s line) obtained on the various samples calcined at 500 °C. The values of the binding energies of Ni2p3/2, Fe2p3/2, Cr2p3/2, and O1s lines, as well as the results of the quantification of the atomic ratios Ni/Fe and Ni/Cr, calculated from the photopeak intensities are gathered in Table 4.

Table 4. Binding energy (eV) and Ni/M atomic ratios (M = Fe or Cr) obtained by XPS.

Catalysts	Binding Energy (eV)			Atomic Ratio [1]	
	Ni	Fe	Cr	Ni/Fe	Ni/Cr
Ni_2Fe-500	854.7	711.1	-	0.7	-
Ni_3Fe-500	854.9	711.6	-	0.9	-
Ni_2Cr-500	855.0	-	576.5	-	2.2
Ni_3Cr-500	855.0	-	576.9	-	3.2

[1] Atomic ratio equal to 2 or 3.

The surface compositions depend on the nature of the used metals. Both chromium-based catalysts (Ni_2Cr-500 and Ni_3Cr-500) show Ni/Cr ratio very close to the nominal bulk composition, suggesting little or negligible surface segregation on these samples. In contrast, Ni_2Fe-500 and Ni_3Fe-500 catalysts show a Ni/Fe atomic ratio lower than expected, highlighting the presence of more iron species on the surface than in the bulk of the catalysts. This excess in iron species on the surface can be correlated to the nature of $NiFe_2O_4$ inverse spinel structure as Fe(III+) species occupy both crystallographic positions: 50% of the ions in the octahedral-[Oh] position and 50% in tetrahedral-[Td] sites.

The decomposition of the spectra for Ni, Cr and O species shows two components, while only one component is observed for Fe (Figures S3–S6). For the latter, the binding energy values are 711.1 eV for Ni_2Fe-500 and 711.6 eV for Ni_3Fe-500 (Figure S3), accompanied by the presence of a satellite peak at higher energy (7.7 eV) vs. the main peak as a clear indication of the presence of Fe(III) species only on the catalyst surface [33]. The *Ni2p3/2* peaks (Figure S4), are composed of the main peak located at ≈855 eV and a relatively intense satellite peak at about 7 eV higher energy. The existence of such a satellite is characteristic of the oxidation state (+II) of nickel [33,34]. According to literature data [31,32], the decomposition of these spectra (Figure S4) shows the presence of Ni (II+) in NiO by the lines located at ≈855. $Ni(OH)_2$ hydroxide (Ni, II+) shows values close to that of NiO oxide (861 and 867 eV), but its presence can be excluded because the calcination is carried out at 500 °C where the total transformation of $Ni(OH)_2$ hydroxide into oxide is ensured. The peaks situated at ≈856 and 862 eV can therefore be attributed to nickel in the spinel structure (Ni in $NiFe_2O_4$ or in $NiCr_2O_4$). Both Ni_2Cr-500 and Ni_3Cr-500 systems show similar *Cr2p* spectra (Figure S5). The binding energy value of the *Cr2p* line is 576.5 and 576.9 eV for Ni_2Cr-500 and Ni_3Cr-500, respectively. These values characterize the presence of Cr^{3+} in our formulations. After the decomposition of the spectra (Figure S5) of *Cr2p*, we note the appearance of a band around 579.1 eV for Ni_2Cr-500 and 579.3 eV for Ni_3Cr-500 which can be associated with Cr^{6+} species [33]. Several studies reported that a fraction of Cr^{3+} ions exposed in the chromium oxide is easily oxidized to Cr^{6+} during the calcination step under an ambient atmosphere [35]. The photopeak 1s of oxygen (Figure S6) reveals two components for all formulations. The first component, corresponding to the lowest binding energy (~530 eV), is associated with the lattice oxygen O^{2-} and the second component of higher binding energy (~ 532 eV), is due to the presence oxygen localized on the outer layer of the solid and belonging to -OH groups or probably to H_2O adsorbed on the surface.

The H_2-TPR profiles are given in Figure 9. The hydrogen consumption displays different profiles depending on both the trivalent cation and the molar ratios used. The amount of consumed H_2 depends significantly on the nature of the trivalent metal (Fe or Cr) and does not depend on the Ni/M ratios (M = Fe, Cr); the amount of consumed H_2 for Ni_RFe-500 (16–17 mmol/g) catalysts is greater with a factor of ≈2 compared to that of Ni_RCr-500 (9–10 mmol/g).

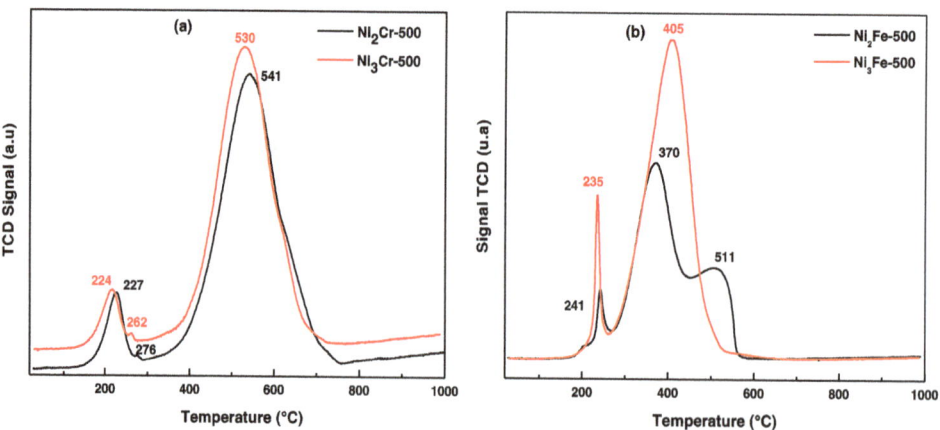

Figure 9. H_2-TPR profiles of Ni_RCr-500 (**a**) and Ni_RFe-500 (**b**).

The Ni$_2$Cr-500 and Ni$_3$Cr-500 catalysts possess a similar TPR with reduction peaks, which shift slightly to higher temperature upon decreasing Ni/Cr ratio. This means that the Ni$_2$Cr-500 catalyst is less reducible and more stable. In Figure 9a, two domains of hydrogen consumption can be observed in the temperature region 200–700 °C, which are related mainly to the reduction of both Ni^{2+} species. For both chromium-based catalysts, the first peak of H$_2$ consumption at 227 for Ni$_2$Cr-500 and at 224 °C for Ni$_3$Cr-500 is correlated to the reduction of surface oxygen species, which can be reduced by hydrogen at low temperatures [36]. The peaks at about 262–276 °C and at 530–541 °C could be attributed to the reduction of Ni^{2+} present in NiO and in the lattice of NiCr$_2$O$_4$ spinel phase detected by XRD and Raman analyses as mentioned above.

In contrast to Cr-based catalysts, the H$_2$-TPR of iron-based catalyst exhibits two different profiles (Figure 9b) according to the Ni/Fe ratio. Ni$_2$Fe-500 catalyst exhibits three reduction peaks centered at 241, 370, and 511 °C. The first and the second peaks (located at 241 and 370 °C) may be attributed to the simultaneous reduction of (i) Ni(II+) present in NiO and in NiFe$_2$O$_4$ and (ii) Fe(III+) in tetrahedral-[Td] sites of the NiFe$_2$O$_4$ spinel phase. The third peak located at 511 °C is assigned to the reduction of Fe^{3+} in the octahedral-[Oh] position of the NiFe$_2$O$_4$ structure. However, the catalyst richer in Ni species (Ni$_3$Fe-500) shows a fairly similar profile compared to NiO oxide [37] in accordance with XRD data, which showed only NiO oxide (Figure 6). The profile reveals two neat reduction peaks centered at 225 and 405 °C. The first of low intensity at ~225 °C and the second with strong intensity at ~405 °C are attributed to the reduction of amorphous α-NiO and clustered β-NiO, respectively.

2.4. Catalytic Properties in CO$_2$-Reforming of Methane

The catalysts obtained after synthesis (LDH) and calcination at 500 °C under air flow (LDO) were tested for DRM. Figures 10 and 11 and Table S2 show the catalytic performances (CH$_4$ conversion and CO$_2$ conversion, H$_2$ selectivity and H$_2$/CO ratio) obtained in temperature-programmed reaction conditions between 400 and 650 °C.

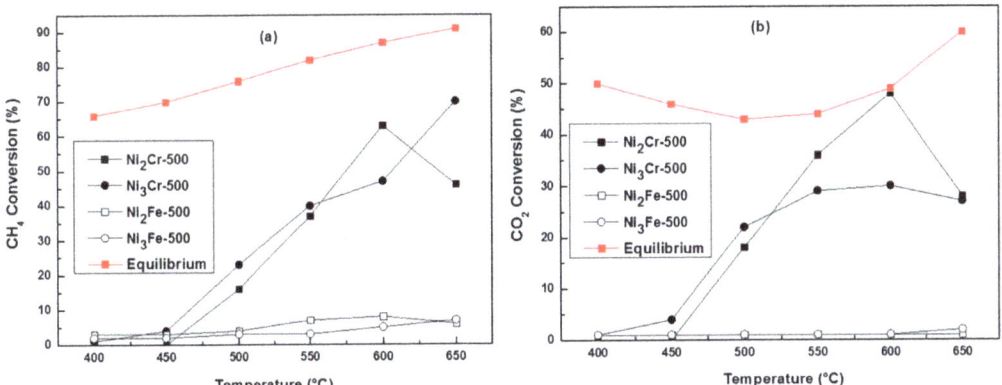

Figure 10. CH$_4$ (a) and CO$_2$ (b) conversions obtained on the fresh Ni$_R$M-500 (M = Cr or Fe, R = 2 or 3) catalysts issued from LDH structure and calcined at 500 °C (CH$_4$ = 20%; CO$_2$ = 20%; 100 mg; F = 100 mL/min).

Both chromium-based catalysts (Ni$_2$Cr-500 and Ni$_3$Cr-500) are catalytically active and selective. The conversions of CH$_4$ and CO$_2$ (Figure 10), H$_2$-selectivity, and H$_2$/CO ratio (Figure 11) show very similar behaviors, suggesting the little effect of Ni/Cr ratios on the catalytic performances for these formulations.

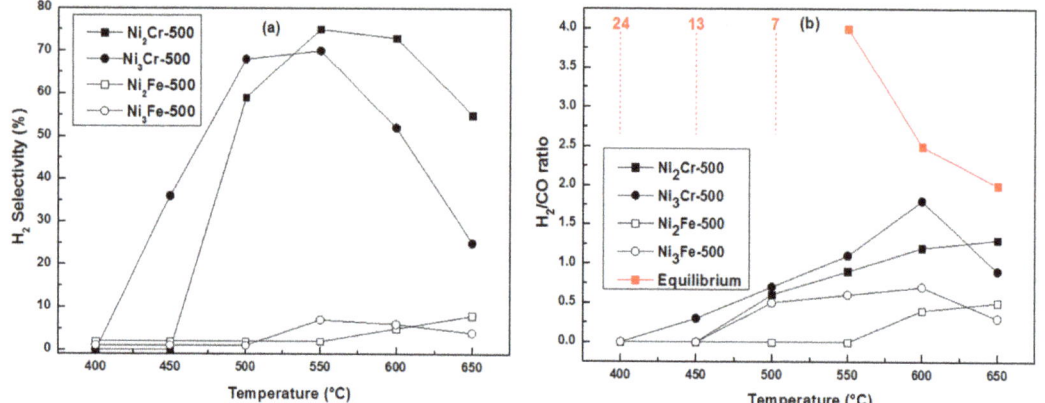

Figure 11. H$_2$ selectivity (**a**) and H$_2$/CO ratios (**b**) obtained on the fresh Ni$_R$M-500 (M = Cr or Fe, R = 2 or 3) catalysts issued from LDH structure and calcined at 500 °C (CH$_4$ = 20%; CO$_2$ = 20%; 100 mg; F = 100 mL/min).

This behavior is not very surprising because both systems, as shown in the characterization section, have similar structural (NiO and NiCr$_2$O$_4$ in their structure) and textural (73–74 m^2/g and Ni/Cr ≈ stoichiometry) properties. CH$_4$ and CO$_2$ conversion remain well below equilibrium values in the full range of temperature explored. In particular, in the 450–550 °C range, thermodynamics should favor CH$_4$ conversion and carbon deposition on one side, and CO$_2$ conversion through RWGS to form water on the other. This would lead to significantly higher methane conversion with respect to CO$_2$ conversion, strong carbon deposition, and a high H$_2$/CO ratio (above 4). The performances observed for Ni$_2$Cr-500 and Ni$_3$Cr-500 samples are very far from the thermodynamic conversions, confirming that the reactivity is effectively governed by the catalytic properties of the materials.

Moreover, if one looks more carefully at the values obtained at 500 °C (Table 5), both chromium-based samples show rather similar behaviors. Conversions of methane are in the range of 16–23% for the two samples and are close to those of CO$_2$. Hydrogen selectivity is high (60–70%), whereas H$_2$/CO is around 0.7. Water is certainly produced, either through a contribution of RWGS reaction or through the reduction of the solid. These results remain exceptional in terms of selectivity in such low temperature ranges.

Table 5. DRM performances at 500 °C in temperature-programmed and isothermal modes.

Catalysts	X% CH$_4$	X% CO$_2$	S% H$_2$	H$_2$/CO
Ni$_2$Cr-500 (TP [1])	16	18	59	0.6
Ni$_3$Cr-500 (TP [1])	23	22	68	0.7
Ni$_2$Fe-500 (TP [1])	4	1	2	-
Ni$_3$Fe-500 (TP [1])	3	1	1	0.5
Ni$_2$Cr-500 (ISO [2], t = 20 min)	30	18	40	1
Ni$_3$Cr-500 (ISO [2], t = 20 min)	16	21	88	0.7

[1] TP: temperature programmed mode (cf. Figures 10 and 11); [2] ISO: isothermal mode (cf. Figure 12).

Above 600 °C, the curves of CH$_4$ and CO$_2$ conversion do not increase anymore as should be expected. On the contrary, the catalysts are progressively deactivated. This catalytic behavior is very different from the results obtained in our previous work [36] on Ni-Cr spinel oxide prepared by the coprecipitation method. These catalysts showed excellent activity both in terms of conversions and selectivity at a high temperature, whereas below 700 °C, carbon deposition mostly occurred even though the Ni content of those catalysts was much lower (Ni/Cr = 0.5 as compared to 2 or 3 in the present catalysts). The carbon deposition is usually attributed to large metallic nickel particles. The results obtained at a low temperature on Ni$_2$Cr-500 and Ni$_3$Cr-500 are therefore particularly interesting.

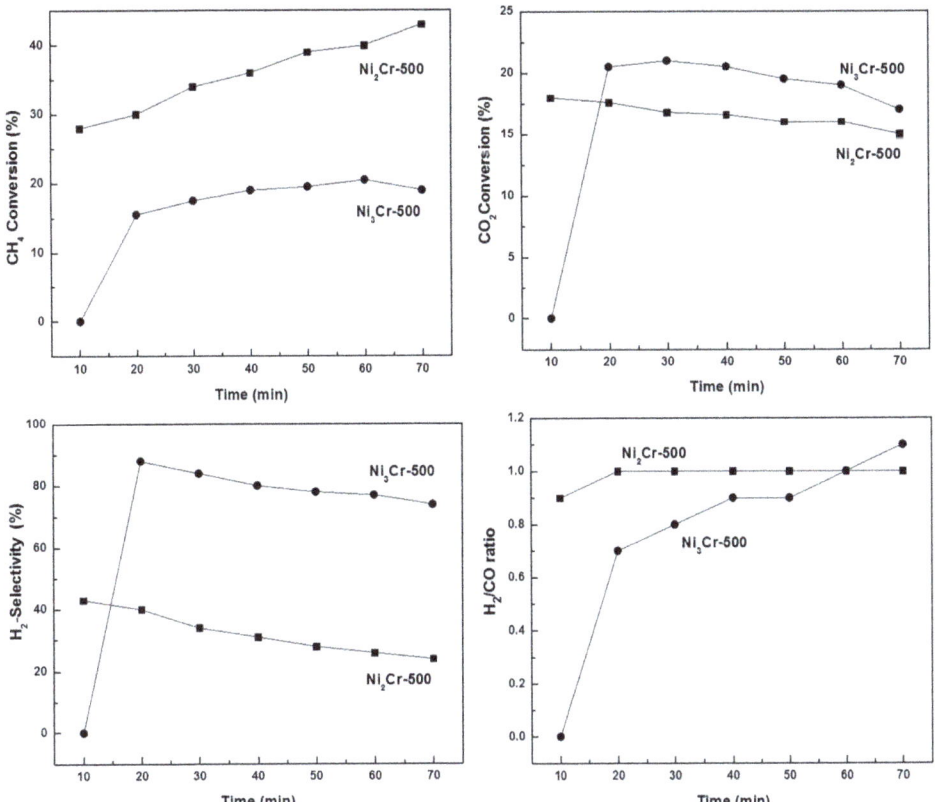

Figure 12. Isothermal test of catalytic performances in terms of conversions (CH$_4$ and CO$_2$), H$_2$ selectivity and H$_2$/CO of fresh Ni$_2$Cr-500 and Ni$_3$Cr-500 catalyst at 500 °C. (CH$_4$ = 20%; CO$_2$ = 20%; 100 mg; F = 100 mL/min).

For Ni$_2$Fe-500 and Ni$_3$Fe-500, a very different catalytic behavior from those of Ni$_R$Cr-500 was noticed. In spite of their high specific surface area (144–160 m^2/g, Table 3) and their low crystallite size (53–69 Å), both samples showed poor catalytic performances at all temperatures in the range of 400–650 °C (Figures 10 and 11). In addition, although Ni$_2$Fe-500 and Ni$_3$Fe-500 samples have different reducibility patterns (Figure 9), the activity remains almost negligible for both catalysts. The low activity of Ni$_2$Fe-500 and Ni$_3$Fe-500 is rather surprising because the amount of Ni species used in departure, which is responsible for DRM reaction, is the major constituent of the catalysts with respect to iron (Ni/Fe = 2 or 3). However, the catalytic behavior of Ni$_2$Fe-500 and Ni$_3$Fe-500 obtained from LDH structure is similar to ferrite spinel nanoparticles prepared by coprecipitation [19], hydrothermal [19] and sol-gel [38] methods. We can assign the poor catalytic performances of Ni$_R$Fe-500 to the presence of excess Fe^{3+} species on the Ni$_R$Fe-500 surface as revealed by XPS (Table 4). The Fe^{3+} species mainly favor RWGS reaction. On the other hand, the low activity can be linked to the disappearance of active Ni-metallic phase related to the formation of Ni-Fe alloy at the expense of Ni° and Fe° reduced species under reaction mixture, as confirmed in our previous works by in situ HT-XRD under flowing H$_2$ [19,38].

To better evaluate the catalytic properties of Ni_2Cr-500 and Ni_3Cr-500 catalysts, the fresh catalysts were heated from RT to reaction temperature in inert gas and then exposed to DRM mixture at 500 °C. Figure 12 shows the evolution of CH_4, CO_2 conversions, H_2 selectivity, and H_2/CO as a function of time. Ni_2Cr-500 shows higher conversion than Ni_3Cr-500, but activity decreases progressively with time, whereas that of Ni_3Cr-500 remains rather stable throughout the period studied (up to 70 min). CO_2 conversions are in the same range and tend to diminish with time on both samples. Although methane conversion increases and CO_2 decreases, the H_2 selectivity tends to decrease with time, while H_2/CO is very stable (Ni_2Cr-500) or increases progressively (Ni_3Cr-500) tending to the optimal stoichiometry of $H_2/CO = 1$. This is rather surprising, especially on Ni_2Cr-500, given that CO_2 conversion is significantly lower than that of CH_4 on this catalyst. This suggests that the presence of Cr^{3+} species probably limits CO_2 activation and the participation of sides reactions such as RWGS. However, in such conditions, higher H_2 selectivity and H_2/CO ratio should be expected. The low values observed can only be explained by simultaneous water production (not quantified), which would need a significant supply of oxygen species. This could be due to the reduction of the catalytic material.

The reactions could not be studied for longer period because after approx. 1 h, the pressure inside the reactor increased, brutally triggering the safety circuit of the setup and stopping the reaction. This could only be caused by an increase in the pressure drop due to severe carbon deposition in the catalyst bed.

The necessary production of water to close the mass balance and the brutal modification of the catalytic behavior after approx. one hour suggest that the catalysts undergo significant modifications during this period. Most probably, the NiO species are reduced to form metallic nickel, which is then responsible for the high carbon deposition. One would nevertheless expect that the activity and selectivity would progressively evolve, with carbon starting to be deposited as soon as nickel particles start being formed. On the contrary, the activity is rather stable during this reduction process, especially on Ni_3Cr-500, which, paradoxically, contains the largest amount of Ni. This suggests that the underlying $NiCr_2O_4$ phase can stabilize the metallic particles before it starts being reduced itself. At that moment, Ni particles may sinter rapidly and provoke sudden catalyst deactivation.

Given the evolution of the material during reaction, this catalytic behavior must be considered transient and cannot be extrapolated straightforwardly to a continuous DRM application. However, the selectivity towards syngas is remarkable at such low temperatures for the Ni_3Cr-500 catalyst. This opens the path for further investigation through adequate process conditions (e.g., by varying the CH_4/CO_2 composition) in order to slow the reduction of the material or through further investigation of material design and synthesis to better stabilize the active species. Another potential route to explore is the use of this material in non-steady state processes such as chemical looping reforming [39–43], or reaction–regeneration cyclic processes [44,45].

3. Materials and Methods

3.1. Chemicals

Nickel (II) nitrate hexahydrate ($Ni(NO_3)_2 \cdot 6H_2O$, $\geq 98\%$, Sigma Aldrich, St. Louis, MO, USA), chromium (III) nitrate nonahydrate ($Cr(NO_3)_3 \cdot 9H_2O$, $\geq 98\%$, Sigma Aldrich), iron (III) nitrate nonahydrate ($Fe(NO_3)_3 \cdot 9H_2O$, $\geq 98\%$, Sigma Aldrich, St. Louis, MO, USA), sodium hydroxide (NaOH, $\geq 98\%$, Sigma Aldrich), and sodium carbonate (Na_2CO_3, $\geq 98\%$, Sigma Aldrich, St. Louis, MO, USA) were used in LDH preparation. All reagents were analytical grade and used without any further purification. Distilled water was used in the synthesis and washing processes.

3.2. Catalyst Preparation

Ni_RFe and Ni_RCr LDH samples were prepared by coprecipitation method at constant pH at 70 °C with a divalent-to-trivalent cation molar ratio R of 2 and 3. These materials were denoted as Ni_2Cr, Ni_3Cr, Ni_2Fe, and Ni_3Fe LDHs. In a typical procedure, $Ni(NO_3)_2 \cdot 6H_2O$

and Fe(NO$_3$)$_3$·9H$_2$O were dissolved in distilled water to prepare a 1 M aqueous solution. Then, under vigorous stirring, Ni(NO$_3$)$_2$ (1M) and Fe(NO$_3$)$_3$ (1M) were dropped simultaneously (with Ni^{2+}/Fe^{3+} molar ratios in solution of 3:1 and 2:1) with an aqueous solution of NaOH on 100 mL of an aqueous solution of sodium carbonate. The pH during precipitation was maintained at a constant value of 10 by dropwise addition of NaOH solution at a temperature of 70 °C. After an aging step at this temperature for 24 h, the precipitates were recovered by filtration, washed several times with distilled water, and finally dried at 100 °C overnight in air. The same procedure was achieved to prepare Ni$_R$Cr LDH. Finally, the dried LDH samples were subjected to calcination at 500 °C (heating rate of 5 °C/min) and held for 4 h. The calcined samples were labeled as Ni$_2$Fe-500, Ni$_3$Fe-500, Ni$_2$Cr-500, and Ni$_3$Cr-500 (500 refers to the applied calcination temperature).

3.3. Catalysts Characterization

Several physicochemical methods were used for the characterization of the catalysts before and after heating treatment.

Powder X-ray powder diffraction (PXRD) was performed using a Bruker AXS D8 Advance diffractometer (Bruker, Billerica, MA, USA) working in Bragg–Brentano geometry using Cu Kα radiation (λ = 1.54 Å), equipped with a LynxEye detector. Patterns were collected at room temperature, in the 2θ = 10–90° range, with a 0.02° step and 96 s counting time per step. The EVA software was used for phase identification. The average crystallite size (CS) is calculated from the line broadening of the most intense peak using Scherer's formula, Cs = (0.9.λ)/(β cosθ), where (CS) is the average crystallite size, β is the half-maximum line width (FWHM), λ is the wavelength of radiation used (1.54056 Å), and θ is the angle of diffraction. X-ray diffraction at variable temperatures (HT-XRD) under an air atmosphere was carried out on the same apparatus equipped with XRK 900 chamber and a LynxEye detector. The patterns were collected every 25 °C, using a 0.1 °C/s heating rate between each temperature. The counting time being chosen to collect a diagram was set to 15 min in the 10–90° 2θ range. The sample was displayed on a platinum sheet. After measurement, the sample was cooled down to room temperature at a 0.3 °C/s cooling rate.

Thermogravimetry analysis (TGA) was performed on a SETARAM TG-92 (KEP Technologies, Caluire, France) thermobalance. The sample was heated at 5 °C/min in airflow conditions from 25 to 1000 °C. The released gases evolved during the analysis were monitored by a mass spectrometer (Pfeiffer Vacuum, Aßlar, Germany).

Laser-Raman spectra were recorded from 200 to 1500 cm^{-1} at room temperature using a FT-Raman spectrometer (Dilor XY Raman, Horiba France, Palaiseau, France) at an excitation wavelength of 647.1 nm, laser power of 3 mW, and spectral resolution of 0.5 cm^{-1}.

Attenuated Total Reflection Infra-Red spectra (IR-ATR) were recorded at room temperature using a Perkin Elmer model 400 (Perkin Elmer Inc., Waltham, MA, USA) in transmission mode, in the range from 350 to 4000 cm^{-1}.

The surface areas and pore size were calculated from N$_2$ adsorption–desorption isotherms measured on an ASAP 2020 (Micromeritics, Norcross, GA, USA) analyzer by Brunauer–Emmett–Teller (B.E.T) and Barret–Joyner–Halenda (B.J.H) methods.

XPS analyses were recorded using a Kratos Analytical Axis UltraDLD spectrometer (Kratos Analytical, Manchester, UK). The excitation was ensured by a monochromatic aluminum Kα source at 1486.6 eV operating at 180 W. The Kratos charge compensation system was applied to neutralize any charging effects. The residual pressure in the analysis chamber was below 5 · 10^{-10} Torr. Survey scans were acquired at a pass energy of 160 eV with a 1 eV step, while core level spectra were acquired at 20 eV pass energy and with a 0.1 eV step. Data were processed using Casa XPS software. All spectra were calibrated using the C1s photoelectron peak corresponding to C-C bonds at 284.8 eV.

The reducible species which exist in the catalysts were profiled by temperature-programmed reduction. Hydrogen temperature-programmed reduction (H$_2$-TPR) was measured on a AutoChem II 2920 (Micromeritics, Norcross, GA, USA) apparatus with a thermal conductivity detector (TCD) to monitor the H$_2$ consumption. After calibration of

H_2 on the TCD, samples were sealed in a U-shaped quartz tube reactor and pre-treated in an argon atmosphere to remove surface impurities. Then, the temperature was raised from 25 to 1000 °C at 5 °C/min in a stream of 5% v/v H_2/Ar.

3.4. Catalytic Reforming Experiments

The tests of catalytic CO_2 reforming of methane were carried out at atmospheric pressure in a fixed-bed U-type quartz reactor. A 100 mg sample of catalyst was thoroughly mixed with SiC powder before loading in the reactor. The gas mixture containing CH_4:CO_2:He:Ar = 20:20:10:50 with a total flow of 100 mL/min was used, and the catalytic reaction was carried out in temperature-programmed mode from room temperature to 650 °C at a 5 °C/min heating rate. The gas flow was continuously monitored online using a Prisma 200 Pfeiffer mass spectrometer. Isothermal reactivity was performed using a new catalyst sample heated to reaction temperature (500 °C) in Argon and then exposed for approx. 1 h in the same reaction conditions.

4. Conclusions

Ni_RM (M = Cr or Fe, R = 2 or 3) hydrotalcite precursors were prepared using the coprecipitation method and were subsequently tested in the dry reforming of methane without any prior H_2 treatment. All the physicochemical characterization confirms the successful formation of the takovite structure. Upon calcination at 500 °C, Ni_RM hydrotalcites yielded stable mixed oxides consisting of a NiO phase and spinel structure ($NiCr_2O_4$ or $NiFe_2O_4$). Surface compositions evaluated by the XPS reveal different surface properties with Fe^{3+} species mainly at the surface of Ni_RFe systems and, in contrast, a balanced surface in Ni^{2+} and Cr^{3+} species for Ni_RCr catalysts. Ni_RCr catalysts are active and selective for DRM compared to Ni_RFe systems, showing the role of the trivalent metal on the structural and textural properties. Despite their high specific surface areas, the activity of Ni_RFe catalysts is low and can be attributed to (i) the localization of Fe^{3+} species on the surface and (ii) the loss of Ni-metal during the catalytic process, due to the formation of the Ni-Fe alloy favoring RWGS reaction. Ni_RCr catalysts show remarkable activity between 450 and 600 °C, in particular in terms of selectivity in such a low-temperature range. The deactivation of the catalysts at higher temperatures or after a long reaction time suggests a transient behavior associated with the reduction of NiO species to metallic Ni particles stabilized by the underlying $NiCr_2O_4$ phase or the presence of Cr_2O_3 oxide. During this process, the Ni particles remain active and selective until the $NiCr_2O_4$ start being reduced, provoking the sintering of the active phase. The remarkable properties of these partially reduced catalysts provide interesting perspectives for the use of these materials in non-steady state (looping or cycling) processes for methane valorization at particularly low temperatures for reforming reactions by CO_2.

Supplementary Materials: The following supporting information can be downloaded at: https://www.mdpi.com/article/10.3390/catal12121507/s1, Figure S1: TG-MS curves of Ni_2Fe and Ni_3Fe LDH precursors performed in air atmosphere; Figure S2: TG-MS curves of Ni_2Cr and Ni_3Cr LDH precursors performed in air atmosphere; Figure S3: XPS spectra of Fe2p3/2 species of (a) Ni_3Fe-500 and (b) Ni_2Fe-500; Figure S4: XPS spectra of Ni2p3/2 species of (a) Ni_3Cr-500, (b) Ni_2Cr-500, (c) Ni_3Fe-500 and (d) Ni_2Fe-500; Figure S5: XPS spectra of Cr2p species of (a) Ni_3Cr-500 and (b) Ni_2Cr-500; Figure S6: XPS spectra of O1s species of (a) Ni_3Cr-500, (b) Ni_2Cr-500, (c) Ni_3Fe-500 and (d) Ni_2Fe-500. Table S1: TGA-MS of Ni_RFe and Ni_RCr LDH precursors. Table S2: Catalytic performances in DRM, temperature-programmed mode.

Author Contributions: Conceptualization, R.B. and A.L.; Data curation, M.H.; Funding acquisition, R.B. and A.L.; Investigation, M.H., R.B., N.F.C., D.L., R.C., K.B., A.R., P.R., R.-N.V. and M.T.; Methodology, R.B. and A.L.; Supervision, R.B. and A.L.; Writing—original draft, M.H.; Writing—review and editing, M.H., R.B., N.F.C., D.L., R.C., K.B., A.R., P.R., R.-N.V., M.T. and A.L. All authors have read and agreed to the published version of the manuscript.

Funding: This work was partially supported by an Algeria–France cooperation program PHC-TASSILI (project N°19MDU206). The Fonds Européen de Développement Régional (FEDER), CNRS, Région Hauts-de-France, Chevreul Institute (FR 2638) and Ministère de l'Education Nationale de l'Enseignement Supérieur et de la Recherche are acknowledged for funding XPS spectrometers and XRD instruments.

Data Availability Statement: Data are available within the article.

Acknowledgments: The authors are grateful to Laurence Burylo, Olivier Gardol, and Nora Djelal, for the technical assistance.

Conflicts of Interest: The authors declare that they have no known competing financial interests or personal relationships that could have appeared to influence the work reported in this paper.

References

1. Tungatarova, S.; Xanthopoulou, G.; Vekinis, G.; Karanasios, K.; Baizhumanova, T.; Zhumabek, M.; Sadenova, M. Ni-Al Self-Propagating High-Temperature Synthesis Catalysts in Dry Reforming of Methane to Hydrogen-Enriched Fuel Mixtures. *Catalysts* **2022**, *12*, 1270. [CrossRef]
2. Pinheiro, A.L.; Pinheiro, A.N.; Valentini, A.; Filho, J.M.; Sousa, F.F.; Sousa, J.R.; Rocha, M.G.C.; Bargiela, P.; Oliveira, A.C. Analysis of coke deposition and study of the structural features of MAl_2O_4 catalysts for the dry reforming of methane. *Catal. Commun.* **2009**, *11*, 11–14. [CrossRef]
3. Crisafulli, C.; Scire, S.; Maggiore, R.; Minico, S.; Galvagno, S. CO_2 reforming of methane over Ni–Ru and Ni–Pd bimetallic catalysts. *Catal. Lett.* **1999**, *59*, 21–26. [CrossRef]
4. Wang, H.Y.; Ruckenstein, E. Carbon dioxide reforming of methane to synthesis gas over supported rhodium catalysts: The effect of support. *Appl. Catal. A-Gen.* **2000**, *204*, 143–152. [CrossRef]
5. Erdohelyi, A. Catalytic Reaction of Carbon Dioxide with Methane on Supported Noble Metal Catalysts. *Catalysts* **2021**, *11*, 159. [CrossRef]
6. Ruckenstein, E.; Hu, Y.H. Carbon dioxide reforming of methane over nickel/alkaline earth metal oxide catalysts. *Appl. Catal. A-Gen.* **1995**, *133*, 149–161. [CrossRef]
7. Rostrup-Nielsen, J.R.; Bak Hansen, J.H. CO_2-reforming of methane over transition metals. *J. Catal.* **1993**, *144*, 38–49. [CrossRef]
8. Bradford, M.C.J.; Vannice, M.A. CO_2 reforming of CH_4. *Catal. Rev.-Sci. Eng.* **1999**, *41*, 1–42. [CrossRef]
9. Rostrup-Nielsen, J.R. Production of synthesis gas. *Catal. Today* **1993**, *18*, 305–324. [CrossRef]
10. Wang, S.B.; Lu, G.Q. Carbon dioxide reforming of methane to produce synthesis gas over metal-supported catalysts: state of the art. *Energy Fuels* **1996**, *10*, 896–904. [CrossRef]
11. Romero, A.; Jobbagy, M.; Laborde, M.; Baronetti, G.; Amadeo, N. Ni(II)–Mg(II)–Al(III) catalysts for hydrogen production from ethanol steam reforming: Influence of the Mg content. *Appl. Catal. A-Gen.* **2014**, *47*, 398–404. [CrossRef]
12. Jin, L.; Xie, T.; Ma, B.; Li, Y.; Hu, H. Preparation of carbon-Ni/MgO-Al_2O_3 composite catalysts for CO_2 reforming of methane. *Int. J. Hydrog. Energy* **2017**, *42*, 5047–5055. [CrossRef]
13. Roussel, H.; Briois, V.; Elkaim, E.; De Roy, A.; Besse, J.P.; Jolivet, J.P. Study of the Formation of the layered double hydroxide [Zn–Cr–Cl]. *Chem. Mater.* **2001**, *13*, 329–337. [CrossRef]
14. You, Y.; Zhao, H.; Vance, G.F. Hybrid organic–inorganic derivatives of layered double hydroxides and dodecylbenzenesulfonate: Preparation and adsorption characteristics. *J. Mater. Chem.* **2002**, *12*, 907–912. [CrossRef]
15. Chatla, A.; Almanassra, I.W.; Kochkodan, V.; Laoui, T.; Alawadhi, H.; Atieh, M.A. Efficient Removal of Eriochrome Black T (EBT) Dye and Chromium (Cr) by Hydrotalcite-Derived Mg-Ca-Al Mixed Metal Oxide Composite. *Catalysts* **2022**, *12*, 1247. [CrossRef]
16. Triantafyllidis, K.S.; Peleka, E.N.; Komvokis, V.G.; Mavros, P.P. Iron-modified hydrotalcite-like materials as highly efficient phosphate sorbents. *J. Colloid Interface Sci.* **2010**, *342*, 427–436. [CrossRef]
17. Frost, R.; Jagannadha-Reddy, B. Thermo-Raman spectroscopic study of the natural layered double hydroxide manasseite. *Spectrochim. Acta A Mol. Biomol. Spectrosc.* **2006**, *65*, 553–559. [CrossRef]
18. Borromeo, L.; Zimmermann, U.; Andò, S.; Coletti, G.; Bersani, D.; Basso, D.; Gentile, P.; Schulz, B.; Garzanti, E. Raman spectroscopy as a tool for magnesium estimation in Mg-calcite. *J. Raman Spectrosc.* **2017**, *48*, 983–992. [CrossRef]
19. Benrabaa, R.; Boukhlouf, H.; Löfberg, A.; Rubbens, A.; Vannier, R.N.; Bordes-Richard, E.; Barama, A. Nickel ferrite spinel as catalyst precursor in the dry reforming of methane: Synthesis, characterization and catalytic properties. *J. Nat. Gas Chem.* **2012**, *21*, 595–604. [CrossRef]
20. Kreisel, J.; Lucazeau, G.; Vincent, H. Raman Spectra and Vibrational Analysis of $BaFe_{12}O_{19}$ Hexagonal Ferrite. *J. Sol. State Chem.* **1998**, *137*, 127–137. [CrossRef]
21. Faid, A.Y.; Barnett, A.O.; Seland, F.; Sunde, S. Ni/NiO nanosheets for alkaline hydrogen evolution reaction: In situ electrochemical-Raman study. *Electrochim. Acta* **2020**, *361*, 137040. [CrossRef]
22. Colomban, P.; Jullian, S.; Parlier, M.; Monge-Cadet, P. Identification of the high-temperature impact/friction of aeroengine blades and cases by micro Raman spectroscopy. *Aerosp. Sci. Technol.* **1999**, *3*, 447–459. [CrossRef]

23. Singh, J.; Kumar, R.; Vermaa, V.; Kumar, R. Role of Ni^{2+} substituent on the structural, optical and magnetic properties of chromium oxide ($Cr_{2-x}Ni_xO_3$) nanoparticles. *Ceram. Int.* **2020**, *46*, 24071–24082. [CrossRef]
24. Olszówka, E.; Karcz, R.; Bielańska, E.; Kryściak-Czerwenka, J.; Napruszewska, D.; Sulikowski, B.; Socha, P.; Gaweł, A.; Bahranowski, K.; Olejniczak, Z.; et al. New insight into the preferred valency of interlayer anions in hydrotalcitelike compounds: The effect of Mg/Al ratio. *Appl. Clay Sci.* **2018**, *155*, 84–94. [CrossRef]
25. Rozov, K.; Berner, U.; Taviot-Gueho, T.; Leroux, F.; Renaudin, G.; Kulil, D.; Diamond, L.W. Synthesis and characterization of the LDH hydrotalcite–pyroaurite solid-solution series. *Cem. Concr. Res.* **2010**, *40*, 1248–1254. [CrossRef]
26. Benito, P.; Labajos, F.M.; Rives, V. Microwave-treated layered double hydroxides containing Ni^{2+} and Al^{3+}: The effect of added Zn^{2+}. *J. Solid State Chem.* **2006**, *179*, 3784–3797. [CrossRef]
27. Kovanda, F.; Rojka, T.; Bezdicka, P.; Jiratova, K.; Obalova, L.; Pacultova, K.; Bastl, Z.; Grygar, T. Effect of hydrothermal treatment on properties of Ni–Al layered double hydroxides and related mixed oxides. *J. Solid State Chem.* **2009**, *182*, 27–36. [CrossRef]
28. Shebanova, O.N.; Lazor, P. Raman spectroscopic study of magnetite ($FeFe_2O_4$): A new assignment for the vibrational spectrum. *J. Solid State Chem.* **2003**, *174*, 424–430. [CrossRef]
29. Graves, P.R.; Johnston, C.; Campaniello, J.J. Raman scattering in spinel structure ferrites. *Mat. Bul Res.* **1988**, *23*, 1651–1660. [CrossRef]
30. D'Ippolito, V.; Andreozzi, G.B.; Bersani, D.; Lotticic, P.P. Raman fingerprint of chromate, aluminate and ferrite spinels. *J. Raman Spectrosc.* **2015**, *46*, 1255–1264. [CrossRef]
31. Hyun-Kim, J.; Soon- Hwang, I. Development of an in situ Raman spectroscopic system for surface oxide films on metals and alloys in high temperature water. *Nucl. Eng. Des.* **2005**, *235*, 1029–1040. [CrossRef]
32. Takehira, K.; Kawabata, T.; Shishido, T.; Murakami, K.; Ohi, T.; Shoro, D.; Honda, M.; Takaki, K. Mechanism of reconstitution of hydrotalcite leading to eggshell-type Ni loading on Mg–Al mixed oxide. *J. Catal.* **2005**, *231*, 92–104. [CrossRef]
33. X-ray Photoelectron Spectroscopy (XPS) Reference Pages. Available online: http://www.xpsfitting.com/ (accessed on 1 September 2022).
34. Biesingera, M.C.; Payne, B.P.; Grosvenor, A.P.; Laua, L.W.M.; Gerson, A.R.; Smart, R.S.C. Resolving surface chemical states in XPS analysis of first row transition metals, oxides and hydroxides: Cr, Mn, Fe, Co and Ni. *Appl. Surf. Sci.* **2011**, *257*, 2717–2730. [CrossRef]
35. Hosseini, S.A.; Alvarez-Galvan, M.C.; Fierro, J.L.G.; Niaei, A.; Salari, D. MCr_2O_4 (M=Co, Cu, and Zn) nanospinels for 2-propanol combustion: Correlation of structural properties with catalytic performance and stability. *Ceram. Int.* **2013**, *39*, 9253–9261. [CrossRef]
36. Rouibah, K.; Barama, A.; Benrabaa, R.; Guerrero-Caballero, J.; Kane, T.; Vannier, R.N.; Rubbens, A.; Löfberg, A. Dry reforming of methane on nickel-chrome, nickel-cobalt and nickel-manganese catalysts. *Int. J. Hydrogen Energy* **2017**, *42*, 29725–29734. [CrossRef]
37. Benrabaa, R.; Aissat, F.; Fodil Cherif, N.; Gouasmia, A.; Yeste, P.; Cauqui, M.A. Catalytic oxidation of carbon monoxide over CeO_2 and La_2O_3 oxides supported nickel catalysts: The effect of the support and NiO loading. *ChemistrySelect* **2022**, *7*, 104–133. [CrossRef]
38. Benrabaa, R.; Löfberg, A.; Rubbens, A.; Bordes-Richard, E.; Vannier, R.N.; Barama, A. Structure, reactivity and catalytic properties of nanoparticles of nickel ferrite in the dry reforming of methane. *Catal. Today* **2013**, *203*, 188–195. [CrossRef]
39. Bhavsar, S.; Najera, M.; Solunke, R.; Veser, G. Chemical looping: To combustion and beyond. *Catal. Today* **2014**, *228*, 96–105. [CrossRef]
40. Bhavsar, S.; Veser, G. Chemical looping beyond combustion: Production of synthesis gas via chemical looping partial oxidation of methane. *RSC Adv.* **2014**, *4*, 47254–47267. [CrossRef]
41. Galvita, V.; Poelman, H.; Detavernier, C.; Marin, G. Catalyst-assisted chemical looping for CO_2 conversion to CO. *Appl. Catal. B Environ.* **2015**, *164*, 184–191. [CrossRef]
42. Löfberg, A.; Guerrero-Caballero, J.; Kane, T.; Rubbens, A.; Jalowiecki-Duhamel, L. Ni/CeO_2 based catalysts as oxygen vectors for the chemical looping dry reforming of methane for syngas production. *Appl. Catal. B Environ.* **2017**, *212*, 159–174. [CrossRef]
43. Löfberg, A.; Guerrero-Caballero, J.; Kane, T.; Jalowiecki-Duhamel, L. Chemical looping dry reforming of methane: Toward shale-gas and biogas valorization. *Chem. Eng. Process. Process Intensif.* **2017**, *122*, 523–529. [CrossRef]
44. Tang, M.; Xu, L.; Fan, M. Progress in oxygen carrier development of methane-based chemical-looping reforming: A review. *Appl. Energy* **2015**, *151*, 143–156. [CrossRef]
45. Assabumrungrat, S.; Charoenseri, S.; Laosiripojana, N.; Kiatkittipong, W.; Praserthdam, P. Effect of oxygen addition on catalytic performance of Ni/SiO_2-MgO toward carbon dioxide reforming of methane under periodic operation. *Int. J. Hydrogen Energy* **2009**, *34*, 6211–6220. [CrossRef]

Article

Catalytic Performance of Alumina-Supported Cobalt Carbide Catalysts for Low-Temperature Fischer–Tropsch Synthesis

Zahra Gholami [1,*], Zdeněk Tišler [1], Eliška Svobodová [1], Ivana Hradecká [1], Nikita Sharkov [1] and Fatemeh Gholami [2]

1 ORLEN UniCRE, a.s., 436 70 Litvínov, Czech Republic
2 New Technologies-Research Centre, University of West Bohemia, 301 00 Plzeň, Czech Republic
* Correspondence: zahra.gholami@orlenunicre.cz; Tel.: +420-731-576-893

Abstract: The determination of the catalyst's active phase helps improve the catalytic performance of the Fischer–Tropsch (FT) synthesis. Different phases of cobalt, including cobalt oxide, carbide, and metal, exist during the reaction. The content of each phase can affect the catalytic performance and product distribution. In this study, a series of cobalt carbide catalysts were synthesized by exposure of Co/Al_2O_3 catalyst to CH_4 at different temperatures from 300 °C to 800 °C. The physicochemical properties of the carbide catalysts (CoC_x/Al_2O_3) were evaluated by different characterization methods. The catalytic performances of the catalysts were investigated in an autoclave reactor to determine the role of cobalt carbides on the CO conversion and product distribution during the reaction. XRD and XPS analysis confirmed the presence of Co_2C in the prepared catalysts. The higher carbidation temperature resulted in the decomposition of methane into hydrogen and carbon, and the presence of graphitic carbon was confirmed by XRD, XPS, SEM, and Raman analysis. The Co_2C also decomposed to metallic cobalt and carbon, and the content of cobalt carbide decreased at higher carbidation temperatures. Higher content of Co_2C resulted in a lower CO conversion and higher selectivity to light alkanes, mainly methane. The higher carbidation temperature resulted in the decomposition of Co_2C to metallic cobalt with higher activity in the FT reaction. The CO conversion increased by increasing the carbidation temperature from 300 °C to 800 °C, due to the higher content of metallic cobalt. In the presence of pure hydrogen, the Co_2C could be converted mainly into hexagonal, close-packed (hcp) Co with higher activity for dissociative adsorption of CO, which resulted in higher catalyst activity and selectivity to heavier hydrocarbons.

Keywords: Fischer-Tropsch; cobalt carbide; active phase; product distribution; metallic cobalt

Citation: Gholami, Z.; Tišler, Z.; Svobodová, E.; Hradecká, I.; Sharkov, N.; Gholami, F. Catalytic Performance of Alumina-Supported Cobalt Carbide Catalysts for Low-Temperature Fischer–Tropsch Synthesis. *Catalysts* 2022, 12, 1222. https://doi.org/10.3390/catal12101222

Academic Editors: Fanhui Meng and Wenlong Mo

Received: 27 September 2022
Accepted: 10 October 2022
Published: 12 October 2022

Publisher's Note: MDPI stays neutral with regard to jurisdictional claims in published maps and institutional affiliations.

Copyright: © 2022 by the authors. Licensee MDPI, Basel, Switzerland. This article is an open access article distributed under the terms and conditions of the Creative Commons Attribution (CC BY) license (https://creativecommons.org/licenses/by/4.0/).

1. Introduction

Fischer–Tropsch (FT) synthesis is a well-known catalytic process for the production of sulfur and aromatic-free fuels and other value-added chemicals from syngas. Through FT synthesis, which is a catalytic polymerization reaction, syngas is converted to a wide range of products, such as paraffins, olefins, and alcohols. The catalytic performance and product distribution are directly affected by the type of catalyst and its chemical composition and structure. The elements of groups 8–10 of the periodic table with good ability for adsorption and dissociation of CO and H_2 are reported as active metals for the FT reaction, whereas iron and cobalt are the most commonly used catalysts for this reaction [1–4]. Cobalt-based catalysts with higher stability, more resistance to deactivation, lower water–gas shift activity, and high hydrocarbon productivity were considered the ideal choice for synthesizing long-chain hydrocarbons at moderate temperatures and pressures [5,6].

During the FT reaction over Co-based catalysts (which typically proceeds at 220–250 °C, 20 bar, and an H_2/CO ratio of two), the chain growth and propagation of hydrocarbons mainly occurred on cobalt metal sites. Cobalt oxide and cobalt carbide, which are formed by re-oxidation and carburization of the metal catalyst during the FT synthesis, are known as inactive cobalt components in FT synthesis, and they are often considered deactivation signs [2,5,7,8]. However, some other groups reported that cobalt carbides could enhance the selectivity to lower olefins [9–11]; they also act as an active site for a water–gas shift reaction and oxygenates production [12–14] or act as an intermediate reacts with hydrogen to form methylene, acting as a chain-growth monomer for the formation of long-chain hydrocarbons [15]. However, unlike the cobalt catalysts, during the FT synthesis over iron-based catalysts, the Hägg carbide (χ-Fe_5C_2), as the real active phase, plays the main role in catalytic performances [8,16]. The presence of cobalt carbide phases during the FT synthesis could affect the CO conversion and product selectivity [17].

However, Co_2C is often known as an inactive component in FT synthesis via Co-based catalysts, but Co_2C content, their different morphologies, and the form of the Co_2C phase could have different effects on the catalytic performance of Co-based catalysts during the FT reaction. The Co_2C(111) facet is ascribed to the nanosphere-like particles with lower activity and higher methane selectivity under mild FT reaction conditions [10,13,18]. Co_2C nanoprisms with specific exposed facets of (101) and (020), as the FTO active phase, showed a higher intrinsic activity for CO hydrogenation and a lower methane selectivity [9,10,18]. The presence of Co^0 resulted in a lower CO_2 and methane selectivity and higher selectivity to longer-chain hydrocarbons. The Co_2C was reported to be relatively stable under FT reaction conditions [5,19]. At the same time, it was reported that the cobalt carbides appear to be unstable and can be decomposed to metallic cobalt and carbon during the FT reaction, especially at higher reaction temperatures and lower H_2/CO ratio [1,3,20].

In the present study, a series of CoC_x/Al_2O_3 catalysts were synthesized by carbidation of Co/Al_2O_3 catalysts using methane at different temperatures from 300 °C to 800 °C. The properties of these catalysts changed according to their carbidation temperatures. Changing the carbidation temperature could result in the formation of different content and types of cobalt phases, which can alter their catalyst activity and product selectivity during the FT reaction. The physicochemical properties of the catalysts were investigated by different characterization methods, including scanning electron microscopy (SEM), X-ray diffraction (XRD), X-ray photoelectron spectroscopy (XPS), and Raman analysis. The catalytic performances of the catalysts were evaluated in an autoclave reactor to determine the role of cobalt carbides on the CO conversion and product distribution during the reaction.

2. Results and Discussion

2.1. Catalyst Characterization

SEM images (Figure 1) of the prepared CoC_x/Al_2O_3 catalysts at different temperatures confirmed the formation of carbon fibers on the surface after the decomposition of methane at higher temperatures. The formation of carbon at lower temperatures was not detected in SEM images, but the carbon fibers can be seen on the surface of catalysts at high temperatures of 700 °C and 800 °C, which was in good accordance with the obtained results in XRD and elemental analysis results.

The peaks belonging to Al_2O_3 at ~37.5°, 46.1°, 56°, and 67° (JCPDS 10-0425) were observed in the XRD patterns of all prepared catalysts (Figure 2a) [21–24]. The peaks at 37.0°, 41.3°, 42.5°, 45.7°, and 56.6° can be assigned to the Co_2C crystalline plane (110), (002), (111), (021), and (112), (JCPDS 65-1457) [3,9,10,19,25,26] respectively. However, the peak at 56.6° was not very visible in the XRD patterns shown in Figure 1a. The cobalt carbide decomposition to graphitic carbon and metallic cobalt ($Co_2C \rightarrow Co + C$) was started at temperatures above 275 °C [5,27,28], and it could be the reason for gradually decreasing the cobalt carbide and increasing the metallic cobalt by increasing the temperature. In addition, in the presence of Co/Al_2O_3 catalyst, methane also can be decomposed directly to hydrogen and carbon ($CH_4 \rightarrow 2H_2 + C$) [29–33]. At high temperatures (above 600 °C), due to the decomposition of methane, the concentration of graphitic carbon also significantly increased, as shown by the XRD spectra of the catalysts. Decomposition of methane on cobalt catalysts led to the formation of fcc cobalt [34]. The peak at 26.3° (graphite, JCPDS 41-

1478) and 51.5° (fcc Co, JCPDS 1-1255) [35,36] were observed only in the sample prepared at a temperature above 600 °C. The peak at 44.3° was very weak for the samples prepared at low temperatures, and the peak's intensity increased by increasing the temperature. These peaks can belong to both fcc Co (111) (JCPDS 1-1255) and graphitic carbon (JCPDS 41-1478) [3,9,36], which could be formed due to the reduction of cobalt oxide to Co (fcc) during the catalysts' carbidation in the presence of 20%CH_4 in H_2. By increasing the carbidation temperature, the intensity of fcc Co increased, whereas the intensity of the $Co_2C(111)$ facet (at 2θ of 42.5°) decreased. The $Co_2C(111)$ facet is responsible for lower CO hydrogenation and higher CH_4 selectivity during the FT synthesis under mild reaction conditions.

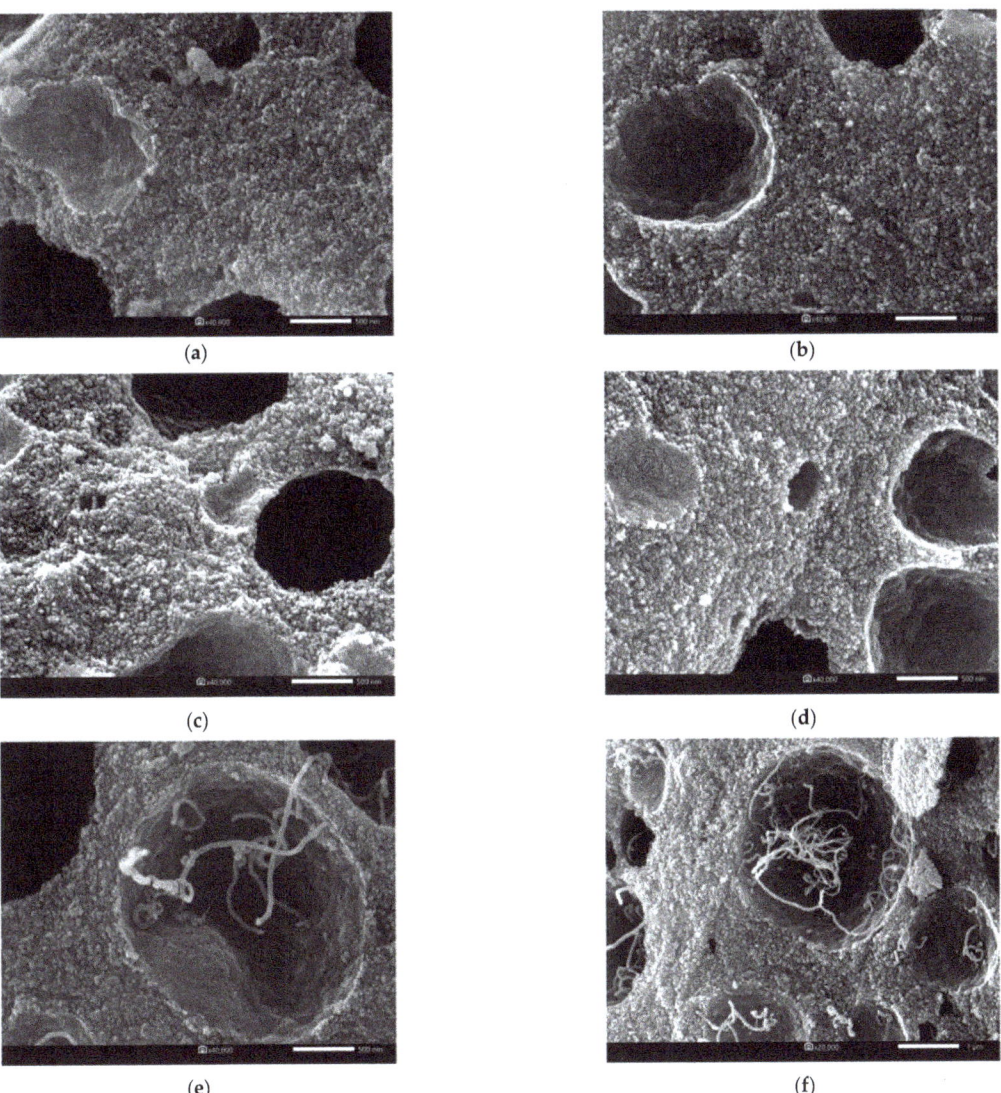

Figure 1. SEM images of the $5CoC_x/Al_2O_3$ catalysts prepared at different temperatures, (**a**) 300 °C, (**b**) 400 °C, (**c**) 500 °C, (**d**) 600 °C, (**e**) 700 °C, (**f**) 800 °C.

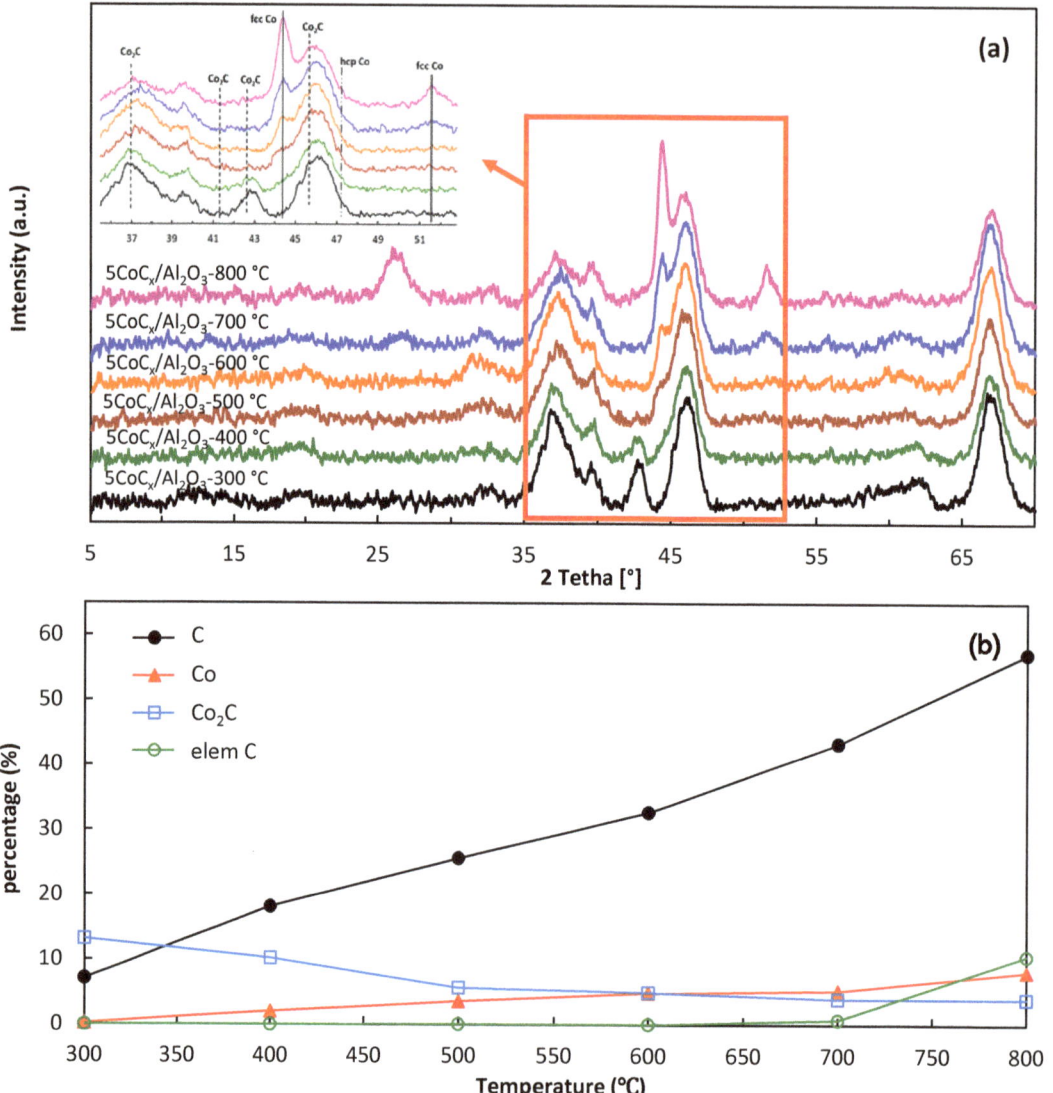

Figure 2. (a) XRD patterns of 5CoC$_x$/Al$_2$O$_3$ and catalysts prepared at different temperatures, (b) Content of individual phases according to SQ performed from XRD spectra (C-carbon, Co-cobalt, Co$_2$C-carbide) and carbon content determined by elemental analysis (elemC).

The contents of phases of C (carbon), Co (cobalt), and Co$_2$C (cobalt carbide) were detected from the XRD results and semi-quantitative (SQ) evaluation, and the carbon content (elemental C) of the prepared samples was also measured using elemental analysis (Figure 2b). Decomposition of methane at higher temperatures increased the carbon content on the surface of the prepared catalysts. By increasing the temperature, the formed Co$_2$C became unstable and decomposed to C and metallic Co, and its content decreased [5,27,28]. Decomposition of cobalt carbide to graphitic carbon and hydrogen is started at temperatures above 275 °C; however, in the presence of hydrogen, it could happen at lower temperatures [5,27,37]. The XRD pattern (Figure 2a) showed that the intensity of the peaks belonging to the Co_2C decreased gradually by increasing the temperature from 300 °C to 800 °C [19]. Elemental analysis of the carbon (Figure 2b) also confirmed that an increase in the temperature led to the formation of more carbon. Cobalt carbides were unstable at high temperatures; they decomposed to metallic cobalt and carbon at high temperatures. However, due to the overlapping of the carbon and cobalt peaks, with the similar cubic structure, it was difficult to accurately distinguish between the two phases in XRD patterns. XPS analysis was performed for further investigation and evaluation of the active phases of the catalysts.

XPS analysis was used to investigate the chemical composition and elemental oxidation states of the catalysts. The survey scan of carbide catalysts is shown in Figure S1, which confirmed the existence of C, O, Co, and Al in these prepared catalysts. The intensity of C 1s peaks increased by increasing the temperature, whereas the intensity of Co 2p peaks decreased. The strong C 1s peak observed at about 284.6 eV was ascribed to sp2 elemental carbon (C-C), such as carbon fibers or carbon nanotubes [38]. The high intensity of this peak in the C 1s spectra (Figure 3a) of the catalysts prepared at higher temperatures (800 °C and 700 °C) was due to the presence of graphitic carbon, as confirmed by SEM, XRD, and Raman analysis. The Co-C peak at around 283.1 eV followed the reverse trend. By increasing the temperature of carbidation, the content of Co-C decreased. In addition to the C-C and Co-C bonds, other bonds at 286.5 eV and 288.5 eV were attributed to the O-C-O and O=C-O bonds, respectively. The $\pi \rightarrow \pi^*$ shake-up satellite peak was also observed at about 291.2 eV. The obtained results are in good agreement with the reported results where the Co-C bonds were C-Co bonds; these were detected, confirming the presence of Co_2C in the prepared catalysts [39,40].

XPS spectra of Co 2p (Figure 3b) showed that the bond at about 778.6 eV was attributed to the Co^{2+} of carbidic Co of the prepared catalysts. The intensity of the Co-C bond decreased by the carbidation temperature, whereas the intensity of peaks belonging to metallic cobalt (Co^0) (at around 782.3 eV) increased at higher temperatures. These findings are in good agreement with the XRD analysis of the catalysts, where the Co_2C content decreased at higher temperatures, and metallic cobalt peaks were detected at higher temperatures. However, by increasing the temperature to above 600 °C and formation of more graphitic carbon on the surface of the catalysts, the intensity of the Co peaks in Co 2p spectra decreased, which is in good accordance with the C 1s spectra of these catalysts, where the content of graphitic carbon increased significantly at high temperatures.

As mentioned earlier, the decomposition of cobalt carbide to cobalt and graphitic carbon started at temperatures above 275 °C [5,27], and this could be the reason for the increase in the Co^0/Co-C and C-C/Co^0 ratios when the carbidation temperature increased (Table 1). However, for the catalyst with the carbidation temperature of 800 °C, the Co^0/Co-C ratios decreased compared with that of 700 °C, which could be due to the deposition of graphitic carbon on the catalyst's surface and blockage of metallic cobalt on the surface. The sharp increase in the C-C/Co^0 at 800 °C was in good agreement with this finding. The XPS analysis of cobalt carbide nanoparticles and nanosheets performed by other research groups also showed similar patterns for cobalt carbides, and the peaks attributed to cobalt carbides were observed at about 283 eV (C 1s) and 778.5 eV (Co 2p 3/2) [26,41,42].

Raman spectroscopy analysis could be beneficial for the investigation of the materials' properties and could provide information about molecular vibrations. Raman spectra of the Co/Al_2O_3 catalysts at different carbidation temperatures are shown in Figure 4. The D and G bands belonging to carbon were not clearly observed for the catalysts prepared at the carbidation temperatures in the range of 300 °C to 600 °C. The D and G bands were clearly detected for the catalysts prepared at 700 °C and 800 °C The peaks at around 1340 cm^{-1} and 1573 cm^{-1} were typical for D and G bands, respectively. The D and G bands of the carbon materials could be observed if amorphous carbon or nanocrystalline graphite existed in a metal carbide matrix [43,44].

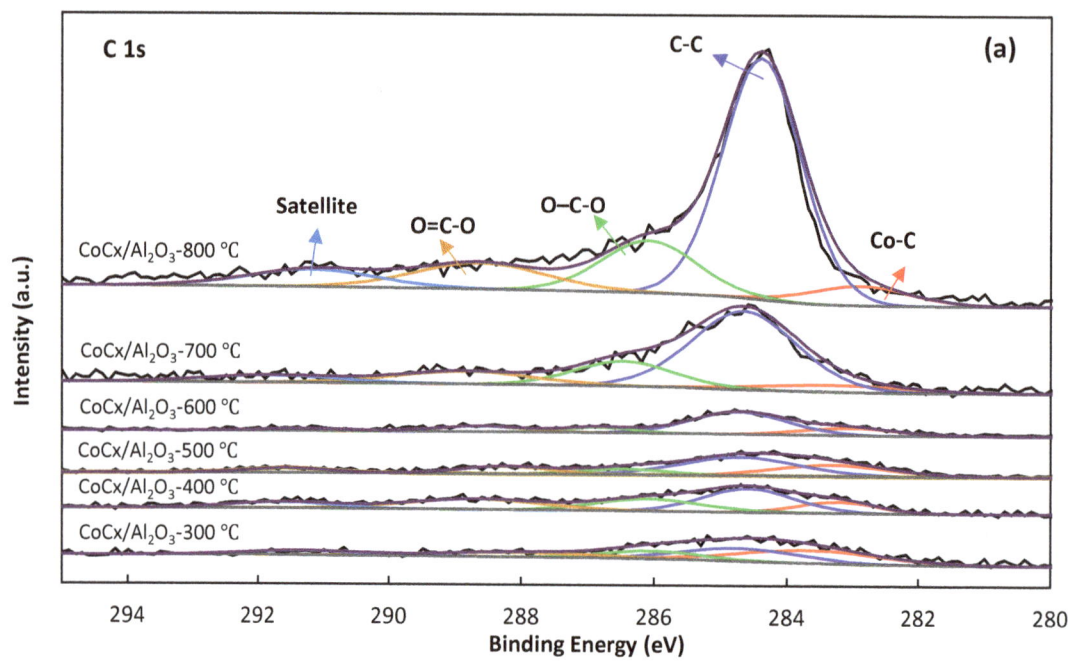

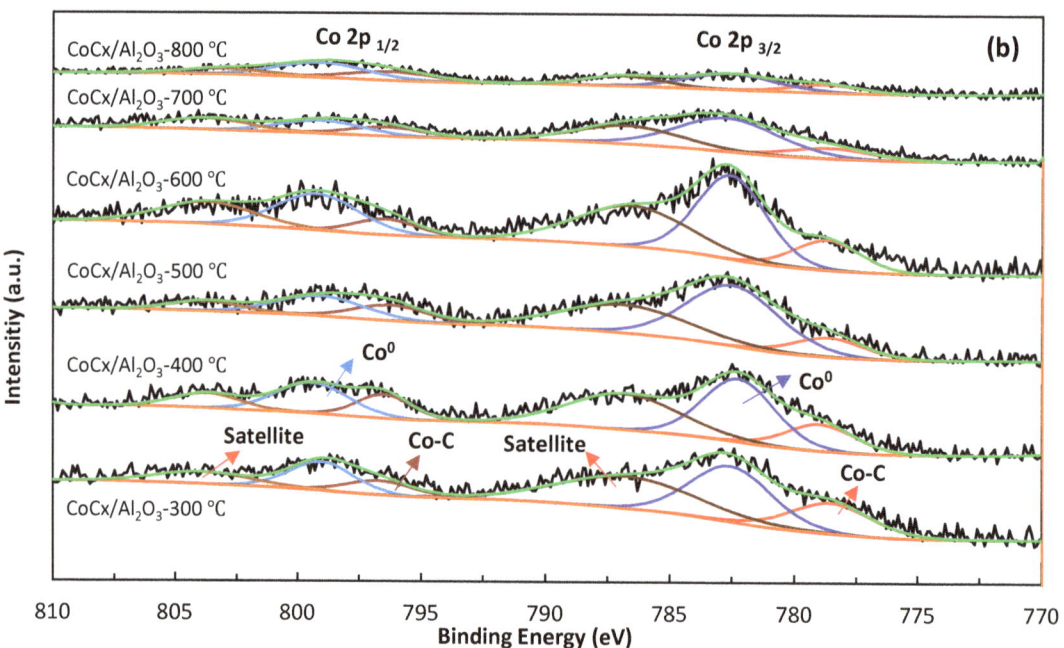

Figure 3. (a) C 1s spectra, and (b) Co 2p spectra of CoC$_x$/Al$_2$O$_3$ catalysts prepared at different temperatures.

Table 1. XPS data analysis of C 1s and Co2p peaks of CoC$_x$/Al$_2$O$_3$ catalysts.

	C 1s			Co 2p	
	Co-C Peak BE (eV)	C-C Peak BE (eV)	C-C/Co-C	Co-C Peak BE (eV)	Co^0/Co-C
CoC$_x$/Al$_2$O$_3$-300 °C	283.2	284.7	0.63	778.5 (2p$_{3/2}$) 796.3 (2p$_{1/2}$)	1.68
CoC$_x$/Al$_2$O$_3$-400 °C	283.3	284.7	1.49	778.6 (2p$_{3/2}$) 796.4 (2p$_{1/2}$)	2.08
CoC$_x$/Al$_2$O$_3$-500 °C	283.1	284.5	0.99	778.7 (2p$_{3/2}$) 796.3 (2p$_{1/2}$)	2.85
CoC$_x$/Al$_2$O$_3$-600 °C	283.2	284.8	2.36	778.7 (2p$_{3/2}$) 796.4 (2p$_{1/2}$)	2.67
CoC$_x$/Al$_2$O$_3$-700 °C	283.3	284.6	6.10	778.5 (2p$_{3/2}$) 796.7 (2p$_{1/2}$)	2.70
CoC$_x$/Al$_2$O$_3$-800 °C	283.3	284.4	14.26	778.6 (2p$_{3/2}$) 796.0 (2p$_{1/2}$)	1.49

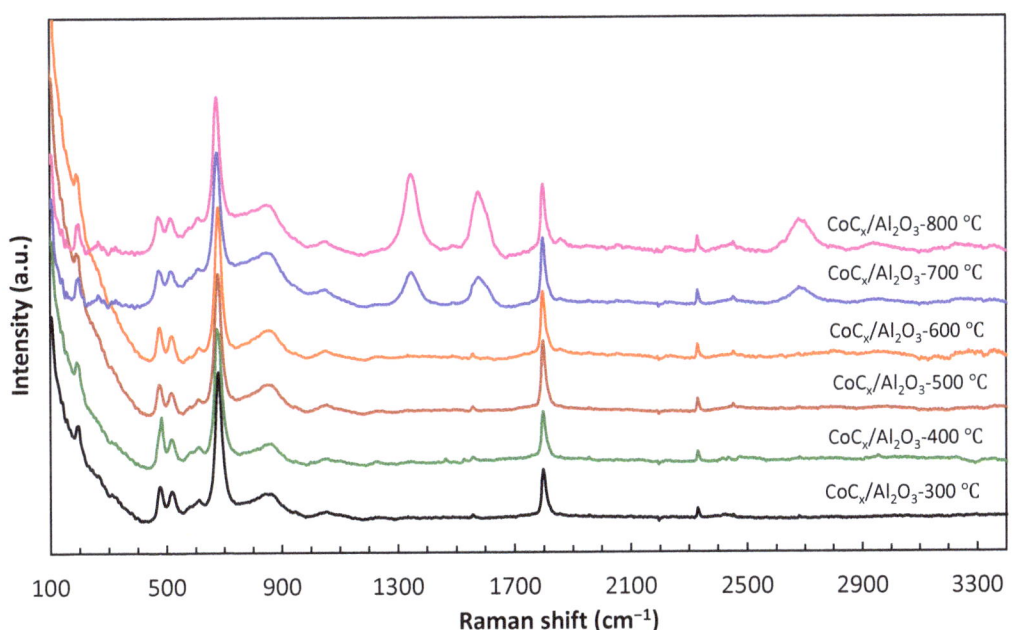

Figure 4. Raman spectrum of the catalysts prepared at different carbidation temperatures.

The D mode is caused by the disordered structure of graphene, and it originates from the A$_{1g}$ breathing vibrations of the sixfold carbon rings involving phonons near the K zone boundary (Figure S2). The G band, which lies in the range of 1500–1600 cm^{-1}, is ascribed to the graphitic carbon with E$_{2g}$ symmetry that involves the in-plane bond-stretching vibration of sp^2 carbon sites. This mode does not require the presence of sixfold rings, so it occurs at all sp^2 sites, not only those in the rings [45,46]. It is worth mentioning that the G mode is due to the relative motion of sp^2 carbon atoms and can also be found in chains.

As mentioned above, the G peak could be observed in the range of 1500–1600 cm^{-1}. By changing the structure of crystallized graphite to nanocrystalline graphite, the peak shifts to 1600 cm^{-1}, and the peak shifts to 1500 cm^{-1} for amorphous carbon when a loss of aromatic bonding appears. Thus, it can be confirmed that the G band positions (1573 cm^{-1})

of the catalysts are close to that of nanocrystalline graphite of nanometer-sized clusters [47]. The degree of crystallinity and graphitization formed carbon on the surface of catalysts after carbidation at 700 °C and 800 °C can be measured by the intensity ratio of I_D/I_G. The higher crystallinity could be observed in catalysts with a lower I_D/I_G ratio (Table 2). In this study, the I_D/I_G decreased from 1.18 to 1.28 by increasing the carbidation temperature from 700 °C to 800 °C, which is not clearly confirming the higher crystallinity of the deposited carbon on the surface of the catalyst prepared at the higher temperature. However, the higher intensities in the catalysts prepared at 800 °C could confirm the higher content of the deposited carbon on the catalyst's surface.

Table 2. Raman spectra for carbonaceous materials observed in the prepared catalysts.

Catalyst	D Peak Position (cm^{-1})	Intensity	G Peak Position (cm^{-1})	Intensity	G' Peak Position (cm^{-1})	Intensity	I_D/I_G
CoC$_x$/Al$_2$O$_3$-300 °C	—	—	—	—	—	—	—
CoC$_x$/Al$_2$O$_3$-400 °C	—	—	—	—	—	—	—
CoC$_x$/Al$_2$O$_3$-500 °C	—	—	—	—	—	—	—
CoC$_x$/Al$_2$O$_3$-600 °C	—	—	—	—	—	—	—
CoC$_x$/Al$_2$O$_3$-700 °C	1340	18.8	1573	16.0	2677	11.1	1.18
CoC$_x$/Al$_2$O$_3$-800 °C	1341	45.0	1573	35.1	2677	19.1	1.28

All kinds of sp^2 carbon materials exhibited a peak in the range 2500–2800 cm^{-1} in the Raman spectra. The G' band (or 2D band) was also observed at 2677 cm^{-1}. The intensity of the G' band in the prepared catalysts increased by increasing the carbidation temperature. The G' band usually helps discriminate the differences between the layers and loading order of the multi-layer graphene system [48]. The single-layer graphene shows a G' band with high intensity. By increasing the graphite layers, the peak becomes broader. In the catalyst prepared at 800 °C, the intensity of the G' band at around 2677 cm^{-1} peak increased due to the higher amount of carbon on the catalyst's surface. The catalysts prepared at lower carbidation temperatures ($\leq$600 °C) did not show any peak belonging to graphitic carbon.

The Raman peaks below 700 cm^{-1} are attributed to the Co-Co stretching mode. The observed peaks for cobalt are very similar to those reported in the Raman spectra of cobalt oxide nanoparticles [49–56]. Five (A_{1g} + E_g + 3 F_{2g}) Raman-active modes were observed and identified. The band at around 187 cm^{-1} was ascribed to the tetrahedral sites' (CoO$_4$) characteristics with the F_{2g} mode (Figure. 3b). The bands at 470, 510, and 607 are attributed to the E_g, F_{2g}, and F_{2g} symmetry, respectively. Generally, the E_g and F_{2g} modes were related to the combined vibrations of the tetrahedral site and octahedral oxygen motions. The band at 671 cm^{-1} with A_{1g} symmetry was attributed to the characteristics of the octahedral sites (CoO$_6$) [49–54]. D'Ippolito and Andreozzi [56] also reported that the Raman vibrational modes of spinel oxides are generally observed in the range of 100–800 cm^{-1}. Diallo et al. [49], Rashad et al. [50], Jiang and Li [51], and Hadjiev et al. [52] also observed the same Raman bands for the Co$_3$O$_4$ nanoparticles. The intensity of the bands in the range of 100–800 cm^{-1} belonging to Co$_3$O$_4$ nanoparticles decreased gradually with increasing carbidation temperature. This phenomenon was more evident in the samples with carbidation temperature above 600 °C, which could be due to the coverage of the surface by deposited carbon. The weak band at around 2330 cm^{-1}, observed in all spectra, could be attributed to the characteristic line of nitrogen from the air [57–59]. The peak at around 1790 cm^{-1} was ascribed to the carbonyl (C=O) stretch bands [60–63], which could be formed after the passivation of the catalyst under oxygen after carbidation.

2.2. Catalyst Evaluation

The results of the FT reaction in the autoclave reactor are shown in Figures 5 and 6. The FT reactions were first performed using the carbide catalysts without prior reduction (WR) by hydrogen (Figure 5a). The catalyst with higher Co$_2$C content (300-WR, the catalyst prepared at the carbidation temperature of 300 °C and used in FT without prior reduction with H_2) showed the lowest CO conversion of about 26%. By increasing the carbidation temperature, the catalyst activity also increased to 74% for the catalyst prepared at 800 °C. As was described earlier, increasing the carbidation temperature resulted in the formation of more metallic cobalt and lower cobalt carbide. Higher content of metallic cobalt could increase the CO conversion. In addition to the cobalt carbide, carbon fibers were also deposited on the catalyst's surface due to the decomposition of methane at higher carbidation temperatures. Deposited carbon could block the active sites and reduce the contact possibility between the reactant gas and the active sites, thus leading to a lower conversion. By increasing the carbidation temperature from 300 °C to 800 °C, the content of products in the form of liquid decreased from 23% to 4%. It seems that the cobalt carbides were acting as an intermediate and reacting with hydrogen to form methylene as a chain-growth monomer for the formation of long-chain hydrocarbons. Therefore, the catalysts with higher cobalt carbide content had a higher selectivity for the heavier hydrocarbons (liquid products) (Figure 5a).

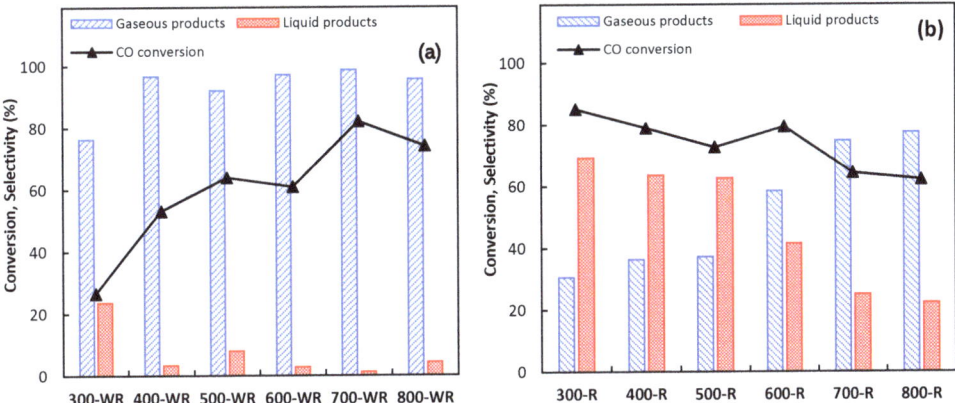

Figure 5. The gaseous and liquid product distributions after FT reaction at 230 °C and 6 h of reaction (**a**) without reduction with H_2, (**b**) reduced with H_2 (WR: without reduction step, R: with reduction step).

During the FT reaction over non-reduced catalysts, the content of formed light alkenes was very low and negligible (maximum 1.3% over 300-WR catalyst), whereas alkanes (56.4%) and CO_2 (42.3%) were the main products of the reaction (Figure 6a-WR). The slightly higher alkene formation for the catalysts with lower carbidation temperature could be attributed to the presence of more Co_2C on the surface of the catalysts [9,64]. By increasing the carbidation temperature, the selectivity to CO_2 decreased and reached 10.7% over the 800-WR catalyst, and alkanes selectivity increased to 89%. Methane was the main produced alkane with the selectivity of 97% for the 300-WR. It was decreased to 73% for 400-WR, then again increased gradually to 97% for the 800-WR catalyst (Figure 6b-WR). Formed alkenes are negligible, and propylene was the main formed alkene in gaseous products of the reaction (Figure 6c-WR). Co_2C with (111) facet and nanosphere-like particles could lead to lower activity and higher methane selectivity [10,13,18]. Moreover, during the FT reaction, some portion of deposited graphitic carbon could also be converted to methane through a hydrogenation reaction. This could be the reason for the increase in the methane content in the catalysts with higher graphitic carbon on their surface. The catalysts prepared at 300 °C and 400 °C with the higher content of $Co_2C(111)$ showed lower CO conversion, whereas increasing the Co^0 content at the catalysts prepared at higher temperatures resulted in a lower CO_2 and alkane (mainly methane) selectivity.

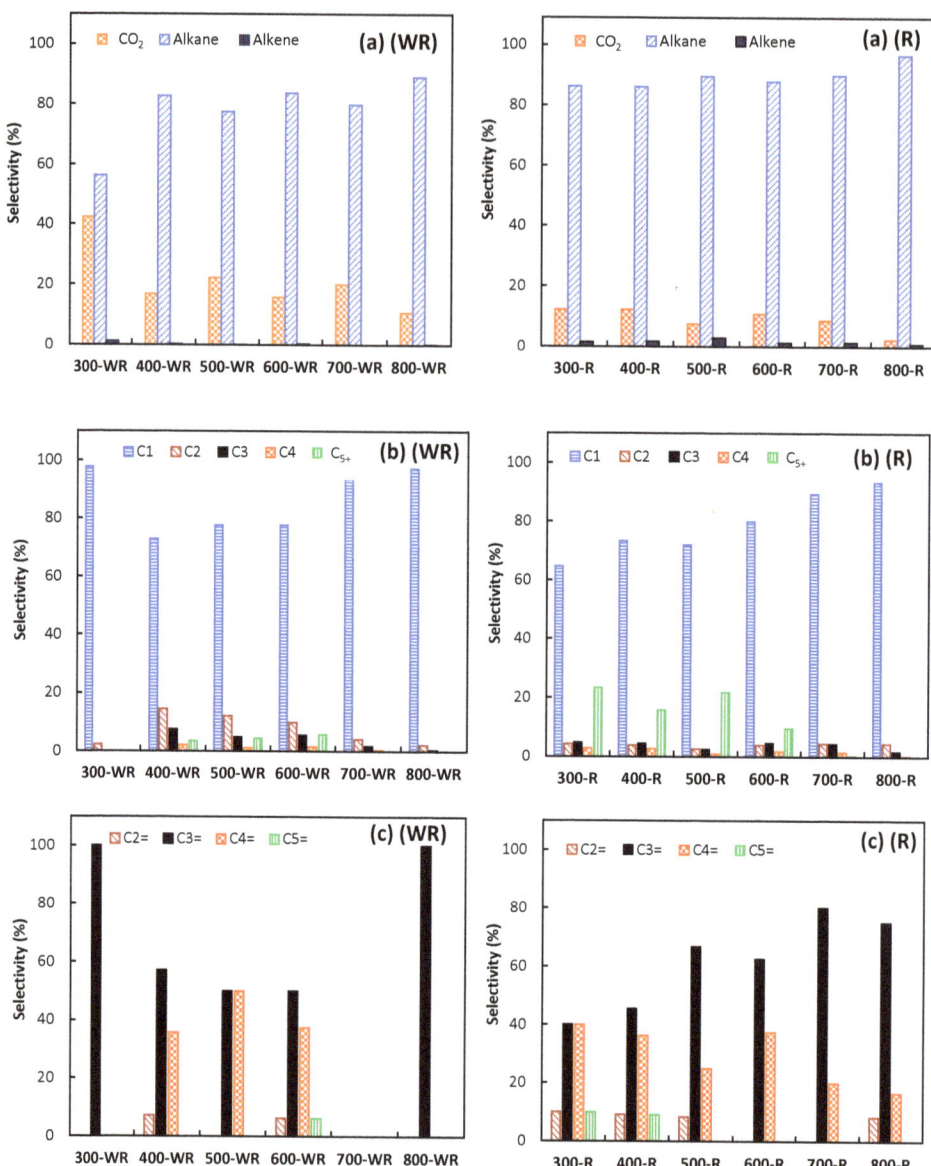

Figure 6. Gaseous product distribution in the FT reaction carbide catalysts: (**a**) hydrocarbon distribution, (**b**) alkanes distribution, and (**c**) alkenes distribution (WR: without reduction step, R: with reduction step).

In the presence of hydrogen, cobalt carbides could be transformed into the face centered-cubic (fcc) and hexagonal close-packed (hcp) Co^0, which are more efficient for the dissociative adsorption of CO during the FT reaction, and hcp cobalt nanoparticles reported to have higher activity compared with fcc cobalt in the FT reaction [65–67]. The reaction route is also reported to be different over these two different Co crystallographic structures, and the direct dissociation of CO (CO* + H → C* + O* + H*) is preferred over hcp Co, whereas H-assisted dissociation (CO* + H* → CHO* → CH* + O*) is the main mechanism over fcc cobalt [67]. In order to investigate the stability of the cobalt carbides in the presence of hydrogen and evaluation of the catalytic performances of catalysts after reduction with hydrogen, the FT reactions in an autoclave reactor were repeated, while the catalysts were first reduced under H_2 at 300 °C for 5 h. Then, the FT reaction was performed at 230 °C for 6 h (Figures 5 and 6). Results of the FT reaction in an autoclave reactor revealed that CO and H_2 conversions gradually decreased by increasing the catalyst carbidation temperature. The distribution of gaseous and liquid products is shown in Figure 5b.

Over the reduced catalysts, the products shifted from heavier hydrocarbons to lighter hydrocarbons by increasing the carbidation temperature, which was attributed to the content of metallic cobalt in these catalysts. The higher catalytic activity over the 300-R catalyst (which means the catalyst was prepared at the carbidation temperature of 300 °C and used in the FT reaction after reducing with H_2) could be due to the formation of more metallic cobalt during the reduction step. This provided more active sites for CO dissociation and the production of heavier hydrocarbons. The content of the products in the form of liquid was higher for the FT reaction over the 300-R catalyst. The catalyst activity and production of heavier hydrocarbons in the liquid phase decreased gradually by increasing the carbidation temperature. Compared with the FT reaction without the reduction step, the FT reactions over the reduced catalysts resulted in the formation of a lower amount of CO_2 and methane (Figure 6a-R).

The formation of methane gradually increased from 64% to 93% by increasing the carbidation temperature from 300 °C to 800 °C, and at the same time, the selectivity to C_{5+} in the gaseous phase decreased (Figure 6b-R). The presence of more cobalt carbides in the catalysts with lower carbidation temperatures resulted in the formation of more metallic cobalt during reduction by hydrogen, which can enhance the CO dissociation and result in higher activity and selectivity to heavier hydrocarbons. Like the non-reduced catalyst, the main formed alkene was propylene over the reduced catalysts, and its content increased by increasing the carbidation temperature (Figure 6c-R). The Co^0 could be carburized in the presence of pure CO and produce Co_2C [68]. However, during the FT reaction, in the presence of CO and H_2, both Co^0 and Co_2C were found to exist. Therefore, both of them can affect the reaction. During the initial reaction stage, the presence of Co^0 in the reduced catalyst led to a higher selectivity for heavier hydrocarbons and a lower CH_4 and CO_2 selectivity. It has been reported that the higher Co_2C content leads to a high methane selectivity and a lower CO conversion [1,69]. Zhong et al. [10] prepared the Co_2C nanospheres to evaluate their activity for the FT reaction. The CO conversion over the Co_2C nanoparticles was about 10%. High selectivity to methane (~80%), C_2-C_4 alkanes of ~18%, and the C_2-C_4 alkenes selectivity of about 2% were obtained during the reaction at 250 °C, 1 bar, and $H_2/CO = 2$. During the FT reaction over the CoMn catalyst, Co^0 was transformed to Co_2C, and the CoMn, which initially existed as $Co_xMn_{1-x}O$, was segregated into Co_2C and MnO. The Co_2C formed during the reaction resulted in higher activity and more light olefins formation. As the reaction proceeded, the formation of C_{5+} and oxygenates decreased.

The evaluation of the liquid phase products for both series of FT reactions (without and with reduction step) showed that the products were a combination of alkanes, alkenes, and a negligible amount of other products such as toluene, and ethylbenzene and propylbenzene were the main products of the FT reaction without the reduction step (Figure 7). The liquid product of the FT reaction with the reduction step consisted of a slightly higher amount of other products (including some oxygenates such as diphenyl ether, acetic anhydride, and dicyclohexyl). The formation of oxygenates was not observed over the catalysts without the reduction step. A small amount of the cyclic products could be due to the reaction of cyclohexane, which was used as the solvent for the FT reaction in the autoclave reactor. The nondissociative adsorption of CO and its insertion into hydrocarbons could result in the formation of oxygenates, which can be converted to alcohol by going through the further hydrogenation process. The FT reaction over carbide catalysts without reduction resulted in the formation of alkanes in the range of C_4-C_{22}, and the lower carbidation temperature led to the formation of slightly lighter hydrocarbons.

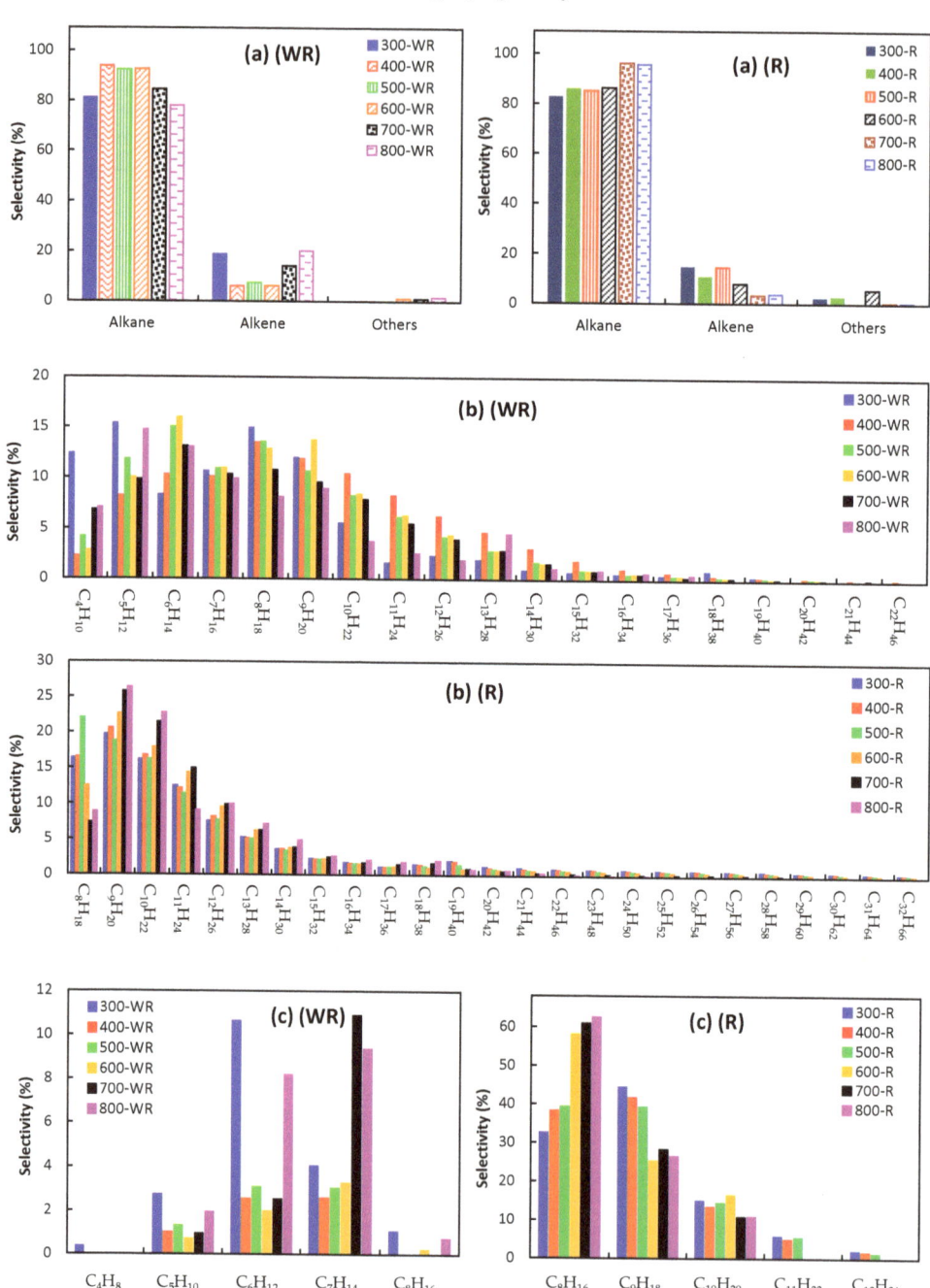

Figure 7. (**a**) Product distribution in liquid phase products, (**b**) distribution of n-alkanes, and (**c**) alkenes in the liquid phase products (WR: without reduction step, R: with reduction step).

By increasing the carbidation temperature, products shifted slightly to heavier hydrocarbons. As expected, the FT reaction over the reduced catalysts resulted in the formation of heavier hydrocarbons in the range of C_8–C_{32}, which could be due to the presence of more active metallic cobalt with higher activity and selectivity to heavy hydrocarbons. The formed alkenes were in the range of C_4–C_8 over carbide catalysts without reduction, whereas heavier alkenes (C_8–C_{12}) were formed over the reaction of reduced catalysts. According to the density functional theory calculations, the active sites for the CO nondissociative adsorption are provided by Co_2C. Co^0 provides the active sites for CO dissociative adsorption; the insertion of CO to hydrocarbon for oxygenates at the interface of Co and Co_2C is facilitated [19]. In the presence of pure H_2, Co_2C could preserve its activity until 150 °C, whereas at higher temperatures, it starts to decompose to the fcc and hcp Co^0 and C. At temperatures above 300 °C, Co_2C is completely decomposed [19]. Due to the presence of both H_2 and CO during the FT reaction, cobalt in the form of both Co^0 and Co_2C existed in the catalyst. The presence of Co^0 could be beneficial for the production of heavier hydrocarbons. The FT reaction over the reduced catalysts prepared at lower carbidation temperature (with more carbides and less carbon formed on the surface) showed a higher selectivity to heavier hydrocarbons.

3. Materials and Methods

3.1. Catalyst Preparation

The 5%Co/Al_2O_3, 5%Fe/Al_2O_3, and 2.5%Fe2.5%Co/Al_2O_3 as precursors were prepared in advance using the incipient wetness impregnation (IWI) method. First, the required amount of cobalt nitrate hexahydrate (Lech-Ner s.r.o., p.a. purity) was dissolved in deionized water. Then, this solution was added dropwise to the Al_2O_3 support (SASOL, alumina balls with 2.5 mm diameter) at room temperature and under continuous stirring for 1 h. The mixture was then dried at 120 °C overnight. The dried samples were used as precursors in the carbidation step. Prior to the carbidation step, the precursors were preheated at 200 °C for 12 h under N_2 atmosphere (75 cm^3/min). For the preparation of carbide catalysts, 4 g of the prepared precursor was exposed to a gas containing 20% CH_4 in H_2 with the flow rate of 300 cm^3/min for 3 h at different temperatures in the range of 300 °C to 800 °C. After the carbidation step, catalysts were purged with N_2 for 30 min and passivated for 2 h under 1% O_2 in Ar with a flow rate of 75 cm^3/h.

3.2. Catalyst Characterizations

The scanning electron microscope (SEM) (JEOL JSM-IT500HR; JEOL Ltd., Tokyo, Japan) was used to study the surface morphology of the prepared catalysts. Representative backscattered electron or secondary electron images of microstructures were taken in high vacuum mode, using an accelerating voltage of 15 kV. An inductively coupled plasma-optical emission spectrometer (ICP-OES; Agilent 725/Agilent Technologies Inc., Santa Clara, CA, USA) was used for the determination of the bulk metal content in the prepared catalysts. X-ray diffraction (XRD) was used to determine the type of phases and crystallinity of the catalysts using a D8 Advance ECO (Bruker AXC GmbH, Karlsruhe, Germany) with CuKα radiation (λ = 1.5406 Å). The step time of 0.5 s and the step size of 0.02° in a 2θ angle ranging from 10° to 70° were used for XRD analysis. X-ray photoelectron spectroscopy (XPS) spectra of the samples were obtained using the XPS instrument in a high-vacuum chamber, which was equipped with a source SPECS X-ray XR50 (SPECS Surface Nano Analysis GmbH, Germany), where an Al anode (1486.6 eV) and a Mg anode (1253.6 eV) were used, and with a hemispherical analyzer SPECS PHOIBOS 100 with a five-channel detector. Thermo Scientific™ Avantage Software was used for XPS data processing. The Raman analysis for the samples was performed using a DXR (Thermo Fischer Scientific, Waltham, MA, USA) Raman microscope. The laser beam was focused onto the catalyst surface using an optical microscope with a magnification of 10×. A 532 nm laser with an exposure time of 7 s and a power of 9 mW was used for sample excitation. Analyses were performed in the area of 100–3400 cm^{-1}.

3.3. Catalyst Evaluation

The FT reaction was performed in a 1 L stainless steel autoclave reactor (Parr instruments). In a typical experiment, 1.5 g of catalyst and 35 mL of cyclohexane as a solvent were added to the reactor vessel. The reactor temperature was increased to 230 °C (3 °C/min) and pressurized to 5 MPa with syngas ($H_2/CO = 2$). The reaction was conducted in batch mode for 6 h under constant stirring of 800 rpm to eliminate the diffusion control region. The conversion rate was measured according to the decrease in pressure during the reaction and the composition of the reactant gas before and after the reaction. For the FT reaction with the in situ catalyst reduction, the reactor was heated to 300 °C (3 °C/min) and then pressurized with H_2 to 5 MPa for 5 h. After reduction, the reactor system was cooled to room temperature, and the gas was switched to syngas ($H_2/CO = 2$). The FT reaction was performed at 230 °C and 5 MPa for 6 h. After the termination of the reaction, the gaseous and liquid products were analyzed using different chromatographic procedures. The resultant gas sample was transferred to a gas bag and analyzed with a gas chromatograph, the Agilent 7890A, with three parallel channels that collected data simultaneously. The channels are equipped with two thermal conductivity detectors (TCD), CO, H_2, N_2, and CO_2 gases, and a flame ionization detector (FID) to detect hydrocarbons. The liquid samples were analyzed using Thermo Scientific ITQ Series GC/MS Ion Trap Mass Spectrometers.

4. Conclusions

Cobalt-based catalysts with high activity and selectivity to long-chain hydrocarbons are widely used for this reaction. Identification of the active phase could be beneficial in enhancing the product yield of the FT reaction. A series of FT reactions were performed in an autoclave reactor to evaluate the catalytic activity of cobalt carbide catalysts and their role in the product distribution during the FT reaction. CoC_x/Al_2O_3 catalysts were prepared by carbidation of Co/Al_2O_3 catalysts at different temperatures and over 20% CH_4 in an H_2 gas stream. Different characterization methods confirmed the presence of Co_2C. By increasing the carbidation temperature, the content of Co_2C decreased, whereas the Co^0 content increased by temperature. The Co/Al_2O_3 catalyst was also active for the methane decomposition to hydrogen and graphitic carbon at higher carbidation temperatures. Cobalt carbides were unstable in the presence of hydrogen at high temperatures and decomposed to metallic cobalt and carbon, which resulted in the increase of metallic cobalt's content. The catalysts with higher cobalt carbide had the lowest CO conversion of 26%, which then increased to 74% for the catalyst with lower Co_2C content prepared at 800 °C. The main products were light alkanes in the range of C_1–C_5 (mainly methane), and the heavier hydrocarbons (in the range of C_4–C_{22}) decreased from 23% to 4%. By increasing the carbidation temperature, the Co_2C content decreased, and the metallic cobalt content increased, which resulted in higher catalytic activity. At the same time, the selectivity to CO_2 decreased. The $Co_2C(111)$ facet with the nanosphere-like particles showed a lower activity and higher methane selectivity under mild FT reaction conditions. The formation of alkenes was negligible over these carbide catalysts, which could be due to the lower content of Co_2C nanoprisms with specific exposed facets of (101) and (020); these are known as the FTO active phase, with higher activity for CO hydrogenation and lower methane selectivity. In addition, a series of FT reactions were performed over the carbide catalyst with a prior reduction step using pure hydrogen at 300 °C. These reduced catalysts showed a different catalytic performance. The main products were the heavier hydrocarbons (C_8–C_{32}) with 69%. By increasing the carbidation temperature to 800 °C, the content of liquid products decreased to 22%, whereas the light hydrocarbons content increased to 78%. The Co conversion decreased by increasing the carbidation temperature in the reduced catalysts. For the catalysts prepared at higher temperatures, the presence of lower Co_2C, which is transformed into hcp cobalt during the reduction with hydrogen, as well as the presence of lower metallic fcc cobalt, resulted in lower CO conversion and less heavy hydrocarbons.

Supplementary Materials: The following supporting information can be downloaded at: https://www.mdpi.com/article/10.3390/catal12101222/s1, Figure S1: Survey XPS spectra scan of carbide catalysts, Figure S2: Carbon motions in the D and G modes [45].

Author Contributions: Conceptualization, Z.G. and Z.T.; methodology, Z.G.; validation, Z.G., Z.T. and F.G.; formal analysis, Z.G., Z.T., E.S., I.H., N.S.; investigation, Z.G., Z.T., E.S., I.H., N.S. and F.G.; data curation, Z.G., Z.T., E.S., I.H., N.S.; writing—original draft preparation, Z.G., Z.T., F.G.; writing—review and editing, Z.G., Z.T., F.G.; visualization, Z.G.; supervision, Z.G., Z.T. All authors have read and agreed to the published version of the manuscript.

Funding: The publication is a result of the project which was carried out within the financial support of the Ministry of Industry and Trade of the Czech Republic with institutional support for long-term conceptual development of research organization. The result was achieved using the infrastructure included in the project Efficient Use of Energy Resources Using Catalytic Processes (LM2018119), which has been financially supported by MEYS within the targeted support of large infrastructures.

Data Availability Statement: MDPI Research Data Policies.

Conflicts of Interest: The authors declare no conflict of interest. The funders had no role in the design of the study; in the collection, analyses, or interpretation of data; in the writing of the manuscript, or in the decision to publish the results.

References

1. Mohandas, J.C.; Gnanamani, M.K.; Jacobs, G.; Ma, W.; Ji, Y.; Khalid, S.; Davis, B.H. Fischer–Tropsch synthesis: Characterization and reaction testing of cobalt carbide. *ACS Catal.* **2011**, *1*, 1581–1588. [CrossRef]
2. Kwak, G.; Woo, M.H.; Kang, S.C.; Park, H.-G.; Lee, Y.-J.; Jun, K.-W.; Ha, K.-S. In situ monitoring during the transition of cobalt carbide to metal state and its application as Fischer–Tropsch catalyst in slurry phase. *J. Catal.* **2013**, *307*, 27–36. [CrossRef]
3. Lin, Q.; Liu, B.; Jiang, F.; Fang, X.; Xu, Y.; Liu, X. Assessing the formation of cobalt carbide and its catalytic performance under realistic reaction conditions and tuning product selectivity in a cobalt-based FTS reaction. *Catal. Sci. Technol.* **2019**, *9*, 3238–3258. [CrossRef]
4. Gholami, Z.; Tišler, Z.; Rubáš, V. Recent advances in Fischer-Tropsch synthesis using cobalt-based catalysts: A review on supports, promoters, and reactors. *Catal. Rev.* **2021**, *63*, 512–595. [CrossRef]
5. Claeys, M.; Dry, M.E.; van Steen, E.; du Plessis, E.; van Berge, P.J.; Saib, A.M.; Moodley, D.J. In situ magnetometer study on the formation and stability of cobalt carbide in Fischer–Tropsch synthesis. *J. Catal.* **2014**, *318*, 193–202. [CrossRef]
6. Yang, X.; Yang, J.; Zhao, T.; Qian, W.; Wang, Y.; Holmen, A.; Jiang, W.; Chen, D.; Ben, H. Kinetic insights into the effect of promoters on Co/Al$_2$O$_3$ for Fischer-Tropsch synthesis. *Chem. Eng. J.* **2022**, *445*, 136655. [CrossRef]
7. Karaca, H.; Safonova, O.V.; Chambrey, S.; Fongarland, P.; Roussel, P.; Griboval-Constant, A.; Lacroix, M.; Khodakov, A.Y. Structure and catalytic performance of Pt-promoted alumina-supported cobalt catalysts under realistic conditions of Fischer–Tropsch synthesis. *J. Catal.* **2011**, *277*, 14–26. [CrossRef]
8. Cheng, J.; Hu, P.; Ellis, P.; French, S.; Kelly, G.; Lok, C.M. Density functional theory study of iron and cobalt carbides for Fischer-Tropsch synthesis. *J. Phys. Chem. C* **2010**, *114*, 1085–1093. [CrossRef]
9. Li, Z.; Zhong, L.; Yu, F.; An, Y.; Dai, Y.; Yang, Y.; Lin, T.; Li, S.; Wang, H.; Gao, P.; et al. Effects of sodium on the catalytic performance of CoMn catalysts for Fischer–Tropsch to olefin reactions. *ACS Catal.* **2017**, *7*, 3622–3631. [CrossRef]
10. Zhong, L.; Yu, F.; An, Y.; Zhao, Y.; Sun, Y.; Li, Z.; Lin, T.; Lin, Y.; Qi, X.; Dai, Y.; et al. Cobalt carbide nanoprisms for direct production of lower olefins from syngas. *Nature* **2016**, *538*, 84–87. [CrossRef]
11. Xiang, Y.; Kruse, N. Tuning the catalytic CO hydrogenation to straight- and long-chain aldehydes/alcohols and olefins/paraffins. *Nat. Commun.* **2016**, *7*, 13058. [CrossRef] [PubMed]
12. Gnanamani, M.K.; Jacobs, G.; Shafer, W.D.; Sparks, D.E.; Hopps, S.; Thomas, G.A.; Davis, B.H. Low temperature water–gas shift reaction over alkali metal promoted cobalt carbide catalysts. *Top. Catal.* **2014**, *57*, 612–618. [CrossRef]
13. Lin, T.; Yu, F.; An, Y.; Qin, T.; Li, L.; Gong, K.; Zhong, L.; Sun, Y. Cobalt carbide nanocatalysts for efficient syngas conversion to value-added chemicals with high selectivity. *Acc. Chem. Res.* **2021**, *54*, 1961–1971. [CrossRef] [PubMed]
14. Lin, T.; Qi, X.; Wang, X.; Xia, L.; Wang, C.; Yu, F.; Wang, H.; Li, S.; Zhong, L.; Sun, Y. Direct production of higher oxygenates by syngas conversion over a multifunctional catalyst. *Angew. Chem. Int. Ed.* **2019**, *58*, 4627–4631. [CrossRef] [PubMed]
15. Chen, W.; Filot, I.A.W.; Pestman, R.; Hensen, E.J.M. Mechanism of cobalt-catalyzed CO hydrogenation: 2. Fischer–Tropsch synthesis. *ACS Catal.* **2017**, *7*, 8061–8071. [CrossRef]
16. Gholami, Z.; Gholami, F.; Tišler, Z.; Hubáček, J.; Tomas, M.; Bačiak, M.; Vakili, M. Production of light olefins via Fischer-Tropsch process using iron-based catalysts: A review. *Catalysts* **2022**, *12*, 174. [CrossRef]
17. Davis, B.H.; Occelli, M.L. *Advances in Fischer-Tropsch Synthesis, Catalysts, and Catalysis*, 1st ed.; CRC press: Boca, Raton, 2009. [CrossRef]

18. Xiao, Y.; Sun, P.; Cao, M. Core–Shell bimetallic carbide nanoparticles confined in a three-dimensional N-doped carbon conductive network for efficient lithium storage. *ACS Nano* **2014**, *8*, 7846–7857. [CrossRef]
19. Pei, Y.-P.; Liu, J.-X.; Zhao, Y.-H.; Ding, Y.-J.; Liu, T.; Dong, W.-D.; Zhu, H.-J.; Su, H.-Y.; Yan, L.; Li, J.-L.; et al. High alcohols synthesis via Fischer–Tropsch reaction at cobalt metal/carbide interface. *ACS Catal.* **2015**, *5*, 3620–3624. [CrossRef]
20. Moya-Cancino, J.G.; Honkanen, A.-P.; van der Eerden, A.M.J.; Oord, R.; Monai, M.; ten Have, I.; Sahle, C.J.; Meirer, F.; Weckhuysen, B.M.; de Groot, F.M.F.; et al. In Situ X-ray Raman scattering spectroscopy of the formation of cobalt carbides in a Co/TiO$_2$ Fischer–Tropsch synthesis catalyst. *ACS Catal.* **2021**, *11*, 809–819. [CrossRef]
21. Osa, A.R.; Lucas, A.; Valverde, J.; Romero, A.; Monteagudo, I.; Coca, P.; Sánchez, P. Influence of alkali promoters on synthetic diesel production over Co catalyst. *Catal. Today* **2011**, *167*, 96–106. [CrossRef]
22. Lu, M.; Fatah, N.; Khodakov, A.Y. Optimization of solvent-free mechanochemical synthesis of Co/Al$_2$O$_3$ catalysts using low- and high-energy processes. *J. Mater. Sci.* **2017**, *52*, 12031–12043. [CrossRef]
23. Peña, D.; Gribovar-Constant, A.; Lecocq, V.; Diehl, F.; Khodakov, A.Y. Influence of operating conditions in a continuously stirred tank reactor on the formation of carbon species on alumina supported cobalt Fischer–Tropsch catalysts. *Catal. Today* **2013**, *215*, 43–51. [CrossRef]
24. Yi, J.-H.; Sun, Y.-Y.; Gao, J.-F.; Xu, C.-Y. Synthesis of crystalline γ-Al$_2$O$_3$ with high purity. *Trans. Nonferrous Met. Soc. China* **2009**, *19*, 1237–1242. [CrossRef]
25. Choi, Y.I.; Yang, J.H.; Park, S.J.; Sohn, Y. Energy storage and CO$_2$ reduction performances of Co/Co$_2$C/C prepared by an anaerobic ethanol oxidation reaction using sacrificial SnO$_2$. *Catalysts* **2020**, *10*, 1116. [CrossRef]
26. Xia, S.-G.; Zhang, Z.; Wu, J.-N.; Wang, Y.; Sun, M.-J.; Cui, Y.; Zhao, C.-L.; Zhong, J.-Y.; Cao, W.; Wang, H.; et al. Cobalt carbide nanosheets as effective catalysts toward photothermal degradation of mustard-gas simulants under solar light. *Appl. Catal. B Environ.* **2021**, *284*, 119703. [CrossRef]
27. Petersen, A.P.; Claeys, M.; Kooyman, P.J.; van Steen, E. Cobalt-based Fischer–Tropsch synthesis: A kinetic evaluation of metal–support interactions using an inverse model system. *Catalysts* **2019**, *9*, 794. [CrossRef]
28. van Ravenhorst, I.K.; Hoffman, A.S.; Vogt, C.; Boubnov, A.; Patra, N.; Oord, R.; Akatay, C.; Meirer, F.; Bare, S.R.; Weckhuysen, B.M. On the cobalt carbide formation in a Co/TiO$_2$ Fischer–Tropsch synthesis catalyst as studied by high-pressure, long-term operando X-ray absorption and diffraction. *ACS Catal.* **2021**, *11*, 2956–2967. [CrossRef]
29. Nuernberg, G.B.; Fajardo, H.V.; Mezalira, D.Z.; Casarin, T.J.; Probst, L.F.D.; Carreño, N.L.V. Preparation and evaluation of Co/Al$_2$O$_3$ catalysts in the production of hydrogen from thermo-catalytic decomposition of methane: Influence of operating conditions on catalyst performance. *Fuel* **2008**, *87*, 1698–1704. [CrossRef]
30. Gao, B.; Wang, I.W.; Ren, L.; Haines, T.; Hu, J. Catalytic performance and reproducibility of Ni/Al$_2$O$_3$ and Co/Al$_2$O$_3$ mesoporous aerogel catalysts for methane decomposition. *Ind. Eng. Chem. Res.* **2019**, *58*, 798–807. [CrossRef]
31. Fakeeha, A.H.; Ibrahim, A.A.; Khan, W.U.; Abasaeed, A.E.; Al-Fatesh, A.S. Hydrogen production by catalytic methane decomposition over Ni, Co, and Ni-Co/Al$_2$O$_3$ catalyst. *Pet. Sci. Technol.* **2016**, *34*, 1617–1623. [CrossRef]
32. Saraswat, S.K.; Pant, K. Progressive loading effect of Co Over SiO$_2$/Al$_2$O$_3$ catalyst for CO$_x$ free hydrogen and carbon nanotubes production via catalytic decomposition of methane. *Int. J. Chem. Mol. Eng.* **2015**, *9*, 485–489. [CrossRef]
33. Koerts, T.; Deelen, M.J.A.G.; van Santen, R.A. Hydrocarbon formation from methane by a low-temperature two-step reaction sequence. *J. Catal.* **1992**, *138*, 101–114. [CrossRef]
34. Narkiewicz, U.; Podsiadły, M.; Jędrzejewski, R.; Pełech, I. Catalytic decomposition of hydrocarbons on cobalt, nickel and iron catalysts to obtain carbon nanomaterials. *Appl. Catal. A Gen.* **2010**, *384*, 27–35. [CrossRef]
35. Pang, H.; Wang, X.; Zhang, G.; Chen, H.; Lv, G.; Yang, S. Characterization of diamond-like carbon films by SEM, XRD and Raman spectroscopy. *Appl. Surf. Sci.* **2010**, *256*, 6403–6407. [CrossRef]
36. Matveev, V.V.; Baranov, D.A.; Yurkov, G.Y.; Akatiev, N.G.; Dotsenko, I.P.; Gubin, S.P. Cobalt nanoparticles with preferential hcp structure: A confirmation by X-ray diffraction and NMR. *Chem. Phys. Lett.* **2006**, *422*, 402–405. [CrossRef]
37. Paterson, J.; Peacock, M.; Ferguson, E.; Purves, R.; Ojeda, M. In situ diffraction of Fischer–Tropsch catalysts: Cobalt reduction and carbide formation. *Chem. Cat. Chem.* **2017**, *9*, 3463–3469. [CrossRef]
38. Kumar, C.V.; Pattammattel, A. Chapter 3-Characterization techniques for graphene. In *Introduction to Graphene*; Kumar, C.V., Pattammattel, A., Eds.; Elsevier: Amsterdam, The Netherlands, 2017; pp. 45–74.
39. Fang, H.; Huang, T.; Sun, Y.; Kang, B.; Liang, D.; Yao, S.; Yu, J.; Dinesh, M.M.; Wu, S.; Lee, J.Y.; et al. Metal-organic framework-derived core-shell-structured nitrogen-doped CoC$_x$/FeCo@C hybrid supported by reduced graphene oxide sheets as high performance bifunctional electrocatalysts for ORR and OER. *J. Catal.* **2019**, *371*, 185–195. [CrossRef]
40. Xiong, J.; Ding, Y.; Wang, T.; Yan, L.; Chen, W.; Zhu, H.; Lu, Y. The formation of Co$_2$C species in activated carbon supported cobalt-based catalysts and its impact on Fischer–Tropsch reaction. *Catal. Lett.* **2005**, *102*, 265–269. [CrossRef]
41. Li, S.; Yang, C.; Yin, Z.; Yang, H.; Chen, Y.; Lin, L.; Li, M.; Li, W.; Hu, G.; Ma, D. Wet-chemistry synthesis of cobalt carbide nanoparticles as highly active and stable electrocatalyst for hydrogen evolution reaction. *Nano Res.* **2017**, *10*, 1322–1328. [CrossRef]
42. Zhang, T.; Wu, J.; Xu, Y.; Wang, X.; Ni, J.; Li, Y.; Niemantsverdriet, J.W. Cobalt and cobalt carbide on alumina/NiAl(110) as model catalysts. *Catal. Sci. Technol.* **2017**, *7*, 5893–5899. [CrossRef]
43. Bayer, B.C.; Bosworth, D.A.; Michaelis, F.B.; Blume, R.; Habler, G.; Abart, R.; Weatherup, R.S.; Kidambi, P.R.; Baumberg, J.J.; Knop-Gericke, A.; et al. In situ observations of phase transitions in metastable nickel (carbide)/carbon nanocomposites. *J. Phys. Chem. C* **2016**, *120*, 22571–22584. [CrossRef] [PubMed]

44. Sarr, M.; Bahlawane, N.; Arl, D.; Dossot, M.; McRae, E.; Lenoble, D. Tailoring the properties of atomic layer deposited nickel and nickel carbide thin films via chain-length control of the alcohol reducing agents. *J. Phys. Chem. C* **2014**, *118*, 23385–23392. [CrossRef]
45. Ferrari, A.C.; Robertson, J. Interpretation of Raman spectra of disordered and amorphous carbon. *Phys. Rev. B* **2000**, *61*, 14095. [CrossRef]
46. Sfyris, D.; Sfyris, G.; Galiotis, C. Stress intrepretation of graphene E-2g and A-1g vibrational modes: Theoretical analysis. *arXiv* **2017**, arXiv:1706.04465. [CrossRef]
47. Tembre, A.; Hénocque, J.; Clin, M. Infrared and Raman spectroscopic study of carbon-cobalt composites. *Int. J. Spectrosc.* **2011**, *2011*, 186471. [CrossRef]
48. Issi, J.-P.; Araujo, P.T.; Dresselhaus, M.S. Electron and phonon transport in graphene in and out of the bulk. In *Physics of Graphene*; Aoki, H., Dresselhaus, M.S., Eds.; Springer International Publishing: Cham, Switzerland, 2014; pp. 65–112. [CrossRef]
49. Diallo, A.; Beye, A.C.; Doyle, T.B.; Park, E.; Maaza, M. Green synthesis of Co_3O_4 nanoparticles via Aspalathus linearis: Physical properties. *Green Chem. Lett. Rev.* **2015**, *8*, 30–36. [CrossRef]
50. Rashad, M.; Rüsing, M.; Berth, G.; Lischka, K.; Pawlis, A. CuO and Co_3O_4 nanoparticles: Synthesis, characterizations, and Raman spectroscopy. *J. Nanomater.* **2013**, *2013*, [CrossRef]
51. Jiang, J.; Li, L. Synthesis of sphere-like Co_3O_4 nanocrystals via a simple polyol route. *Mater. Lett.* **2007**, *61*, 4894–4896. [CrossRef]
52. Hadjiev, V.G.; Iliev, M.N.; Vergilov, I.V. The Raman spectra of Co_3O_4. *J. Phys. C Solid State Phys.* **1988**, *21*, L199–L201. [CrossRef]
53. Maaz, K. *Cobalt*; IntechOpen: London, UK, 2017; Available online: https://www.intechopen.com/books/6133 (accesses on 21 July 2022).
54. Rivas-Murias, B.; Salgueiriño, V. Thermodynamic CoO-Co_3O_4 crossover using Raman spectroscopy in magnetic octahedron-shaped nanocrystals. *J. Raman Spectrosc.* **2017**, *48*, 837–841. [CrossRef]
55. Xiao, T.-C.; York, A.P.E.; Al-Megren, H.; Williams, C.V.; Wang, H.-T.; Green, M.L.H. Preparation and characterisation of bimetallic cobalt and molybdenum carbides. *J. Catal.* **2001**, *202*, 100–109. [CrossRef]
56. D'Ippolito, V.; Andreozzi, G.B. Linking crystal chemistry and physical properties of natural and synthetic spinels: An UV-VIS-NIR and Raman study. Ph.D. Thesis, Sapienza Universita di Roma, Rome, Italy, 2013.
57. Mashkovtsev, R.I.; Thomas, V.G. Nitrogen atoms encased in cavities within the beryl structure as candidates for qubits. *Appl. Magn. Reson.* **2005**, *28*, 401–409. [CrossRef]
58. Zhao, Y.; Yamaguchi, Y.; Liu, C.; Sekine, S.; Dou, X. Quantitative detection of ethanol/acetone in complex solutions using Raman spectroscopy based on headspace gas analysis. *Appl. Spectrosc.* **2018**, *72*, 280–287. [CrossRef] [PubMed]
59. Hoekstra, J.; Beale, A.M.; Soulimani, F.; Versluijs-Helder, M.; Geus, J.W.; Jenneskens, L.W. Base metal catalyzed graphitization of cellulose: A combined Raman spectroscopy, Temperature-Dependent X-ray Diffraction and High-Resolution Transmission Electron Microscopy Study. *J. Phys. Chem. C* **2015**, *119*, 10653–10661. [CrossRef]
60. Armenta, S.; Garrigues, S.; de la Guardia, M. Determination of iprodione in agrochemicals by infrared and Raman spectrometry. *Anal. Bioanal. Chem.* **2007**, *387*, 2887–2894. [CrossRef]
61. Perelygin, I.S.; Itkulov, I.G. Spontaneous raman spectroscopic study of the association of liquid γ-butyrolactone molecules. *J. Struct. Chem.* **1996**, *37*, 918–932. [CrossRef]
62. Ostrovskii, D.; Edvardsson, M.; Jacobsson, P. Weak polymer–electrolyte interaction revealed by Fermi resonance perturbed Raman bands. *J. Raman Spectrosc.* **2003**, *34*, 40–49. [CrossRef]
63. Thombare, J.V.; Lohar, G.M.; Shinde, S.K.; Dhasade, S.S.; Rath, M.C.; Fulari, V.J. Synthesis, characterization and surface wettability study of polypyrrole films: Effect of applied constant current density. *Electron. Mater. Lett.* **2015**, *11*, 266–270. [CrossRef]
64. Dai, Y.; Zhao, Y.; Lin, T.; Li, S.; Yu, F.; An, Y.; Wang, X.; Xiao, K.; Sun, F.; Jiang, Z.; et al. Particle size effects of cobalt carbide for Fischer–Tropsch to olefins. *ACS Catal.* **2019**, *9*, 798–809. [CrossRef]
65. ten Have, I.C.; Weckhuysen, B.M. The active phase in cobalt-based Fischer-Tropsch synthesis. *Chem. Catal.* **2021**, *1*, 339–363. [CrossRef]
66. Lyu, S.; Wang, L.; Zhang, J.; Liu, C.; Sun, J.; Peng, B.; Wang, Y.; Rappé, K.G.; Zhang, Y.; Li, J.; et al. Role of active phase in Fischer–Tropsch synthesis: Experimental evidence of CO activation over single-phase cobalt catalysts. *ACS Catal.* **2018**, *8*, 7787–7798. [CrossRef]
67. Liu, J.-X.; Su, H.-Y.; Sun, D.-P.; Zhang, B.-Y.; Li, W.-X. Crystallographic dependence of CO activation on cobalt catalysts: HCP versus FCC. *J. Am. Chem. Soc.* **2013**, *135*, 16284–16287. [CrossRef] [PubMed]
68. Chen, P.-P.; Liu, J.-X.; Li, W.-X. Carbon monoxide activation on cobalt carbide for Fischer–Tropsch synthesis from first-principles theory. *ACS Catal.* **2019**, *9*, 8093–8103. [CrossRef]
69. Weller, S.; Hofer, L.J.E.; Anderson, R.B. The role of bulk cobalt carbide in the Fischer–Tropsch synthesis. *J. Am. Chem. Soc.* **1948**, *70*, 799–801. [CrossRef]

Article

Effect of MnO₂ Crystal Type on the Oxidation of Furfural to Furoic Acid

Xu Wu [1,2], Heqin Guo [1,*], Litao Jia [1,3], Yong Xiao [1], Bo Hou [1] and Debao Li [1,3,*]

1. State Key Laboratory of Coal Conversion, Institute of Coal Chemistry, Chinese Academy of Sciences, Taiyuan 030001, China; wuxu18@mails.ucas.ac.cn (X.W.)
2. University of Chinese Academy of Sciences, Beijing 100049, China
3. Dalian National Laboratory for Clean Energy, Chinese Academy of Sciences, Dalian 116023, China
* Correspondence: guoheqin@sxicc.ac.cn (H.G.); dbli@sxicc.ac.cn (D.L.)

Abstract: The base-free oxidation of furfural by non-noble metal systems has been challenging. Although MnO₂ emerges as a potential catalyst application in base-free conditions, its catalytic efficiency still needs to be improved. The crystalline form of MnO₂ is an important factor affecting the oxidation ability of furfural. For this reason, four crystalline forms of MnO₂ (α, β, γ, and δ-MnO₂) were selected. Their oxidation performance and surface functional groups were analyzed and compared in detail. Only δ-MnO₂ exhibited excellent activity, achieving 99.04% furfural conversion and 100% Propo.$_{FA}$ (Only furoic acid was detected by HPLC in the product) under base-free conditions, while the furfural conversion of α, β, and γ-MnO₂ was below 10%. Characterization by XPS, IR, O₂-TPD and other means revealed that δ-MnO₂ has the most abundant active oxygen species and surface hydroxyl groups, which are responsible for the best performance of δ-MnO₂. This work achieves the green and efficient oxidation of furfural to furoic acid over non-noble metal catalysts.

Keywords: base-free; MnO₂; crystal type; furfural oxidation to furoic acid

1. Introduction

With the decrease in global fossil resources, the production of biochemical products from biomass feedstock, such as agricultural waste and forestry waste, is important for sustainable development and CO₂ emission reduction. Furthermore, as a value-added biochemical derived from lignin, furoic acid is widely used in the pharmaceutical, agrochemical, and fragrance industries. Most importantly, furoic acid can be used to synthesize 2,5-Furandicarboxylicacid (FDCA), which is a key monomer for the synthesis of polyethylene furanoate (PEF), a substitute for polyethylene terephthalate (PET) [1–3].

Currently, there are three main methods for preparing furoic acid: the Cannizaro method, the base-free esterification method [4–6] and the catalytic oxidation method. Cannizaro method needs to consume a lot of bases, and the maximum yield is only 50%. The base-free esterification method consumes a lot of organic solvents, and the steps are tedious. Compared with the above two methods, the direct catalytic oxidation method is a green process that can oxidize furfural to furfural acid in one step. Two types of catalysts, including noble metal and non-noble metal, have been applied in this technique. The noble metal catalysts mainly focused on Au-based [7–11] and Pt-based and Pd-based systems [12–14]. These catalytic systems can efficiently convert furfural under low or base-free conditions, but the use of noble metals makes the catalysts too expensive. Non-noble metal catalysis mainly includes copper oxide [15], cobalt [16], etc., as active components. This catalytic system needs to be carried out under high alkali conditions. In addition, a large amount of acid neutralization is required for product separation, which is not environmentally friendly. Consequently, the creation of non-noble and base-free furfural oxidation catalysts remains the research focus.

As a potential oxidation catalyst, MnO$_2$ can activate oxygen to produce abundant active oxygen species at low temperatures [17,18], which can activate the aldehyde group to the carboxyl group occur at low temperatures. Thus, MnO$_2$ is widely used in aldehyde reactions, such as the oxidation of formaldehyde [19–21] and the oxidation of 5-hydroxymethylfurfural [22,23]. For example, in furfural oxidation, it is reported by Camila Palombo [24] et al. found that MnO$_2$ exhibited 55% furfural conversion and 25% furoic acid selectivity in base-free conditions, indicating that the non-noble metal MnO$_2$ has potential application in furfural oxidation. However, the furoic acid yield is much lower than that in a strongly alkaline environment [15,16].

The inherent properties such as morphology, crystalline surface, and crystalline shape of MnO$_2$ strongly influence catalytic performance [23,25–27], in which crystalline shape plays a crucial role in the oxidation properties. For example, Xiao et al. [28] found that product selectivity in the ammonization of alcohols reaction varies significantly on different crystals of MnO$_2$. For further study, they found that the hydroxyl group on MnO$_2$ plays an essential role in the activation of nitrile, and hydroxyl-rich MnO$_2$ has higher amide selectivity. Furthermore, Hayashi et al. [23] found that the oxidation activity of 5-hydroxymethylfurfural varies greatly on different crystalline forms of MnO$_2$. They found that the vacancy formation energies in different crystalline forms of MnO$_2$ are different, and low vacancy formation energies can promote the activation of oxygen.

In summary, the oxygen vacancy formation energy and surface groups vary greatly on different types of MnO$_2$, resulting in different abilities for producing reactive oxygen species and activating reactants, which can significantly affect the catalyst activity and product selectivity. In the furfural oxidation reaction, the activation of furfural and oxygen are both essential for furfural conversion [1,12,29]. Thus, it inspired that the furfural conversion and furoic acid yield might be significantly enhanced by adjusting the crystalline shape of MnO$_2$ in base-free conditions.

Based on the above analysis, four MnO$_2$ with different crystalline structures were synthesized in the present study. The obtained MnO$_2$ was characterized by XRD, IR, XPS, O$_2$-TPD and other techniques. Furthermore, the catalytic activity was evaluated in the oxidation of furfural to furoic acid. And the factors affecting the catalytic performance over different MnO$_2$ are discussed.

2. Results

2.1. Furfural Oxidation Properties of Different Crystalline MnO$_2$

The catalytic performance of different MnO$_2$ in the furfural oxidation to furoic acid is shown in Table 1. In the reaction temperature range of 60–120 °C, the furoic acid selectivity of all four crystalline MnO$_2$ types was close to 100%, but δ-MnO$_2$ shows much higher furfural conversion than α-, β-, and γ-MnO$_2$. For example, the furfural conversion on α-, β-, and γ-MnO$_2$ does not exceed 10% at 120° (Entry 4, Entry 8, Entry 12). While δ-MnO$_2$ shows significantly higher furfural conversions, which can reach more than 69.68% at 120 °C (Entry 16).

Table 1. Furfural oxidation properties of different crystalline MnO$_2$.

Catalyst Name	Temperature /(°C)	O$_2$ Pressure/(MPa)	Reaction Time/(h)	Catalyst Weight/(g)	Conver.[1] /(%)	Propo.$_{FA}$[2]/(%)	Productivity/ (mmol$_{furoic\ acid}$/g$_{cat}$/h)	Carbon Balance[3]/(%)
α-MnO$_2$	60	1	1	0.2	3.48	100	0.09	96.19
α-MnO$_2$	80	1	1	0.2	3.37	100	0.09	98.47
α-MnO$_2$	100	1	1	0.2	4.44	100	0.12	95.57
α-MnO$_2$	120	1	1	0.2	5.02	100	0.13	91.64

Table 1. Cont.

Catalyst Name	Temperature /(°C)	O_2 Pressure/(MPa)	Reaction Time/(h)	Catalyst Weight/(g)	Conver.[1] /(%)	Propo.$_{FA}$[2]/(%)	Productivity/ (mmol$_{furoic\ acid}$/ g$_{cat}$/h)	Carbon Balance [3]/(%)
β-MnO$_2$	60	1	1	0.2	3.43	100	0.09	91.26
β-MnO$_2$	80	1	1	0.2	3.74	100	0.10	90.20
β-MnO$_2$	100	1	1	0.2	4.48	100	0.12	94.70
β-MnO$_2$	120	1	1	0.2	5.89	100	0.15	91.22
γ-MnO$_2$	60	1	1	0.2	5.79	100	0.15	96.67
γ-MnO$_2$	80	1	1	0.2	5.80	100	0.15	93.68
γ-MnO$_2$	100	1	1	0.2	6.76	100	0.18	96.13
γ-MnO$_2$	120	1	1	0.2	9.23	100	0.24	99.05
δ-MnO$_2$	60	1	1	0.2	17.03	100	0.44	98.44
δ-MnO$_2$	80	1	1	0.2	39.09	100	1.02	85.99
δ-MnO$_2$	100	1	1	0.2	50.07	100	1.42	85.77
δ-MnO$_2$	120	1	1	0.2	69.68	100	1.81	88.44
δ-MnO$_2$	100	1	3	0.2	63.52	100	0.55	94.47
δ-MnO$_2$	100	1	6	0.2	67.06	100	0.29	88.83
δ-MnO$_2$	100	1	12	0.2	78.37	100	0.17	91.81
δ-MnO$_2$	100	1	24	0.2	88.54	100	0.10	75.61
δ-MnO$_2$	100	1	12	0.05	22.23	100	0.19	96.40
δ-MnO$_2$	100	1	12	0.2	78.37	100	0.17	91.81
δ-MnO$_2$	100	1	12	0.3	99.04	100	0.14	89.56

[1] Conver. = (n$_{FF\ in}$ − n$_{FF\ out}$(by HPLC))/n$_{FF\ in}$ × 100%, n$_{FF\ in}$: Furfural feed molarity, n$_{FF\ out}$: Molarity of furfural in the product detected by HPLC. [2] Propo.$_{FA}$ = n$_{FA}$ (by HPLC)/n$_{product}$ (by HPLC) × 100%, n$_{FA}$ represents the molar amount of furoic acid in the product detected by HPLC, and n$_{product}$ represents the sum of the molar amounts of all products detected by HPLC. [3] Carbon balance = (n$_{FF\ out}$ (by HPLC) + n$_{FA\ out}$ (by HPLC))/n$_{FF\ in}$ × 100%, n$_{FF\ in}$: Furfural feed molarity, n$_{FF\ out}$: Molarity of furfural in the product, n$_{FA\ out}$: Molarity of furoic acid in the product. Reaction conditions: The amount of furfural added is 50 mg, the volume of the reactor is 100 mL, and the stirring speed is 500 rpm.

Further, the reaction conditions were optimized for the δ-MnO$_2$ catalyst, and the results are shown in Table 1. When only the reaction time was varied, the conversion kept increasing with the increase of reaction time (Entry 15, Entry 17–Entry 20). However, the carbon balance decreased sharply when the reaction time reached 24 h, which may be related to the increased reaction time to generate the polymeric compounds (Entry 20) [14,24]. When only the catalyst dosage was changed, the furfural conversion increased continuously with the increase of catalyst amount, but the carbon balance data decreased slightly (Entry 21–Entry 23).

Under the optimized reaction conditions, the conversion of furfural on δ-MnO$_2$ can even reach 99.04% without by-products (Entry 23). And this data is much higher than the reported 55% conversion and 25% selectivity on MnO$_2$ [24]. The stability of the δ-MnO$_2$ in the reaction was further investigated. After three cycles, the δ-MnO$_2$ still maintains a high furoic acid yield of 96.80%, showing great potential for application (see Figure 1).

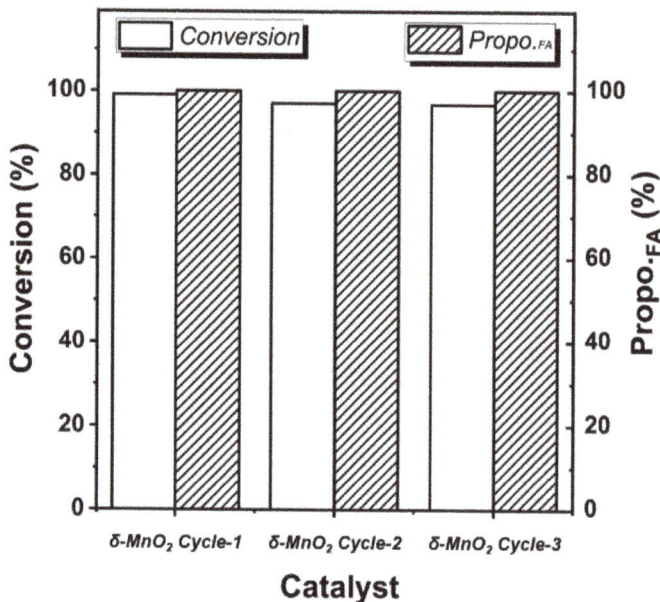

Figure 1. Recyclability of δ-MnO$_2$ in furfural oxidation reaction (the reaction condition is the same as Entry 23 in Table 1, Propo.$_{FA}$ represents the percentage of furoic acid in the product detected by HPLC).

2.2. Structural Properties of Different Crystalline MnO$_2$

To further explore the reasons for the excellent performance of δ-MnO$_2$, the physical and chemical properties of the four MnO$_2$ were characterized. The XRD spectrum of MnO$_2$ with different crystalline structures is shown in Figure 2a. A check of the XRD standard PDF cards shows that the spectrum corresponds well with the standard α-type (JCPDS 44-0141), β-type (JCPDS 24-0735), γ-type (JCPDS 14-0644), and δ-type (JCPDS 80-1098) MnO$_2$ cards (see Table S2), indicating the successful synthesis of four crystalline types of MnO$_2$. PDF cards and previous studies have shown that α-MnO$_2$ of Hollandite-type exhibits a one-dimensional (1 × 1) tunneling structure, β-MnO$_2$ of pyrolusite-type exhibits a one-dimensional (2 × 2) tunneling structure, γ-MnO$_2$ of Nsutite-type exhibits a one-dimensional (1 × 1) tunneling structure, and δ-MnO$_2$ of Birnessite-type exhibits a two-dimensional lamellar structure. And these structural differences can be seen clearly in SEM images (Figure 2b). The α-MnO$_2$, β-MnO$_2$, and γ-MnO$_2$ are assembled by fibers, while δ-MnO$_2$ is assembled by sheets. All of which are composed of six-coordinated [MnO$_6$] basic units [23,30].

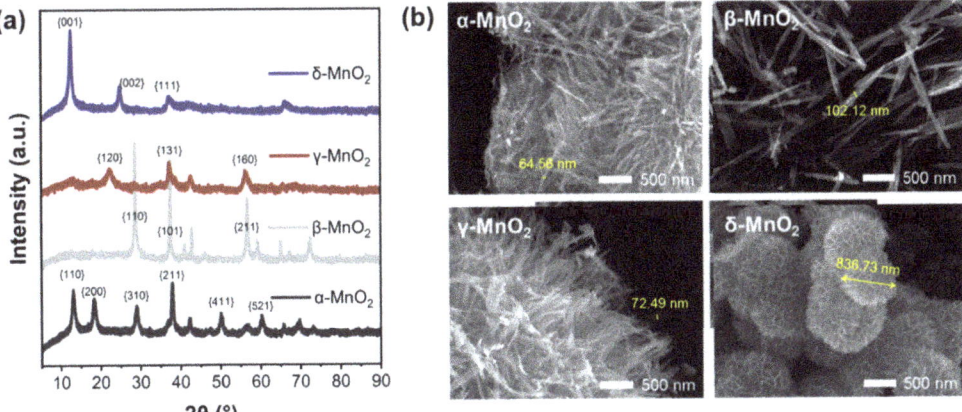

Figure 2. (a) XRD spectra of different crystalline MnO_2, (b) SEM morphology of different crystalline MnO_2.

The functional groups on MnO_2 with different crystalline structures were characterized by the infrared technique (see Figure 3). As can be seen, three peaks at 3186 cm^{-1} and 3557 cm^{-1} attributed to hydroxyl groups can be found on δ-MnO_2 [31–33]. While these peaks are very weak on α-MnO_2, β-MnO_2 and γ-MnO_2. These phenomena indicated that the δ-MnO_2 has significantly more structural hydroxyl groups than the other three MnO_2.

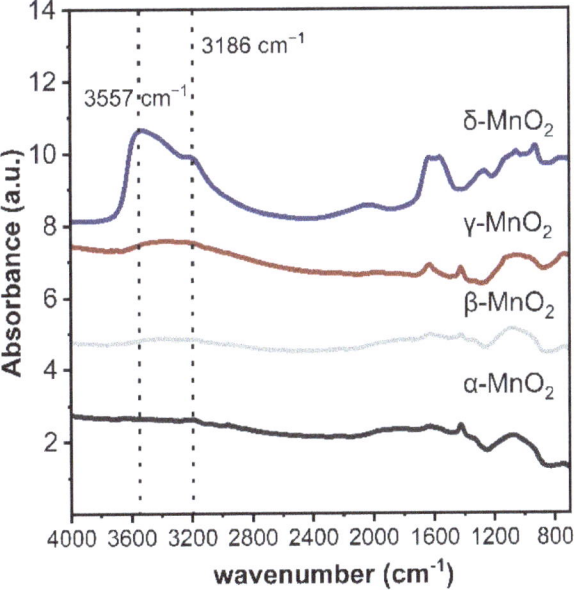

Figure 3. Ex situ DRIFTS spectra of different crystalline MnO_2.

The strength of the Mn-O bond of α-MnO$_2$, β-MnO$_2$, γ-MnO$_2$ and δ-MnO$_2$ was characterized by Raman spectroscopy, and the results are shown in Figure 4. The Raman vibration peaks in 565–580 cm^{-1} and 635–650 cm^{-1} are attributed to the Mn-O bond in the [MnO$_6$] octahedron [22,34,35]. As can be seen, the vibrational peaks appear near 575 cm^{-1} on α-, β- and γ-MnO$_2$, which shifts to 565 cm^{-1} on δ-MnO$_2$. From Hooke's law (1), it follows that:

$$\omega = \frac{1}{2\pi c}\sqrt{\frac{k}{\mu}} \quad (1)$$

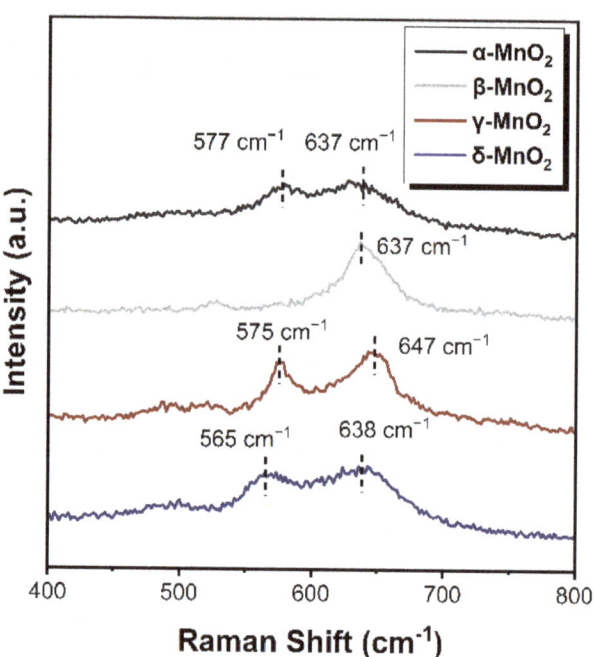

Figure 4. Raman spectrum of different crystalline MnO$_2$.

The mechanical constant k of the Mn-O bond is positively correlated with the Raman shift ω. Therefore, the Mn-O bond is weaker as the corresponding peak shift to a low wavenumber [25]. Thus, the above results indicated that the Mn-O bond in δ-MnO$_2$ is weaker than those in α-, β- and γ-MnO$_2$, which leads to the easier breakthrough of the Mn-O bond and might produce more active oxygen species.

To further characterize the structural stability of MnO$_2$ with different crystalline forms, thermogravimetry characterization was carried out (see Figure 5a). For α-, β-, and γ-MnO$_2$, two weight loss peaks at 300–550 °C and 650–800 °C appeared, which correspond to the transformation of MnO$_2$ to Mn$_2$O$_3$ and Mn$_2$O$_3$ to Mn$_3$O$_4$, respectively [35]. While for δ-MnO$_2$, two weight loss peaks appeared at 50–200 °C and 650–800 °C, respectively. To gain further insight into the cause of the weight loss observed at 50–200 °C, a TG-MS analysis was conducted on δ-MnO$_2$ (see Figure 5b). As can be seen, a large amount of water and oxygen was detected in the temperature range of 50–200 °C, and the peak shapes and emerging temperature correspond well. Combining the IR results indicating that δ-MnO$_2$ has a significantly higher concentration of hydroxyl groups and the Raman results showing that δ-MnO$_2$ has the weakest Mn-O bond, the weight loss in 50–200 °C region on δ-MnO$_2$ might be related to the release of oxygen from structural hydroxyl groups (2-OH→1/2O$_2$ + H$_2$O), which might enhance the easier phase transformation from MnO$_2$ to Mn$_2$O$_3$. Meanwhile, δ-MnO$_2$ contains a high amount of H-bonded hydroxyl likely evolving from water and other surface hydroxyls as shown by the broad 3000–3700 cm^{-1} band, also by adsorbed water in 1625 cm^{-1} [36] (see Figure 3). Therefore, TGA from 80–200 °C should be a combination of weakly adsorbed water.

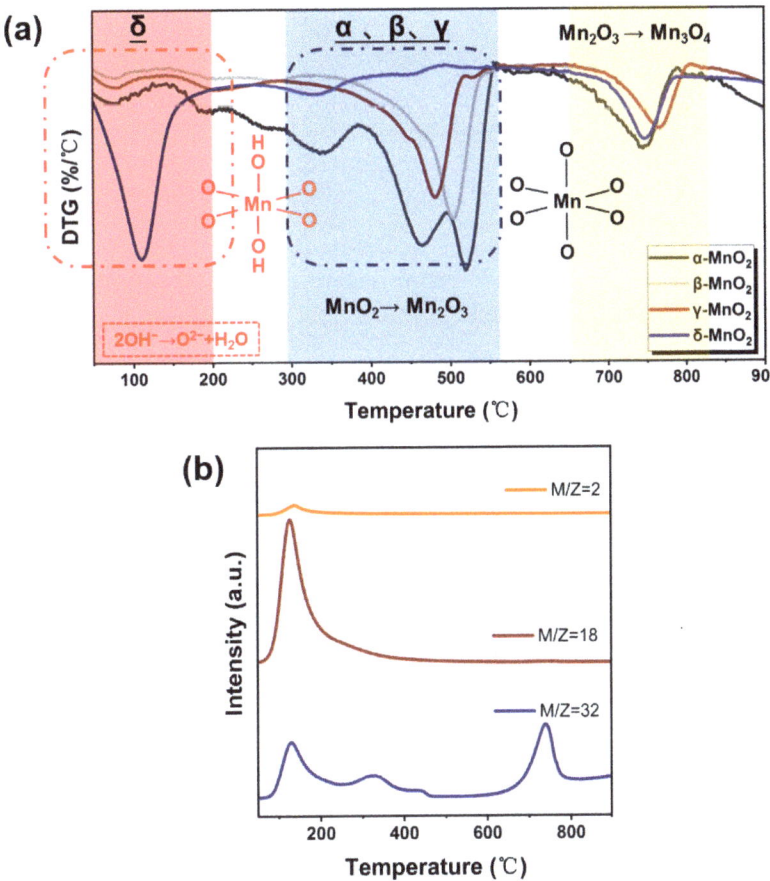

Figure 5. (a) DTG curves of different crystalline MnO_2 (catalyst weight range: ~10 mg, gas type: nitrogen, flow rates: 30 mL/min, temperature ramp: 5 °C/min.), (b) TG-MS spectra of different crystalline MnO_2.

2.3. Oxidation Capacity of Different Crystalline MnO_2

The oxidation ability of α-, β-, γ-MnO_2 and δ-MnO_2 were characterized by H_2-TPR and O_2-TPD techniques. (see Figure 6a,b). Figure 6a shows that the main H_2 consumption peaks of α, β, and γ-MnO_2 appeared near 330 °C and 500 °C, respectively. While the main H_2 consumption peak on δ-MnO_2 is at 328–340 °C with a shoulder peak near 278 °C. These results indicated that the oxygen in δ-MnO_2 is more easily utilized. And this deduction was further supported by the O_2-TPD result (see Figure 6b). The O_2 desorption peaks of α-, β- and γ-MnO_2 mainly appeared near 500 °C and 750 °C. While, for δ-MnO_2, the desorption peak of O_2 shifted to 140 °C, 330 °C and 730 °C. It was reported that the region lower than 400 °C corresponds to the active oxygen species [37,38], while that higher than 400 °C corresponds to the lattice oxygen (O_{latt}). Thus, the above result indicated that δ-MnO_2 has a stronger ability to generate active oxygen species than α-, β- and γ-MnO_2, consistent with the Raman and TG results.

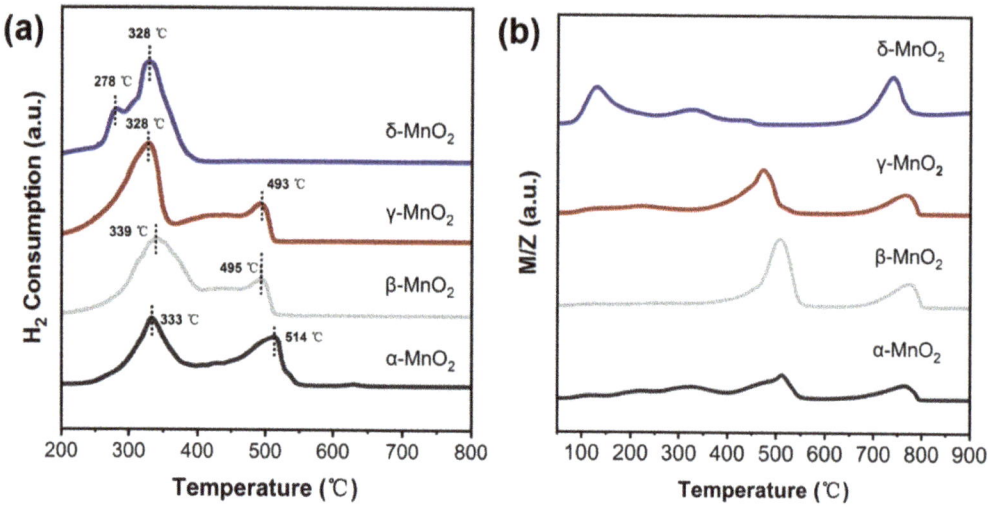

Figure 6. (a) H_2-TPR spectra of different crystalline MnO_2 (catalyst weight: 50 mg, gas type: 10% H_2/Ar, flow rates: 30 mL/min); (b) O_2-TPD spectra of different crystalline MnO_2 (catalyst weight: 50 mg, gas type: 5% O_2/Ar, flow rates: 30 mL/min).

To verify the strong ability to produce active oxygen species on δ-MnO_2, the binding energy of Mn was further analyzed (see Figure 7a), and the average oxidation state (AOS) of Mn was calculated by an empirical formula (AOS = 8.956 − 1.126 × ΔE) [39,40]. Where the lower the AOS of Mn, the more electrons can be transferred to the adsorbed oxygen, resulting in more active oxygen species. Compared with α-MnO_2 (AOS = 3.76), β-MnO_2 (AOS = 3.66), and γ-MnO_2 (AOS = 3.64), the δ-MnO_2 exhibits lower AOS of 3.52 for Mn, indicating δ-MnO_2 has a stronger ability to activate oxygen and is more likely to produce active oxygen species than α-, β-, and γ-MnO_2.

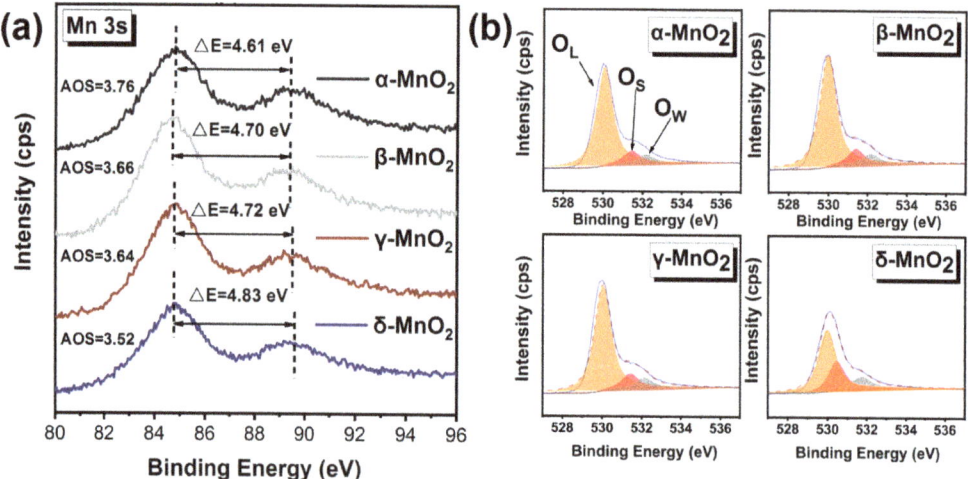

Figure 7. Characterization of the oxidation capacity of different crystalline MnO_2. (a) XPS spectrum of Mn 3 s; (b) XPS spectrum of O 1 s.

And this deduction was further verified by the binding energy of O 1 s (see Figure 7b and Table 2). The O 1 s spectra show that the binding energy of O is at 528–534 eV, which can be divided into three peaks at 530 eV, 531.5 eV, and 533 eV corresponding to lattice oxygen, active oxygen species, and water or oxygen adsorbed on the surface, respectively [39,41]. Based on the peak areas, the percentages of surface-active oxygen species on different MnO_2 have been calculated (see Table 2). The percentage of active oxygen species on δ-MnO_2 was a staggering 27.70%, which is significantly higher than those on α-, β-, and γ-MnO_2. It demonstrates that δ-MnO_2 is capable of producing more active oxygen species.

Table 2. XPS O 1S peak splitting data of different MnO_2.

Catalyst	Oxygen Species Type	Position/Ev	Area Percentage/%
α-MnO_2	O_L [1]	530.06	82.6034
	O_S [2]	531.44	10.5468
	O_W [3]	532.15	6.8498
β-MnO_2	O_L	529.95	81.7358
	O_S	531.38	11.3416
	O_W	532.14	6.9227
γ-MnO_2	O_L	530.02	81.0782
	O_S	531.42	11.3644
	O_W	532.13	7.5575
δ-MnO_2	O_L	529.96	58.3633
	O_S	530.43	27.6973
	O_W	531.70	13.9394

[1] O_L: lattice oxygen, [2] O_S: active oxygen species, [3] O_W: water or oxygen adsorbed on the surface.

2.4. Discussion

The above results show that δ-MnO_2 has a more abundant hydroxyl group and a stronger ability to activate oxygen compared with α-, β- and γ-MnO_2. Therefore, the effect of the hydroxyl group and the oxygen activation ability of MnO_2 on the oxidation of furfural are discussed in detail below.

2.4.1. The Role of the Hydroxyl Group of MnO_2

For the oxidation reaction of furfural, it is generally acknowledged that the attack of the aldehyde group by OH^- or H^+ is essential for the activation of furfural. Under homogeneous alkaline conditions, the free OH^- can attack the aldehyde group leading to opening the C=O bond of furfural. As there is no free OH^- under base-free conditions over MnO_2, it is speculated that the attack of the aldehyde group of furfural might be due to the inherent hydroxyl group of MnO_2 (see Figure 8a). To verify this deduction, the furfural-TPD was carried out over α-, β-, γ- and δ-MnO_2 (see Figure 8b). On δ-MnO_2, a big desorption peak corresponding to furfural appeared at 160 °C, but almost no furfural desorption peak was observed on α-, β-, and γ-MnO_2. And this trend is consistent with the changing trend of hydroxyl amount over different MnO_2 in IR results. Thus, it can be deduced that the inherent hydroxyl group on MnO_2 plays an active role in furfural activation. And this deduction was supported by the formaldehyde oxidation over CeO_2 or TiO_2 catalyst in the literature [42,43].

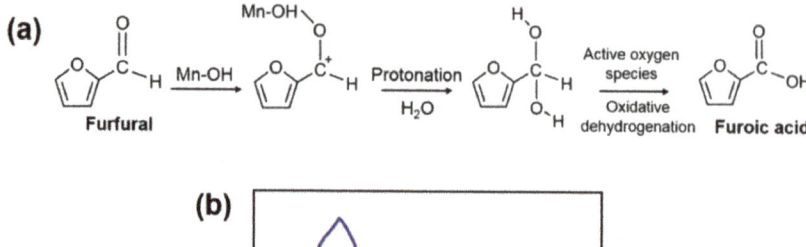

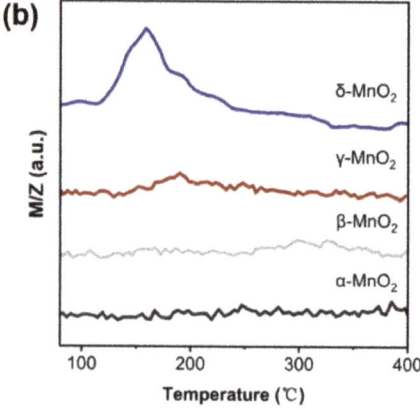

Figure 8. (a) Schematics of the oxidation of furfural to furoic acid; (b) Furfural-TPD spectrum of different crystalline MnO₂ (m/z corresponds to the mass signal of furfural).

To verify the above deductions, the infrared spectra of different MnO$_2$ before and after the adsorption of furfural were characterized by the furfural-IR technique (see Figure 9). The peaks at 3186 cm^{-1} and 3557 cm^{-1} are attributed to hydroxyl groups [33,44,45]. Before furfural adsorption, the hydroxyl group was only found on γ-MnO$_2$ and δ-MnO$_2$, with the corresponding peak intensity stronger on δ-MnO$_2$ than γ-MnO$_2$. However, after the adsorption of furfural, the hydroxyl peaks in both γ-MnO$_2$ and δ-MnO$_2$ are much weaker compared with the pure γ-MnO$_2$ and δ-MnO$_2$ without any change of other groups, which indicated that the inherent hydroxyl group on MnO$_2$ may act as an attacking agent for aldehyde group of furfural.

2.4.2. The Role of Active Oxygen Species of MnO$_2$

It is generally realized that oxidation ability and the amount of activated oxygen play an important role in the activation of oxygen and facilitation of intermediates in aldehyde oxidation reactions [22,42]. The H$_2$-TPR and O$_2$-TPD results in the present study show that the δ-MnO$_2$ has both the strongest ability to activate oxygen and the most abundant active oxygen species (desorption peaks below 400 °C), which might be beneficial to active oxygen and promote the conversion of the geminal diol intermediates in the furfural oxidation reaction.

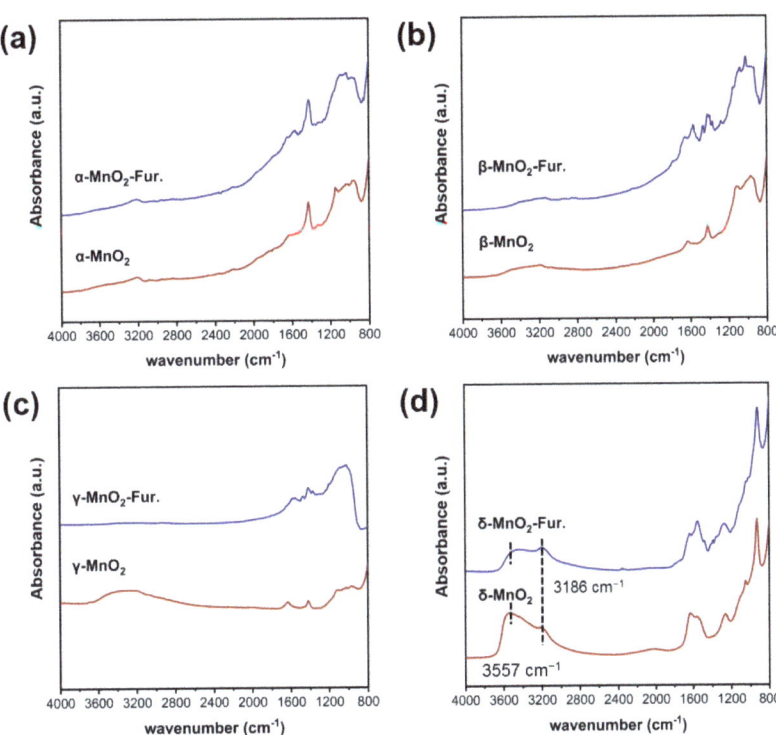

Figure 9. Ex situ DRIFTS Furfural-IR spectrum of different crystalline MnO_2 ((**a**–**d**) are the infrared spectra measured at room temperature for α-, β-, γ- and δ-MnO_2, respectively).

3. Conclusions

In summary, δ-MnO_2 has the weakest Mn-O bond, which can be easily broken to produce the most abundant active oxygen species, giving rise to the beneficial oxidation of intermediates. Moreover, the plentiful inherent hydroxyl groups on δ-MnO_2 can activate the aldehyde group of furfural and favors furfural conversion. The activation behavior of the hydroxyl group on the aldehyde group allows the reaction to occur under base-free conditions. Thus, a 99.04% furfural conversion and 100% furoic acid selectivity can be obtained on δ-MnO_2. Moreover, the δ-MnO_2 shows super stability after three reaction cycles.

4. Materials and Methods

Ammonium persulfate (($NH_4)_2S_2O_8$, 99.99%, 7727-54-0), ammonium sulfate (($NH_4)_2SO_4$, 99.99%, 7783-20-2), manganese sulfate monohydrate ($MnSO_4·H_2O$, 99.99%, 10034-96-5), potassium permanganate ($KMnO_4$, 99%, 7722-64-7) and Furfural ($C_5H_4O_2$, 99%, 98-01-1) were used in the catalyst synthesis process, all of which were purchased from Aladdin (Shanghai, China).

4.1. Preparation of MnO_2 with Different Crystalline Forms

The synthesis steps of α-MnO_2 were as follows: 0.056 mol of $(NH_4)_2S_2O_8$, 0.056 mol of $MnSO_4·H_2O$, and 0.14 mol of $(NH_4)_2SO_4$ were added into 140 mL of deionized water and stirred for 30 min until fully dissolved, then transferred to 200 mL of PTFE liner and hydrothermally heated at 140 °C for 12 h. After cooling, the obtained precipitate was filtered and washed three times with deionized water and finally dried in a static air oven at 80 °C for 12 h. The obtained catalyst was recorded as α-MnO_2.

The synthesis steps of β-MnO$_2$ were as follows: 0.056 mol of (NH$_4$)$_2$S$_2$O$_8$, 0.056 mol of MnSO$_4$·H$_2$O were added into 140 mL of deionized water and stirred for 30 min until fully dissolved, then transferred to 200 mL of PTFE liner and hydrothermally heated at 140 °C for 12 h. After cooling, the obtained precipitate was filtered and washed three times with deionized water and finally dried in a static air oven at 80 °C for 12 h. The obtained catalyst was recorded as β-MnO$_2$.

The synthesis steps of γ-MnO$_2$ were as follows: 0.056 mol of (NH$_4$)$_2$S$_2$O$_8$, 0.056 mol of MnSO$_4$·H$_2$O were added into 140 mL of deionized water and stirred for 30 min until fully dissolved, then transferred to 200 mL of PTFE liner and hydrothermally heated at 90 °C for 12 h. After cooling, the obtained precipitate was filtered and washed three times with deionized water and finally dried in a static air oven at 80 °C for 12 h. The obtained catalyst was recorded as γ-MnO$_2$.

The synthesis steps of δ-MnO$_2$ were as follows: 0.024 mol of KMnO$_4$ and 0.004 mol of MnSO$_4$·H$_2$O were added into 140 mL of deionized water and stirred for 30 min until fully dissolved, then transferred to 200 mL of PTFE liner and hydrothermally heated at 160 °C for 12 h. After cooling, the obtained precipitate was filtered and washed three times with deionized water and finally dried in a static air oven at 80 °C for 12 h. The obtained catalyst was recorded as δ-MnO$_2$.

4.2. Evaluation of Catalysts

The evaluation of the catalyst was carried out in a 100 mL stainless-steel magnetic stirring batch reactor. In general, 50 mg of furfural, 10 mL of deionized water, and 200 mg of catalyst were loaded into the batch reactor, replaced with pure oxygen five times, charged with 1 Mpa of pure oxygen, and raised to 100 °C for 1 h. The experimental parameters were changed with additional explanations in the text. The reactor was cooled immediately after the completion of the reaction, and the resulting mixture was filtered through a syringe filter equipped with a 22 μm PTEF membrane. The filtrate was diluted and analyzed by high-performance liquid chromatography with the following chromatographic detection conditions. The mobile phase ratio was acetonitrile: water = 1:1, and the detection wavelength was 220 nm. The carbon equilibrium of the reaction was calculated by standard solutions. The equations for each parameter in Table 1 are as follows:

$$\text{Conver.} = (n_{FF\ in} - n_{FF\ out}(\text{by HPLC}))/n_{FF\ in} \times 100\%$$

$n_{FF\ in}$: Furfural feed molarity, $n_{FF\ out}$ Molarity of furfural in the product detected by HPLC.

$$\text{Propo.}_{FA} = n_{furoic\ acid}(\text{by HPLC})/n_{product}(\text{by HPLC}) \times 100\%$$

$n_{furoic\ acid}$ represents the molar amount of furoic acid in the product detected by HPLC, and $n_{product}$ represents the sum of the molar amounts of all products detected by HPLC.

$$\text{Carbon balance} = (n_{FF\ out}(\text{by HPLC}) + n_{FA\ out}(\text{by HPLC}))/n_{FF\ in}$$

$n_{FF\ in}$: Furfural feed molarity, $n_{FF\ out}$: Molarity of furfural in the product, $n_{FA\ out}$: Molarity of furoic acid.

4.3. Catalyst Reuse

The used catalyst was filtered out, soaked in 100 mL 50% ethanol/50% sodium hydroxide (1M) mixture for 6 h, then warmed up to 80 °C and stirred for 2 h, and finally filtered and washed five times, dried at 60 °C for 1 h and used in the reaction.

4.4. Catalyst Characterization

H_2-TPR: The H_2-TPR was tested on a Tianjin Xianquan TP-5080 Automatic Multi-use Adsorption Instrument (Xianquan Industrial and Trading Co., Ltd., Tianjing, China) with 50 mg catalyst (20–40 mesh), purged under a reducing gas (5% H_2/95% Ar) atmosphere for 30 min to a smooth baseline under room temperature with a gas flow rate of 30 mL/min. The temperature was programmed to 800 °C from room temperature at a ramp rate of 10 °C/min, and the H_2 consumption was measured by TCD.

O_2-TPD: The O_2-TPD procedure was as follows, using 50 mg of catalyst (20–40 mesh) in quartz tubes purged at 50 °C for 30 min under an Ar atmosphere, followed by the introduction of 5% O_2/95% Ar (30 mL/min) without pretreatment for 1 h to saturate the O_2 on the sample surface, and then switching to pure argon for 40 min to a smooth baseline. Finally, the temperature was increased from 50 °C to 900 °C at a 10 °C/min rate in an argon atmosphere. The m/z was used to monitor O_2 and H_2O by a mass spectrometer (IPI GAM200, Pfeiffer Vacuum GmbH (Hessen, Germany)).

XRD: The crystalline phase of the catalyst was tested using an X-ray diffraction (XRD) analyzer (PANalytical X'pert3, Malvern Panalytical Ltd., Overijssel, The Netherlands) with the parameters Cu target, Kα (λ = 1.54056 Å), 40 kV tube voltage, and 40 mA tube current.

SEM: Morphological observation of the catalyst using a scanning electron microscope (JSM-7001F, JEOL Ltd., Beijing, China). The sample powder adheres to the conductive adhesive with an operating voltage of 10 KV.

XPS: XPS analysis was performed using an AXIS ULTRADLD X-ray photoelectron spectrometer (Kratos Ltd., Manchester, UK). The spectra were charge-corrected using a C1s signal located at 284.5 eV. The spectra of all samples were fitted using the XPS PEAK41 software (Kratos Ltd., UK), and the peaks were resolved using an 80% Gaussian/20% Lorentzian model function.

Raman: Raman spectroscopy was performed on a LabRAM HR Evolution Raman instrument (Horiba Ltd., Kyoto, Japan). The excitation source was an argon ion laser 532 nm visible Raman spectral line (laser power:100 mW, Acq. time: 100 s, Accumulations: 1, Hole:150). The instrument was calibrated with silica before testing.

Ex situ DRIFTS: Diffuse reflectance infrared experiments were performed on a Nicolet iS10 infrared spectrometer (Thermo Fisher Scientific Inc., Shanghai, China, spectra resolution: 8 cm^{-1}, scans averaged: 64).

TG: Thermogravimetric curves of the catalysts were measured on a Setaram TGA-92 platform (Setaram Ltd., Bourges, France). About 10 mg of the sample was purged under N_2 at room temperature for 30 min and then ramped up from room temperature to 900 °C under N_2 at a ramp rate of 5 °C/min.

Ex situ DRIFTS Furfural-IR: After ~50 mg catalyst acquisition of ordinary diffuse reflectance IR, the spectrum was acquired by adding 5 μL of pure furfural dropwise on the catalyst. The specific details are purging it for 10 min under an argon atmosphere (60 mL/min) by warming up to 80 °C in the in-situ pool (spectra resolution: 8 cm^{-1}, scans averaged: 64).

Furfural-TPD: 50 mg of catalyst was loaded into the reaction furnace, furfural was placed in the wash bottle ice bath thermostat (0 °C), and Ar was passed through the furfural wash bottle at a flow rate of 30 mL/min to bring the furfural vapor into the reaction furnace. Next, the product was adsorbed for 30 min until saturation (The mass spectral signal of furfural is not changing) at 80 °C, and then argon purge (30 mL/min) was switched to for 40 min until the baseline was smooth. Finally, the product was desorbed under an Ar atmosphere at a rate of 10 °C/min up to 400 °C, and the desorbed product was detected using mass spectrometry (InProcess Instruments Ltd., Beijing, China).

Supplementary Materials: The following supporting information can be downloaded at: https://www.mdpi.com/article/10.3390/catal13040663/s1, Table S1: BET data for different crystalline MnO_2. Table S2: JCPDS main peaks data of different crystalline MnO_2. Figure S1: XPS S 2p binding energy of MnO_2 with different crystal types. Figure S2: Ex situ DRIFTS spectra of pure furfural. Figure S3: Reactor for the oxidation of furfural to furoic acid.

Author Contributions: Investigation, X.W., H.G. and D.L.; writing—original draft preparation, X.W.; writing—review and editing, X.W., H.G., L.J. and D.L.; supervision, L.J., Y.X., B.H. and D.L.; project administration, H.G., L.J., Y.X., B.H. and D.L.; funding acquisition, H.G., L.J. and D.L. All authors have read and agreed to the published version of the manuscript.

Funding: This research was funded by financial support from Shanxi Science and Technology Major Project, grant number 202005D121002 and "The Natural Science Foundation of Shanxi Province, grant number 20210302123007". And "The Shanxi Key R&D Programme 202102090301022".

Data Availability Statement: Not applicable.

Conflicts of Interest: The authors declare no conflict of interest.

References

1. Mariscal, R.; Maireles-Torres, P.; Ojeda, M.; Sádaba, I.; López Granados, M. Furfural: A renewable and versatile platform molecule for the synthesis of chemicals and fuels. *Energy Environ. Sci.* **2016**, *9*, 1144–1189. [CrossRef]
2. Banerjee, A.; Dick, G.R.; Yoshino, T.; Kanan, M.W. Carbon dioxide utilization via carbonate-promoted C–H carboxylation. *Nature* **2016**, *531*, 215–219. [CrossRef] [PubMed]
3. Jiang, L.; Gonzalez-Diaz, A.; Ling-Chin, J.; Malik, A.; Roskilly, A.P.; Smallbone, A.J. PEF plastic synthesized from industrial carbon dioxide and biowaste. *Nat. Sustain.* **2020**, *3*, 761–767. [CrossRef]
4. Pinna, F.; Olivo, A.; Trevisan, V.; Menegazzo, F.; Signoretto, M.; Manzoli, M.; Boccuzzi, F. The effects of gold nanosize for the exploitation of furfural by selective oxidation. *Catal. Today* **2013**, *203*, 196–201. [CrossRef]
5. Menegazzo, F.; Signoretto, M.; Pinna, F.; Manzoli, M.; Aina, V.; Cerrato, G.; Boccuzzi, F. Oxidative esterification of renewable furfural on gold-based catalysts: Which is the best support? *J. Catal.* **2014**, *309*, 241–247. [CrossRef]
6. Menegazzo, F.; Fantinel, T.; Signoretto, M.; Pinna, F.; Manzoli, M. On the process for furfural and HMF oxidative esterification over Au/ZrO_2. *J. Catal.* **2014**, *319*, 61–70. [CrossRef]
7. Ferraz, C.P.; Zieliński, M.; Pietrowski, M.; Heyte, S.; Dumeignil, F.; Rossi, L.M.; Wojcieszak, R. Influence of Support Basic Sites in Green Oxidation of Biobased Substrates Using Au-Promoted Catalysts. *ACS Sustain. Chem. Eng.* **2018**, *6*, 16332–16340. [CrossRef]
8. Douthwaite, M.; Huang, X.; Iqbal, S.; Miedziak, P.J.; Brett, G.L.; Kondrat, S.A.; Edwards, J.K.; Sankar, M.; Knight, D.W.; Bethell, D.; et al. The controlled catalytic oxidation of furfural to furoic acid using $AuPd/Mg(OH)_2$. *Catal. Sci. Technol.* **2017**, *7*, 5284–5293. [CrossRef]
9. Ferraz, C.P.; Navarro-Jaén, S.; Rossi, L.M.; Dumeignil, F.; Ghazzal, M.N.; Wojcieszak, R. Enhancing the activity of gold supported catalysts by oxide coating: Towards efficient oxidations. *Green Chem.* **2021**, *23*, 8453–8457. [CrossRef]
10. Papanikolaou, G.; Lanzafame, P.; Perathoner, S.; Centi, G.; Cozza, D.; Giorgianni, G.; Migliori, M.; Giordano, G. High performance of Au/ZTC based catalysts for the selective oxidation of bio-derivative furfural to 2-furoic acid11This paper in honor of Professor James G. Goodwin, Jr., in the occasion of his 75th birthday, to celebrate his outstanding contribution to catalysis sciences and technology. *Catal. Commun.* **2021**, *149*, 106234. [CrossRef]
11. Daniel, S.; Subbiah, V.P.; Ramaganthan, B.; Kanthapazham, R.; Vanaraj, R. Upgrading the Strategy of Multistage Torrefaction Liquid by the Selective Oxidation Reaction Route Using a Reusable MgO-Based Au/Al_2O_3 Catalyst. *Energy Fuel* **2021**, *35*, 15831–15841. [CrossRef]
12. Ren, Z.; Yang, Y.; Wang, S.; Li, X.; Feng, H.; Wang, L.; Li, Y.; Zhang, X.; Wei, M. Pt atomic clusters catalysts with local charge transfer towards selective oxidation of furfural. *Appl. Catal. B Environ.* **2021**, *295*, 120290. [CrossRef]
13. Gupta, K.; Rai, R.K.; Dwivedi, A.D.; Singh, S.K. Catalytic Aerial Oxidation of Biomass-Derived Furans to Furan Carboxylic Acids in Water over Bimetallic Nickel–Palladium Alloy Nanoparticles. *Chemcatchem* **2017**, *9*, 2760–2767. [CrossRef]
14. Al Rawas, H.K.; Ferraz, C.P.; Thuriot-Roukos, J.; Heyte, S.; Paul, S.; Wojcieszak, R. Influence of Pd and Pt Promotion in Gold Based Bimetallic Catalysts on Selectivity Modulation in Furfural Base-Free Oxidation. *Catalysts* **2021**, *11*, 1226. [CrossRef]
15. Tian, Q.; Shi, D.; Sha, Y. CuO and Ag_2O/CuO Catalyzed Oxidation of Aldehydes to the Corresponding Carboxylic Acids by Molecular Oxygen. *Molecules* **2008**, *13*, 948. [CrossRef]
16. Zhou, H.; Xu, H.; Wang, X.; Liu, Y. Convergent production of 2,5-furandicarboxylic acid from biomass and CO_2. *Green Chem.* **2019**, *21*, 2923–2927. [CrossRef]
17. Yu, K.; Lou, L.-L.; Liu, S.; Zhou, W. Asymmetric Oxygen Vacancies: The Intrinsic Redox Active Sites in Metal Oxide Catalysts. *Adv. Sci.* **2020**, *7*, 1901970. [CrossRef] [PubMed]

18. Yu, K.; Liu, Y.; Lei, D.; Jiang, Y.; Wang, Y.; Feng, Y.; Lou, L.-L.; Liu, S.; Zhou, W. $M^{3+}O(-Mn^{4+})_2$ clusters in doped MnO_x catalysts as promoted active sites for the aerobic oxidation of 5-hydroxymethylfurfural. *Catal. Sci. Technol.* **2018**, *8*, 2299–2303. [CrossRef]
19. Zhou, J.; Qin, L.; Xiao, W.; Zeng, C.; Li, N.; Lv, T.; Zhu, H. Oriented growth of layered-MnO_2 nanosheets over α-MnO_2 nanotubes for enhanced room-temperature HCHO oxidation. *Appl. Catal. B Environ.* **2017**, *207*, 233–243. [CrossRef]
20. Wang, J.; Li, J.; Jiang, C.; Zhou, P.; Zhang, P.; Yu, J. The effect of manganese vacancy in birnessite-type MnO_2 on room-temperature oxidation of formaldehyde in air. *Appl. Catal. B Environ.* **2017**, *204*, 147–155. [CrossRef]
21. Wang, J.; Zhang, P.; Li, J.; Jiang, C.; Yunus, R.; Kim, J. Room-Temperature Oxidation of Formaldehyde by Layered Manganese Oxide: Effect of Water. *Environ. Sci. Technol.* **2015**, *49*, 12372–12379. [CrossRef] [PubMed]
22. Liu, H.; Jia, W.; Yu, X.; Tang, X.; Zeng, X.; Sun, Y.; Lei, T.; Fang, H.; Li, T.; Lin, L. Vitamin C-Assisted Synthesized Mn–Co Oxides with Improved Oxygen Vacancy Concentration: Boosting Lattice Oxygen Activity for the Air-Oxidation of 5-(Hydroxymethyl)furfural. *ACS Catal.* **2021**, *11*, 7828–7844. [CrossRef]
23. Hayashi, E.; Yamaguchi, Y.; Kamata, K.; Tsunoda, N.; Kumagai, Y.; Oba, F.; Hara, M. Effect of MnO_2 Crystal Structure on Aerobic Oxidation of 5-Hydroxymethylfurfural to 2,5-Furandicarboxylic Acid. *J. Am. Chem. Soc.* **2019**, *141*, 890–900. [CrossRef]
24. Ferraz, C.P.; Da Silva, A.G.; Rodrigues, T.S.; Camargo, P.H.; Paul, S.; Wojcieszak, R. Furfural Oxidation on Gold Supported on MnO_2: Influence of the Support Structure on the Catalytic Performances. *Appl. Sci.* **2018**, *8*, 1246. [CrossRef]
25. Rong, S.; Zhang, P.; Liu, F.; Yang, Y. Engineering Crystal Facet of α-MnO_2 Nanowire for Highly Efficient Catalytic Oxidation of Carcinogenic Airborne Formaldehyde. *ACS Catal.* **2018**, *8*, 3435–3446. [CrossRef]
26. Saputra, E.; Muhammad, S.; Sun, H.; Ang, H.M.; Tade, M.O.; Wang, S. Different crystallographic one-dimensional MnO_2 nanomaterials and their superior performance in catalytic phenol degradation. *Environ. Sci. Technol.* **2013**, *47*, 5882–5887. [CrossRef] [PubMed]
27. Qiu, G.; Huang, H.; Dharmarathna, S.; Benbow, E.; Stafford, L.; Suib, S.L. Hydrothermal Synthesis of Manganese Oxide Nanomaterials and Their Catalytic and Electrochemical Properties. *Chem. Mater.* **2011**, *23*, 3892–3901. [CrossRef]
28. Wang, H.; Luo, Q.; Wang, L.; Hui, Y.; Qin, Y.; Song, L.; Xiao, F.-S. Product selectivity controlled by manganese oxide crystals in catalytic ammoxidation. *Chin. J. Catal.* **2021**, *42*, 2164–2172. [CrossRef]
29. Verdeguer, P.; Merat, N.; Gaset, A. Lead/platinum on charcoal as catalyst for oxidation of furfural. Effect of main parameters. *Appl. Catal. A Gen.* **1994**, *112*, 1–11. [CrossRef]
30. Wang, X.; Li, Y. Synthesis and Formation Mechanism of Manganese Dioxide Nanowires/Nanorods. *Chem. A Eur. J.* **2003**, *9*, 300–306. [CrossRef]
31. Luo, J.; Huang, A.; Park, S.H.; Suib, S.L.; O'Young, C.-L. Crystallization of Sodium−Birnessite and Accompanied Phase Transformation. *Chem. Mater.* **1998**, *10*, 1561–1568. [CrossRef]
32. Yang, D.S.; Wang, M.K. Syntheses and Characterization of Well-Crystallized Birnessite. *Chem. Mater.* **2001**, *13*, 2589–2594. [CrossRef]
33. Xu, Z.; Yu, J.; Jaroniec, M. Efficient catalytic removal of formaldehyde at room temperature using AlOOH nanoflakes with deposited Pt. *Appl. Catal. B Environ.* **2015**, *163*, 306–312. [CrossRef]
34. Zhang, J.; Li, Y.; Wang, L.; Zhang, C.; He, H. Catalytic oxidation of formaldehyde over manganese oxides with different crystal structures. *Catal. Sci. Technol.* **2015**, *5*, 2305–2313. [CrossRef]
35. Jia, J.; Zhang, P.; Chen, L. The effect of morphology of α-MnO_2 on catalytic decomposition of gaseous ozone. *Catal. Sci. Technol.* **2016**, *6*, 5841–5847. [CrossRef]
36. Hadjiivanov, K. Chapter Two—Identification and Characterization of Surface Hydroxyl Groups by Infrared Spectroscopy. In *Advances in Catalysis*; Jentoft, F.C., Ed.; Academic Press: Cambridge, MA, USA, 2014; Volume 57, pp. 99–318.
37. Gu, H.; Liu, X.; Liu, X.; Ling, C.; Wei, K.; Zhan, G.; Guo, Y.; Zhang, L. Adjacent single-atom irons boosting molecular oxygen activation on MnO_2. *Nat. Commun.* **2021**, *12*, 5422. [CrossRef]
38. Li, X.; Wang, Y.; Chen, D.; Li, N.; Xu, Q.; Li, H.; He, J.; Lu, J. A highly dispersed Pt/copper modified-MnO_2 catalyst for the complete oxidation of volatile organic compounds: The effect of oxygen species on the catalytic mechanism. *Green Energy Environ.* **2021**, *8*, 538–547. [CrossRef]
39. Xu, T.; Zhang, P.; Zhang, H. Ultrathin δ-MnO_2 nanoribbons for highly efficient removal of a human-related low threshold odorant—Acetic acid. *Appl. Catal. B Environ.* **2022**, *309*, 121273. [CrossRef]
40. Zhu, G.; Zhu, W.; Lou, Y.; Ma, J.; Yao, W.; Zong, R.; Zhu, Y. Encapsulate α-MnO_2 nanofiber within graphene layer to tune surface electronic structure for efficient ozone decomposition. *Nat. Commun.* **2021**, *12*, 4152. [CrossRef]
41. Chen, D.; Zhang, G.; Wang, M.; Li, N.; Xu, Q.; Li, H.; He, J.; Lu, J. Pt/MnO_2 Nanoflowers Anchored to Boron Nitride Aerogels for Highly Efficient Enrichment and Catalytic Oxidation of Formaldehyde at Room Temperature. *Angew. Chem. Int. Ed.* **2021**, *60*, 6377–6381. [CrossRef]
42. Zhang, L.; Bao, Q.; Zhang, B.; Zhang, Y.; Wan, S.; Wang, S.; Lin, J.; Xiong, H.; Mei, D.; Wang, Y. Distinct Role of Surface Hydroxyls in Single-Atom Pt_1/CeO_2 Catalyst for Room-Temperature Formaldehyde Oxidation: Acid–Base Versus Redox. *JACS Au* **2022**, *2*, 1651–1660. [CrossRef] [PubMed]
43. Chen, X.; He, G.; Li, Y.; Chen, M.; Qin, X.; Zhang, C.; He, H. Identification of a Facile Pathway for Dioxymethylene Conversion to Formate Catalyzed by Surface Hydroxyl on TiO_2-Based Catalyst. *ACS Catal.* **2020**, *10*, 9706–9715. [CrossRef]

44. Rong, S.; Zhang, P.; Wang, J.; Liu, F.; Yang, Y.; Yang, G.; Liu, S. Ultrathin manganese dioxide nanosheets for formaldehyde removal and regeneration performance. *Chem. Eng. J.* **2016**, *306*, 1172–1179. [CrossRef]
45. Yang, X.; Makita, Y.; Liu, Z.-h.; Sakane, K.; Ooi, K. Structural Characterization of Self-Assembled MnO_2 Nanosheets from Birnessite Manganese Oxide Single Crystals. *Chem. Mater.* **2004**, *16*, 5581–5588. [CrossRef]

Disclaimer/Publisher's Note: The statements, opinions and data contained in all publications are solely those of the individual author(s) and contributor(s) and not of MDPI and/or the editor(s). MDPI and/or the editor(s) disclaim responsibility for any injury to people or property resulting from any ideas, methods, instructions or products referred to in the content.

Article

DFT Investigations of the Reaction Mechanism of Dimethyl Carbonate Synthesis from Methanol and CO on Various Cu Species in Y Zeolites

Yuan Zhou [1,2], Guoqiang Zhang [1], Ya Song [1], Shirui Yu [1], Jingjing Zhao [1] and Huayan Zheng [1,*]

1. Department of Food Science and Engineering, Moutai Institute, Renhuai 564502, China
2. Experimental Training Teaching Center, Moutai Institute, Renhuai 564502, China
* Correspondence: andyzheng1109@163.com

Abstract: In this study, a density functional theory method is employed to investigate the reaction mechanisms of dimethyl carbonate (DMC) formation, through oxidative carbonylation of methanol, on four types of Y zeolites doped with Cu^+, Cu^{2+}, Cu_2O and CuO, respectively. A common chemical route is found for these zeolites and identified as, first, the adsorbed CH_3OH is oxidized to CH_3O species; subsequently, CO inserts into CH_3O to CH_3OCO, which reacts with CH_3O to form DMC rapidly; and finally, the adsorbed DMC is released into the gas phase. The rate-limiting step on $Cu^{2+}Y$ zeolite is identified as oxidation of CH_3OH to CH_3O with activation barrier of 66.73 kJ·mol^{-1}. While for Cu^+Y, Cu_2O-Y and CuO-Y zeolites, the rate-limiting step is insertion of CO into CH_3O, and the corresponding activation barriers are 63.73, 60.01 and 104.64 kJ·mol^{-1}, respectively. For Cu^+Y, $Cu^{2+}Y$ and Cu_2O-Y zeolites, adsorbed CH_3OH is oxidized to CH_3O with the presence of oxygen, whereas oxidation of CH_3OH on CuO-Y is caused by the lattice oxygen of CuO. The order of catalytic activities of these four types of zeolites with different Cu states follows $Cu^+Y \approx Cu_2O$-Y > $Cu^{2+}Y$ > CuO-Y zeolite. Therefore, CuY catalysts with Cu^+ and Cu_2O as dominated Cu species are beneficial to the formation of DMC.

Keywords: dimethyl carbonate; Y zeolite; Cu states; density functional theory; reaction mechanism

Citation: Zhou, Y.; Zhang, G.; Song, Y.; Yu, S.; Zhao, J.; Zheng, H. DFT Investigations of the Reaction Mechanism of Dimethyl Carbonate Synthesis from Methanol and CO on Various Cu Species in Y Zeolites. *Catalysts* **2023**, *13*, 477. https://doi.org/10.3390/catal13030477

Academic Editor: C. Heath Turner

Received: 30 January 2023
Revised: 22 February 2023
Accepted: 24 February 2023
Published: 26 February 2023

Copyright: © 2023 by the authors. Licensee MDPI, Basel, Switzerland. This article is an open access article distributed under the terms and conditions of the Creative Commons Attribution (CC BY) license (https:// creativecommons.org/licenses/by/ 4.0/).

1. Introduction

Dimethyl carbonate (DMC), which is considered as one of the environmentally benign chemicals, has been used as a low toxicity solvent and fuel additive. Its production and utilization have recently drawn much attention [1–7]. Meanwhile, DMC synthesis by oxidative carbonylation of methanol is suggested since phosgene is not produced during the process [7–11]. CuO and Cu_2O are p-type semiconductors with a direct band gap of 1.2 and 2.0 eV, respectively, which has been widely used as sensors and active centers in various catalytic reactions due to their unique electronic structure [12–14]. Similarly, Cu-exchanged zeolite catalysts, as the chloride-free catalysts, have been considered as one of the most attractive catalysts for DMC synthesis in recent years [15–19], due to the high catalytic activity and selectivity. CuY zeolite catalyst is one of them [17–23].

The presence of different Cu states in CuY zeolites results in distinct catalytic activities and is achieved using different methods. King [20] reported that Cu^+Y zeolite prepared by solid-state ion exchange showed a satisfying catalytic activity in the oxidative carbonylation reaction, while ion-exchanged $Cu^{2+}Y$ zeolite exhibited a poor performance. Cu^+ and CuO based Cu-FAU catalysts were prepared by Kieger et al. via ion-exchanged method and incipient-wetness-impregnation, respectively. After characterized by UV-VIS, IR, TPR and NH_3-TPD, it was suggested that Cu^+ and CuO were formed in Cu-FAU by ion-exchange method and incipient-wetness impregnation, respectively, and Cu^+ exhibited a better catalytic activity than CuO [24]. Richter and co-workers showed that CuO was formed in CuY zeolite when the Cu loading was above 10 wt% during incipient-wetness impregnation [25]. They pointed out that, due to the formation of CuO_x particles, oxidative carbonylation of methanol proceeded with and without oxygen, meanwhile, CuO_x enhanced the formation of DMC [26,27]. Our study showed that with increasing the exchange degree, different Cu states were produced, such as Cu^{2+}, Cu^+, Cu_2O, CuO and CuY zeolites, leading to different catalytic performance [28].

A number of studies investigated the possible reaction schemes of oxidative carbonylation of methanol to DMC on Cu-exchanged zeolite [16,20,22,25,29–31], Pd-exchanged zeolite [9,32], Cu/AC [33], γ-$Cu_2Cl(OH)_3$ [34], CuCl [35] and Cu_2O [8,36] catalysts. Generally, the molecularly adsorbed methanol is first oxidized by oxygen to methoxide or di-methoxide species. Then the formation of DMC follows two distinct reaction pathways. The first starts with the insertion of CO into methoxide to produce CH_3OCO, which subsequently reacts with CH_3OH to form DMC. The second involves the CO addition to di-methoxide species. Experimental investigation by Engeldinger et al. [26,27] showed that CuO_x aggregates were formed in CuY catalyst when the Cu loading was above 11 wt%, which promoted oxidation and oxocarbonylation reactions of methanol and enhanced the formation of DMC. They suggested that the reaction was closely related to the CH_3OCOOH (MMC), which was produced through participation of lattice oxygen from CuO_x of the catalyst. Cu was reoxidized by gas phase oxygen according to the Mars–van Krevelen mechanism [37]. Although the role of CuO_x has been identified, there is little information on the detailed reaction mechanism which addresses the Cu_2O, CuO and Cu^{2+} species of Y zeolite during oxidative carbonylation of methanol.

In this work, the reaction mechanisms governing oxidative carbonylation of methanol to DMC were studied with Cu^+, Cu^{2+}, Cu_2O and CuO species in Y zeolites using density functional theory (DFT). An appropriate size of CuY cluster was constructed as the stable configuration, reflecting different Cu states in Y zeolite. Then, the reaction mechanisms for DMC formation on four types of Cu species were investigated, and the order of catalytic activity of different Cu states in Y zeolite was characterized. It is expected that these results will provide a theoretical clue to prepare CuY catalyst with better catalytic activity for the DMC synthesis.

2. Results

A faujasite type structure with various cationic sites and different crystallographic oxygen positions is shown in Figure 1. As reactant molecule, CO is very difficult to diffuse inside the sodalite cages and hexagonal prisms (2.3 Å) [38] of Y zeolite because of the large dynamic diameter (3.76 Å), while they easily enter enter supercages (7.4 Å) [38], suggesting that only copper species located at sites II and III are accessible to CO adsorption and act as the active sites for the oxidative carbonylation of methanol to DMC. Based on the previous studies [30,31,39,40], copper cations in site II are more stable than site III, therefore site II was selected to represent the location of active center Cu species in this study.

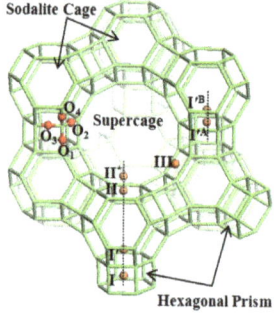

Figure 1. Faujasite type structure with cationic sites (orange balls) and different crystallographic oxygen positions (red balls). Here, site I is at the middle of hexagonal prism; site I'B is in the sodalite cage adjacent to 6MR which is shared by both hexagonal prism and sodalite cage; site I'A is similar to site I'B, but away from the sodalite cage; site II is in the supercage close to the six-membered ring (6 MR) shared by supercage and sodalite cage; site II* is similar to site II, but located towards the supercage; site III is in the supercage that is next to four-membered rings (4 MR) of sodalite cage.

According to the literatures [30,41,42], the local conformation and interactions of molecule can be described using the cluster model. The dangling bonds were saturated by H atoms [41,43]. The terminal H atoms were oriented along the bond direction of Y zeolite. The bond length of O-H was set to 1.0 Å, respectively. During numerical optimization, the local structure of Y zeolite was kept unchanged for Y^{n-}, Y, CuY cluster models. The compensating charges, Al atoms and adjacent SiO_4 units were relaxed, while other atoms were fixed. For the adsorbate-CuY cluster system, the compensating charges, the absorbed molecules and the 6 MR occupied by the active center Cu^+ species were relaxed.

In order to find the appropriate cluster size, five different sized clusters, consisting of 6T, 12T, 24T, 42T, and 60T atoms (T represents an Al or Si atom) (see Figure 2), were constructed.

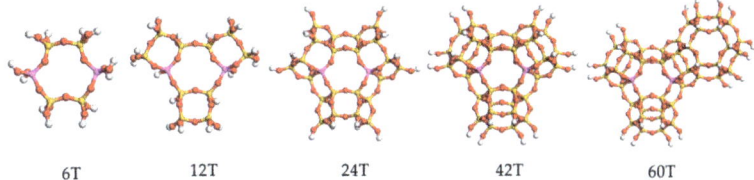

6T   12T   24T   42T   60T

Figure 2. The cluster geometries of Y zeolite with different sizes. Red, yellow, pink and white balls stand for O, Si, Al and H atoms, respectively.

The binding energies of Cu^{2+} in these five Y clusters and the adsorption energies of CO on CuY zeolite with these clusters were calculated, as shown in Table 1.

Table 1. The interaction energies (E_{int}) of Cu^{2+} and the adsorption energies (E_{ads}) of CO on the clusters of Cu^{2+}Y zeolites with different sizes.

Y Zeolite with Different Size	E_{int}/kJ·mol^{-1}	E_{ads}/kJ·mol^{-1}
6T	2713.62	91.45
12T	2588.76	83.33
24T	2529.59	83.08
42T	2454.44	83.27
60T	2442.61	82.88

The interaction energy (E_{int}) between Cu^{2+} and Y^{2-} zeolite was defined as [44]

$$E_{int} = E_{Cu^{2+}} + E_{Y^{2-}} - E_{MY}$$

where $E_{Cu^{2+}}$ is the total energy of Cu^{2+}, $E_{Y^{2-}}$ is the total energy of Y^{2-}, and E_{MY} is the total energy of MY, respectively. Here, a larger E_{int} represents a more stable structure of Cu^{2+}Y system. It can be found from Table 1 that the effect of the cluster size on the adsorption energies of CO is negligible, while the Cu^{2+} interaction energies are significantly influenced by the cluster size. Therefore, a very small cluster model cannot fully reflect the structure of Y zeolite. Comparison of the Cu^{2+} interaction energies suggests that the difference between 24T and 60T cluster is within the allowable error range (<80 kJ/mol). The 24T cluster model was selected in this study to reduce the computing cost.

To better represent the structure of Y zeolite in experiments, four Si atoms of the Y zeolite cluster were substituted by four Al atoms according to the Lowenstein–Dempsey rules. The Y cluster with four Si atoms replaced by Al atoms was denoted as Y^{4-}. Based on the 24T cluster model of Y zeolite, the most stable configurations of Y^{4-} cluster was obtained by evaluating the substitution energy [45] and binding energy of Y^{4-}, which are defined as

$$E_{sub} = E_{Y^{4-}} + 4E_{Si} - E_Y - 4E_{Al}$$

$$E_{bind} = 24E_H + 60E_O + 4E_{Al} + 20E_{Si} - E_{Y^{4-}}$$

where, E_{sub} and E_{bind} are the substitution energy and the binding energy of Y^{4-}, respectively. $E_{Y^{4-}}$ and E_Y are the total energies of Y^{4-} cluster and the Y cluster without the replacement of Si, respectively. E_{Si}, E_{Al}, E_H and E_O are the energies of single Si, Al, H and O atoms, respectively. With these definitions, a smaller E_{sub} indicates an easier replacement of Si by Al and larger E_{bind} means a more stable Y cluster.

Table 2 lists the calculated E_{sub} and E_{bind} for different distribution of Al atoms.

Table 2. The substitution energies (E_{sub}) and binding energies (E_{bind}) for the Y^{4-} cluster with different distribution of Al atoms.

The Distribution of Al Atoms	E_{sub}/kJ·mol^{-1}	E_{bind}/Ha
1-11-12-22	27.31	21.0454
2-11-12-22	31.16	21.0440
3-11-12-22	45.15	21.0387
4-11-12-22	0.28	21.0557
5-11-12-22	37.72	21.0415
8-11-12-22	31.96	21.0436
9-11-12-22	122.11	21.0093
11-12-14-22	107.60	21.0148
11-12-17-22	210.52	20.9756
11-12-20-22	129.81	21.0064
11-12-22-24	125.49	21.0080

Due to its least substitution energy and largest binding energy of Y^{4-}, Y^{4-} cluster with the distribution of 4 Al atoms, denoted as 4-11-12-22, is the most stable structure of Y^{4-} cluster (see Figure 3). Negative charges, which are introduced when Si atoms are replaced by Al atoms, are usually compensated by protons associated with crystallographic oxygen atoms adjacent to the Al atoms.

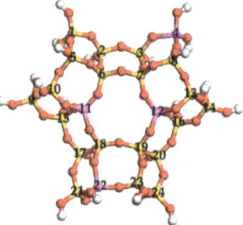

Figure 3. The stable cluster geometry of Y^{4-} zeolite with 24T atoms. See Figure 2 for the color coding.

Based on the stable structure of Y^{4-} cluster, the configurations to reflect the different Cu states in Y zeolite were constructed. According to the literatures [44,46], a majority of charge-compensating protons locate at O1 sites, while the others occupy O3 sites to avoid the formation of -OH$_2$ group. In this study, for Y zeolite with five Al atoms, three charge-compensating protons locate at O1 sites, and two protons are at O3 sites (see Figure 3). For Cu^{2+}Y zeolite, Cu^{2+} is used to balance the negative charge of Al11 and Al12, and charge-compensating protons are located at the O1 site to balance the negative charges of Al4 and Al22, respectively (see Figure 4a). For Cu$^+$Y zeolite, Cu$^+$ balances the negative charge of Al12, when three protons located at the O1 site compensate the negative charges of Al4, Al11 and Al22 (see Figure 4b). For Cu$_2$O-Y and CuO-Y zeolite, all negative charges of Al are compensated by four charge-compensating protons located at O1 and O3 (see Figure 4c,d).

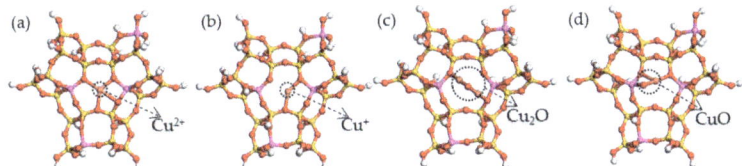

Figure 4. The stable cluster geometries of (**a**) Cu^{2+}Y, (**b**) Cu^+Y, (**c**) Cu_2O-Y and (**d**) CuO-Y zeolites. Red, yellow, pink, white and orange balls stand for O, Si, Al, H and Cu atoms, respectively.

3. Discussion

In this section, the notation (X)* and (X)(Y)* are referred to active center Cu states, such as Cu^+, Cu^{2+}, Cu_2O and CuO interacting with species X and X and Y, respectively. The optimized geometries of reactants, transition states and products for different reaction pathways of DMC formation were calculated.

3.1. The Desorption and Dissociation of CH_3OH

The processes of desorption and dissociation of CH_3OH on these four types of zeolites and the corresponding transition states TS1 are shown in Figure 5.

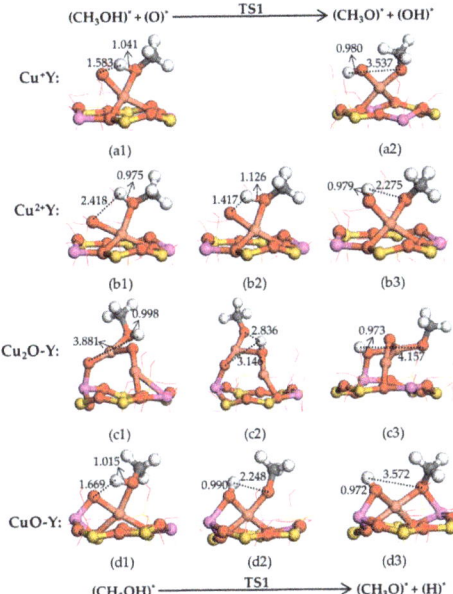

Figure 5. The structures of reactants, products and transition states on Cu^+Y, Cu^{2+}Y, Cu_2O-Y and CuO-Y zeolite for the oxidation of CH_3OH to CH_3O (unit: Å). (**a1**) $(CH_3OH)^*/(O)^*$ on Cu^+Y, (**a2**) $(CH_3O)^*/(OH)^*$ on Cu^+Y, (**b1**) $(CH_3OH)^*/(O)^*$ on Cu^{2+}Y, (**b2**) TS1 on Cu^{2+}Y, (**b3**) $(CH_3O)^*/(OH)^*$ on Cu^{2+}Y, (**c1**) $(CH_3OH)^*/(O)^*$ on Cu_2O-Y, (**c2**) TS1 on Cu_2O-Y, (**c3**) $(CH_3O)^*/(OH)^*$ on Cu_2O-Y, (**d1**) $(CH_3OH)^*/(O)^*$ on CuO-Y, (**d2**) TS1 on CuO-Y and (**d3**) $(CH_3O)^*/(OH)^*$ on CuO-Y. See Figure 4 for the color coding.

As shown in Figure 5, adsorbed CH_3OH on the four types of zeolites are bound to different kinds of active center Cu species through O atom. The adsorption of CH_3OH on Cu^+Y, Cu^{2+}Y, Cu_2O-Y and CuO-Y zeolites is exothermic with the energy release of 65.59, 85.81, 122.70 and 94.07 kJ·mol^{-1}, respectively. Subsequently, for Cu^+Y zeolite, with the presence of oxygen, the O-H bond of CH_3OH breaks to form the co-adsorbed $(CH_3O)^*(OH)^*$ configuration (see Figure 5(a2)). Since no TS state has been found, molecularly adsorbed CH_3OH

is converted rapidly to CH_3O species. This demonstrates that the presence of adsorbed O on Cu^+Y zeolite exhibits a high surface reactivity toward the formation of CH_3O. The results are in good agreement with early reported experimental observations [15,20,22,28].

For $Cu^{2+}Y$ zeolite, adsorbed CH_3OH is oxidized by adsorbed O to form CH_3O species via a transition state (TS1), as shown in Figure 5(b2). The O-H distance in CH_3OH increases from initial 0.975 Å to 1.126 Å of TS1, and finally to 2.275 Å, showing that the O-H bond in CH_3OH is destroyed. Meanwhile, the distance between adsorbed O atom and H atom decreases drastically from initial 2.418 Å to 1.417 Å of TS1, then to 0.979 Å of $(CH_3O)^*/(OH)^*$ (see Figure 5(b3)), revealing that a new O-H bond forms on $Cu^{2+}Y$ zeolite. Similar changes are found on Cu_2O-Y zeolite. For these two zeolites, the oxidation of adsorbed CH_3OH with the presence of oxygen needs to overcome activation barriers of 66.73 and 23.56 kJ·mol^{-1}, respectively (see Table 3).

Table 3. The activation barriers for individual reaction steps based on two proposed reaction mechanisms (/kJ·mol^{-1}).

Catalyst	$(CH_3OH)^*$ + $O^* \to (CH_3O)^*(OH)^*$	$(CH_3O)^*$ + $CO^* \to (CH_3OCO)^*$	$(CH_3OCO)^*$ + $(CH_3O)^* \to (DMC)^*$	$(CH_3O)^*(OH)^*$ + $CH_3OH \to (CH_3O)_2^*$ + H_2O	$(CH_3O)_2^*$ + $CO \to (DMC)^*$	Ref.
Cu_2O	–	161.9	98.8	68.3	308.5	[36]
Cu_2O-Y	23.56	60.01	40.90	116.38	253.96	This study
Cu^+Y	–	63.73	28.27	93.86	201.68	This study
$Cu^{2+}Y$	66.73	64.45	37.95	89.49	164.95	This study
CuO-Y	39.94	104.64	15.95	115.29	210.74	This study
CuO	–	114.5	200.9	25.7	109.1	[47]

It is interesting to note that for CuO-Y zeolite, adsorbed CH_3OH is oxidized to CH_3O species without the presence of O, which is attributed to the presence of lattice oxygen from CuO species. Experimental studies by Engeldinger et al. [26,27] suggested that the formation of methoxy species from the adsorbed CH_3OH proceeded with and without oxygen, indicating that lattice oxygen of CuO_x was able to participate in the oxidation process. According to the Mars–van Krevelen mechanism [37], gas phase oxygen can re-oxidize Cu to CuO species [26,27]. These structures in Figure 5 further prove that for the oxidation reaction of CH_3OH to CH_3O, oxygen is needed for Cu_2O-Y zeolite but is not essential for CuO-Y zeolite. The oxidation reaction of CH_3OH on CuO-Y zeolite is exothermic (85.36 kJ·mol^{-1}) and exhibits an activation barrier of 39.94 kJ·mol^{-1}, as shown in Table 3.

3.2. Insertion of CO into CH_3O (Path I)

Figure 6 shows the processes of inserting CO into CH_3O to form CH_3OCO on these four types of zeolites and the corresponding transition states TS2.

For Cu^+Y zeolite, the distance between the C atom of CO and the O atom of CH_3O decreases from initially 2.787 Å of $(CO)^*/(CH_3O)^*$ to 1.961 Å of TS2, which suggests the formation of a new C-O bond. It is also seen from Figure 6 that insertion of CO into Cu-OCH$_3$ elongates the Cu-O bond from initially 1.901 Å to 2.455 Å of TS2 and then to 2.840 Å of $(CH_3OCO)^*$, indicating that at the final product CH_3OCO adsorbs on the Cu^+ via the C atom of CO (this C atom is denoted as C' for the further analysis). The CO insertion reaction exhibits an activation barrier of 63.73 kJ·mol^{-1} via TS2, which agrees with the results calculated by Zheng et al. [30] (see Table 3). Similar changes from the initial geometries to TS2 states then to the final products happen on the three other types of zeolites. The CO insertion reaction on $Cu^{2+}Y$, Cu_2O-Y and CuO-Y zeolites needs to overcome activation barriers of 64.45, 60.01 and 104.64 kJ·mol^{-1}, respectively (see Table 3).

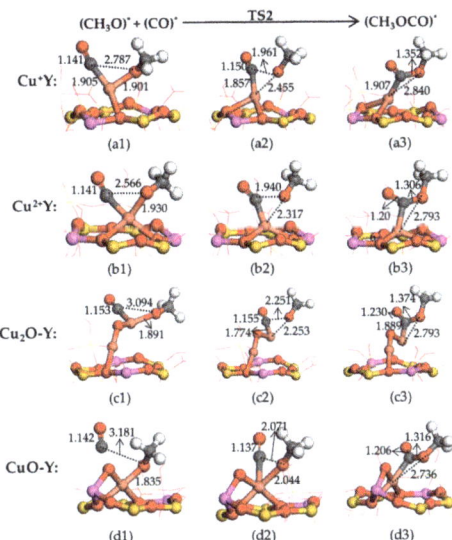

Figure 6. The structures of reactants, products and transition states on Cu^+Y, $Cu^{2+}Y$, Cu_2O-Y and CuO-Y zeolite for the formation of CH_3OCO (unit: Å). (**a1**) $(CH_3O)^*/(CO)^*$ on Cu^+Y, (**a2**) TS2 on Cu^+Y, (**a3**) $(CH_3OCO)^*$ on Cu^+Y, (**b1**) $(CH_3O)^*/(CO)^*$ on $Cu^{2+}Y$, (**b2**) TS2 on $Cu^{2+}Y$, (**b3**) $(CH_3OCO)^*$ on $Cu^{2+}Y$, (**c1**) $(CH_3O)^*/(CO)^*$ on Cu_2O-Y, (**c2**) TS2 on Cu_2O-Y, (**c3**) $(CH_3OCO)^*$ on Cu_2O-Y, (**d1**) $(CH_3O)^*/(CO)^*$ on CuO-Y, (**d2**) TS2 on CuO-Y and (**d3**) $(CH_3OCO)^*$ on CuO-Y. See Figure 4 for the color coding.

3.3. CH_3O Reacts with CH_3OCO to Form DMC (Path I)

$(CH_3OCO)^*$ adsorbed on these four types of zeolites can react with another $(CH_3O)^*$ to form DMC, via a transition state TS3 (see Figure 7).

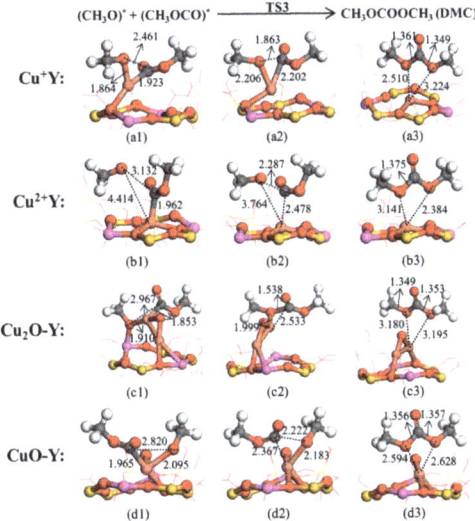

Figure 7. The structures of reactants, products and transition states on Cu^+Y, $Cu^{2+}Y$, Cu_2O-Y and CuO-Y zeolite for the formation of DMC via path I (unit: Å). (**a1**) $(CH_3O)^*/(CH_3OCO)^*$ on Cu^+Y, (**a2**) TS3 on Cu^+Y, (**a3**) DMC on Cu^+Y, (**b1**) $(CH_3O)^*/(CH_3OCO)^*$ on $Cu^{2+}Y$, (**b2**) TS3 on $Cu^{2+}Y$, (**b3**) DMC on $Cu^{2+}Y$, (**c1**) $(CH_3O)^*/(CH_3OCO)^*$ on Cu_2O-Y, (**c2**) TS3 on Cu_2O-Y, (**c3**) DMC on Cu_2O-Y, (**d1**) $(CH_3O)^*/(CH_3OCO)^*$ on CuO-Y, (**d2**) TS3 on CuO-Y and (**d3**) DMC on CuO-Y. See Figure 4 for the color coding.

For Cu$^+$Y zeolite, the distance between C′ atom of CH$_3$OC′O and the O atom of the second CH$_3$O (this O atom is denoted as O′ for the later analysis) decreases from initially 2.461 Å of (CH$_3$OC′O)*/(CH$_3$O′)* to 1.863 Å of TS3, and finally the C′-O′ bond in DMC of 1.361 Å. In addition, the bonds of Cu-C′ and Cu-O′ are elongated to 3.433 (not shown in Figure 7) and 2.510 Å, respectively, suggesting the weak (physical) adsorption of DMC on Cu$^+$Y zeolite. Similar changes from the initial geometries to TS3 states then to the final products happen on the three other types of zeolites. The reaction of CH$_3$O with CH$_3$OCO on Cu$^+$Y, Cu^{2+}Y, Cu$_2$O-Y and CuO-Y zeolites exhibits activation barriers of 28.27, 37.95, 40.90 and 15.95 kJ·mol^{-1} via TS3, respectively (see Table 3), and the exothermic energies are 164.16, 315.11, 313.18 and 312.58 kJ·mol^{-1}, respectively for these four types of zeolites.

3.4. Formation of (CH$_3$O)$_2$ Species (Path II)

The second pathway to form DMC suggests that (CH$_3$O)*/(OH)* reacts with CH$_3$OH, which results in co-adsorption of (CH$_3$O)$_2$* and H$_2$O. Figure 8 shows these adsorption configurations of (CH$_3$O)*(OH)*/CH$_3$OH and (CH$_3$O)$_2$*/H$_2$O on these four types of zeolites, via a transition state TS4.

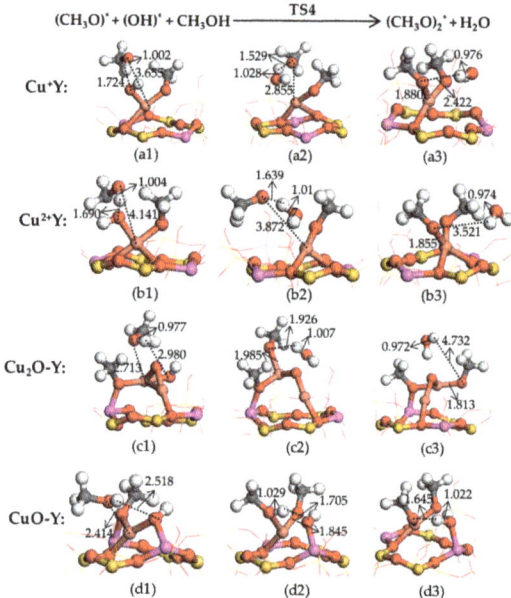

Figure 8. The structures of reactants, products and transition states on Cu$^+$Y, Cu^{2+}Y, Cu$_2$O-Y and CuO-Y zeolite for the formation of (CH$_3$O)$_2$ (unit: Å). (**a1**) (CH$_3$O)*/(OH)*/CH$_3$OH on Cu$^+$Y, (**a2**) TS4 on Cu$^+$Y, (**a3**) (CH$_3$O)$_2$*/H$_2$O on Cu$^+$Y, (**b1**) (CH$_3$O)*/(OH)*/CH$_3$OH on Cu^{2+}Y, (**b2**) TS4 on Cu^{2+}Y, (**b3**) (CH$_3$O)$_2$*/H$_2$O on Cu^{2+}Y, (**c1**) (CH$_3$O)*/(OH)*/CH$_3$OH on Cu$_2$O-Y, (**c2**) TS4 on Cu$_2$O-Y, (**c3**) (CH$_3$O)$_2$*/H$_2$O on Cu$_2$O-Y, (**d1**) (CH$_3$O)*/(OH)*/CH$_3$OH on CuO-Y, (**d2**) TS4 on CuO-Y and (**d3**) (CH$_3$O)$_2$*/H$_2$O on CuO-Y. See Figure 4 for the color coding.

For Cu$^+$Y zeolite, the O-H distance in CH$_3$OH increases from initially 1.002 Å to 1.529 Å of TS3, indicating this O-H bond tends to break. Meanwhile, the distance between the H atom of OH in CH$_3$OH and the O atom of (OH)* decreases from initially 1.724 Å of (CH$_3$O)*(OH)*/CH$_3$OH to 1.028 Å of TS4, demonstrating the migration of the H atom away from CH$_3$OH and towards the O atom of OH. This leads to the formation of additional (CH$_3$O)* and H$_2$O. Moreover, the distance between the O atom of CH$_3$OH and Cu$^+$ decreases from 2.855 Å of TS4 to 1.880 Å of (CH$_3$O)$_2$*/H$_2$O, which suggests the formation of (CH$_3$O)$_2$*. This reaction on Cu$^+$Y, Cu^{2+}Y, Cu$_2$O-Y and CuO-Y zeolites exhibits activation barriers of 93.86, 89.49, 116.38 and 115.29 kJ·mol^{-1} via TS4, respectively (see Table 3).

3.5. Insertion of CO into (CH$_3$O)$_2$ to Form DMC (Path II)

The processes of inserting CO into (CH$_3$O)$_2$ to form DMC on these four types of zeolites and the corresponding transition states TS5 are shown in Figure 9.

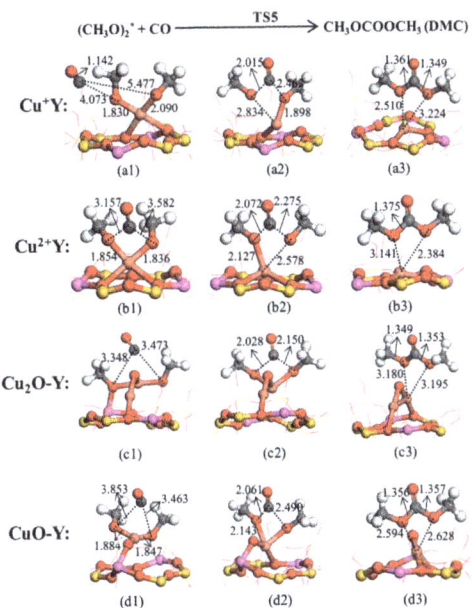

Figure 9. The structures of reactants, products and transition states on Cu$^+$Y, Cu^{2+}Y, Cu$_2$O-Y and CuO-Y zeolite for the formation of DMC via path II (unit: Å). (**a1**) (CH$_3$O)$_2$*/CO on Cu$^+$Y, (**a2**) TS5 on Cu$^+$Y, (**a3**) DMC on Cu$^+$Y, (**b1**) (CH$_3$O)$_2$*/CO on Cu^{2+}Y, (**b2**) TS5 on Cu^{2+}Y, (**b3**) DMC on Cu^{2+}Y, (**c1**) (CH$_3$O)$_2$*/CO on Cu$_2$O-Y, (**c2**) TS5 on Cu$_2$O-Y, (**c3**) DMC on Cu$_2$O-Y, (**d1**) (CH$_3$O)$_2$*/CO on CuO-Y, (**d2**) TS5 on CuO-Y and (**d3**) DMC on CuO-Y. See Figure 4 for the color coding.

As shown in Figure 9, on Cu$^+$Y zeolite the distance of Cu-CO of (CH$_3$O)$_2$*/CO configuration is 5.477 Å, and the C-O bond (1.142 Å) of CO is similar to that (1.143 Å) of CO in gas phase. This suggests that two CH$_3$O (i.e., (CH$_3$O)$_2$*) molecules adsorbed at the active center Cu effectively inhibit the adsorption of CO, which agrees with the stronger adsorption of CH$_3$O than CO (139.59 kJ·mol^{-1} vs. 125.25 kJ·mol^{-1}). Starting from the adsorption configuration of (CH$_3$O)$_2$*/CO, the formation of DMC goes through a transition state TS5 (see Figure 9). On Cu$^+$Y zeolite, the distance of the C atom of CO (denoted as C″) and the O atom of the nearest CH$_3$O (denotes as O″) decreases from initially 4.073 Å to 2.015 Å of TS5. The distance between C″ and the O atom of the furthest CH$_3$O (denotes as O‴) decreases from initially 5.477 Å of (CH$_3$O)$_2$*/CO to 2.462 Å of TS5. Furthermore, in TS5, the distance of Cu-O1 is elongated to 2.834 Å from 1.830 Å of (CH$_3$O)$_2$*/CO to accommodate the insertion of CO. Similar calculated results are found on the other three types of zeolites. The reaction step on Cu$^+$Y, Cu^{2+}Y, Cu$_2$O-Y and CuO-Y zeolites is significantly exothermic by 256.59, 372.06, 232.24 and 323.77 kJ·mol^{-1}, and the corresponding activation barriers are 201.68, 164.95, 253.96 and 210.74 kJ·mol^{-1} via TS5, respectively (see Table 3).

3.6. Desorption of DMC

Desorption of DMC from Cu$^+$Y, Cu^{2+}Y, Cu$_2$O-Y and CuO-Y zeolites is endothermic with the energy input of 24.30, 72.18, 41.31 and 65.69 kJ·mol^{-1}, respectively. These energies needed are compensable by the exothermic reactions of DMC formation on respective zeolites.

3.7. Rate-Limiting Reactions of DMC Formation

The potential energy curves for two reaction paths are plot in Figure 10.

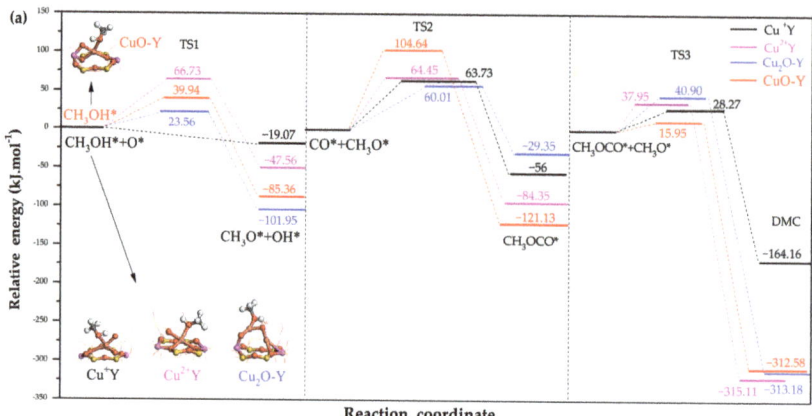

Figure 10. The reaction mechanism for the formation of DMC over Cu^+Y, $Cu^{2+}Y$, Cu_2O-Y and CuO-Y zeolites via path I (**a**) and path II (**b**).

For the path I of DMC formation on these zeolites, insertion of CO into CH_3O is followed by the formation of DMC. On Cu^+Y zeolite, the corresponding activation barriers of these two reactions are 63.73 and 28.27 kJ·mol^{-1}, respectively, which suggests the insertion reaction of CO into CH_3O is rate-limiting. On the other hand, for path II of DMC formation, insertion of CO into $(CH_3O)_2$ is followed by the formation of DMC. On Cu^+Y zeolite, the rate-limiting step for path II is insertion of CO into $(CH_3O)_2$ with an activation barrier of 201.68 kJ·mol^{-1}. The comparison between the rate-limiting reactions of two paths (63.73 vs. 201.68 kJ·mol^{-1}) suggests the path I is favorable for DMC formation on Cu^+Y zeolite. Similar to these processes on Cu^+Y zeolite, DFT calculations further confirm that path I is the favorable process of DMC formation over CuO-Y and Cu_2O-Y zeolites, with the rate-limiting step of inserting CO into CH_3O. For $Cu^{2+}Y$ zeolite, the favorable pathway of DMC formation is also path I, while oxidation of the absorbed CH_3OH to CH_3O becomes rate-limiting. Zhang et al. [36] found the favorable pathway of DMC formation on $Cu_2O(111)$ follows path I, which agrees with the finding in this study. However, they found that path II was favorable for the formation of DMC over

CuO(111) and the insertion of CO into (CH$_3$O)$_2$ was considered as rate-limiting step [47]. Comparison of the activation barriers of the rate-limiting steps on Cu$_2$O-Y (this study) and Cu$_2$O(111) (in the literature [36]) (60.01 kJ·mol^{-1} and 161.9 kJ·mol^{-1}, respectively) suggest the Cu$_2$O species in Y zeolite should exhibit a better catalytic activity than the carrier-free Cu$_2$O crystalline surface (see Table 3). These results indicate that the carrier significantly affects the activation barriers and even the reaction pathways.

Based on the aforementioned analyses, the following reaction route of DMC formation on different Cu states in Y zeolites was proposed. First, the adsorbed CH$_3$OH is oxidized to CH$_3$O species on zeolites. Then CO inserts to CH$_3$O to form CH$_3$OCO, which subsequently reacts with CH$_3$O to form DMC at a relative high reaction rate. Finally, adsorbed DMC is released into the gas phase. A distinction exists for these four types of zeolites investigated in this study. It is found that the rate-limiting step on Cu^{2+}Y zeolite is oxidation of CH$_3$OH to CH$_3$O, while for Cu$^+$Y, Cu$_2$O-Y and CuO-Y zeolites, the rate-limiting step is insertion of CO into CH$_3$O. Moreover, oxidation of CH$_3$OH to form CH$_3$O requires a presence of oxygen on Cu$^+$Y, Cu^{2+}Y and Cu$_2$O-Y zeolites, while on CuO-Y zeolite the adsorbed CH$_3$OH is oxidized by the lattice oxygen of CuO. The latter agrees with experimental findings by Engeldinger et al. [26,27].

The activation barriers of insertion of CO into CH$_3$O over Cu$^+$Y, Cu^{2+}Y, Cu$_2$O-Y and CuO-Y zeolites are found to be 63.73, 64.45, 60.01 and 104.64 kJ·mol^{-1}, respectively. Worthwhile to notice, oxidation of CH$_3$OH to CH$_3$O on Cu^{2+}Y zeolite exhibits an activation barrier of 66.73 kJ·mol^{-1}, while oxidation of CH$_3$OH on Cu$^+$Y zeolite is a barrier free reaction, suggesting that Cu$^+$Y zeolite possess a better catalytic activity than Cu^{2+}Y zeolite. As a result, the order of catalytic activities of these four types of zeolites is derived as Cu$_2$O-Y ≈ Cu$^+$Y > Cu^{2+}Y > CuO-Y, which agrees with a previous experimental study [28].

4. Methodology

Density functional theory calculations were performed using the DMol^3 program package of Materials Studio 8.0 [48]. The generalized-gradient approximation (GGA) with the Perdew–Burke–Ernzerhof (PBE) exchange-correction functional was used in all calculations [49]. The double numerical plus polarization (DNP) basis set [50], which is equivalently accurate to the commonly used 6-31G** Gaussian basis set, was employed to describe the Si-O-H-Al-Cu system. In this approach, for the non-metal Si, O, H and C atoms were treated with the all-electron basis sets, which considers all valence orbitals, while the inner electrons of the Al and Cu atoms were kept frozen and replaced by an effective core potential (ECP), which is attributed to that the metal atom participated into the reaction mainly occurs by the outer valence electron orbitals. The convergence criteria of DFT calculations were set to 2 × 10^{-5} Ha for energy, 4 × 10^{-3} Ha/Å for force, 0.005 Å for displacement. Complete linear synchronous transit (LST) and quadratic synchronous transit (QST) were used to determine the transition states (TS).

For the reaction A + B → AB on CuY zeolite, the reaction enthalpy (ΔH) and activation energy (E_a) were calculated by

$$\Delta H = E_{AB/CuY} - E_{A+B/CuY}$$

$$E_a = E_{TS/CuY} - E_{A+B/CuY}$$

where $E_{AB/CuY}$ is the total energy for the product AB on CuY zeolite, $E_{A+B/CuY}$ is the total energies of the co-adsorbed A and B on CuY zeolite and $E_{TS/CuY}$ is the total energy of the transition state (TS) on CuY zeolite, respectively. The negative ΔH represents an exothermic reaction.

The adsorption energy (E_{ads}) of the adsorbate-cluster system is defined as

$$E_{ads} = E_{adsorbate} + E_{CuY} - E_{adsorbate/CuY}$$

where $E_{\text{adsorbate/CuY}}$ is the total energy of adsorbate-CuY substrate system in the equilibrium state, E_{CuY} and $E_{\text{adsorbate}}$ are the total energies of CuY substrate and free adsorbate alone, respectively. From this definition, the large adsorption energy indicates a strong interaction between the absorbate and CuY zeolite.

5. Conclusions

In this work, the DFT method was employed to investigate the reaction mechanisms of DMC formation on four types of zeolites doped with Cu^+, Cu^{2+}, Cu_2O and CuO, respectively, based on two proposed reaction pathways. The calculation results reveal that path I is dominant for the formation of DMC since the activation barriers of rate-limiting steps for path II are much higher than that of path I. Moreover, the calculation results also suggest the following route to describe DMC formation on these zeolites. First, CH_3OH is adsorbed and oxidized to CH_3O species. Then CO inserts into CH_3O to form CH_3OCO, which reacts with CH_3O to product DMC. Lastly, adsorbed DMC is released into the gas phase. It is found that for Cu^+Y, $Cu^{2+}Y$ and Cu_2O-Y zeolites, adsorbed CH_3OH is oxidized to CH_3O with a presence of oxygen, whereas oxidation of CH_3OH on CuO-Y utilizes the lattice oxygen of CuO. The rate-limiting step on $Cu^{2+}Y$ zeolite is oxidation of CH_3OH to CH_3O, while on three other types of zeolites, the rate-limiting step is insertion of CO into CH_3O, and the corresponding activation barriers of these rate-limiting steps for $Cu^{2+}Y$, Cu^+Y, Cu_2O-Y and CuO-Y zeolites are 66.73, 63.73, 60.01 and 104.64 kJ·mol^{-1}, respectively. Based on above mentioned, the catalytic activities of these four types of zeolites with different Cu states exhibit the order of Cu_2O-Y $\approx$ Cu^+Y > $Cu^{2+}Y$ > CuO-Y. These findings are expected to guide the selection and preparation of CuY catalysts with the best catalytic activity for DMC synthesis.

Author Contributions: H.Z. and S.Y. outlined the work plan; Y.Z. and Y.S. conducted the computations; Y.Z. and G.Z. drew the figures and drafted the manuscript. H.Z. and J.Z. revised the drafted manuscript. All authors have read and agreed to the published version of the manuscript.

Funding: This work was supported by the National Natural Science Foundation of China (22262020), Zunyi Technology and Big data Bureau, Moutai institute Joint Science and Technology Research and Development Project ([2021]328) and Research Foundation for Scientific Scholars of Moutai Institute (mygccrc[2022]081 and [2022]080).

Data Availability Statement: All the relevant data used in this study have been provided in the form of figures and tables in the published article, and all data provided in the present manuscript are available to whom they may concern.

Acknowledgments: The authors are grateful to Zhong Li, John Z. Wen and Nilesh Narkhede for their kindly academic discussion and language help.

Conflicts of Interest: The authors declare no conflict of interest.

References

1. Kohli, K.; Sharma, B.K.; Panchal, C.B. Dimethyl Carbonate: Review of Synthesis Routes and Catalysts Used. *Energies* **2022**, *15*, 5133. [CrossRef]
2. Shi, D.; Heyte, S.; Capron, M.; Paul, S. Catalytic processes for the direct synthesis of dimethyl carbonate from CO_2 and methanol: A review. *Green Chem.* **2022**, *24*, 1067–1089. [CrossRef]
3. Raza, A.; Ikram, M.; Guo, S.; Baiker, A.; Li, G. Green Synthesis of Dimethyl Carbonate from CO_2 and Methanol: New Strategies and Industrial Perspective. *Adv. Sustain. Syst.* **2022**, *6*, 2200087. [CrossRef]
4. Huo, L.; Wang, T.; Xuan, K.; Li, L.; Pu, Y.; Li, C.; Qiao, C.; Yang, H.; Bai, Y. Synthesis of Dimethyl Carbonate from CO_2 and Methanol over Zr-Based Catalysts with Different Chemical Environments. *Catalysts* **2021**, *11*, 710. [CrossRef]
5. Ohno, H.; Ikhlayel, M.; Tamura, M.; Nakao, K.; Suzuki, K.; Morita, K.; Kato, Y.; Tomishige, K.; Fukushima, Y. Direct dimethyl carbonate synthesis from CO_2 and methanol catalyzed by CeO_2 and assisted by 2-cyanopyridine: A cradle-to-gate greenhouse gas emission study. *Green Chem.* **2021**, *23*, 457–469. [CrossRef]
6. Liu, K.; Liu, C. Synthesis of dimethyl carbonate from methanol and CO_2 under low pressure. *RSC Adv.* **2021**, *11*, 35711–35717. [CrossRef] [PubMed]
7. Huang, S.; Yan, B.; Wang, S.; Ma, X. Recent advances in dialkyl carbonates synthesis and applications. *Chem. Soc. Rev.* **2015**, *44*, 3079–3116. [CrossRef]

8. Wang, J.; Fu, T.; Meng, F.; Zhao, D.; Chuang, S.S.C.; Li, Z. Highly active catalysis of methanol oxidative carbonylation over nano Cu_2O supported on micropore-rich mesoporous carbon. *Appl. Catal. B Environ.* **2022**, *303*, 120890. [CrossRef]
9. Wang, C.; Liu, B.; Liu, P.; Huang, K.; Xu, N.; Guo, H.; Bai, P.; Ling, L.; Liu, X.; Mintova, S. Elucidation of the reaction mechanism of indirect oxidative carbonylation of methanol to dimethyl carbonate on Pd/NaY catalyst: Direct identification of reaction intermediates. *J. Catal.* **2022**, *412*, 30–41. [CrossRef]
10. Al-Rabiah, A.A.; Almutlaq, A.M.; Bashth, O.S.; Alyasser, T.M.; Alshehri, F.A.; Alofai, M.S.; Alshehri, A.S. An Intensified Green Process for the Coproduction of DMC and DMO by the Oxidative Carbonylation of Methanol. *Processes* **2022**, *10*, 2094. [CrossRef]
11. Almusaiteer, K.A.; Al-Mayman, S.I.; Mamedov, A.; Al-Zeghayer, Y.S. In Situ IR Studies on the Mechanism of Dimethyl Carbonate Synthesis from Methanol and Carbon Dioxide. *Catalysts* **2021**, *11*, 517. [CrossRef]
12. Zhang, Z.; Che, H.; Wang, Y.; Gao, J.; Ping, Y.; Zhong, Z.; Su, F. Template-free synthesis of $Cu@Cu_2O$ core–shell microspheres and their application as copper-based catalysts for dimethyldichlorosilane synthesis. *Chem. Eng. J.* **2012**, *211–212*, 421–431. [CrossRef]
13. Ai, Z.; Zhang, L.; Lee, S.; Ho, W. Interfacial Hydrothermal Synthesis of $Cu@Cu_2O$ Core−Shell Microspheres with Enhanced Visible-Light-Driven Photocatalytic Activity. *J. Phys. Chem. C* **2009**, *113*, 20896–20902. [CrossRef]
14. Teng, F.; Yao, W.; Zheng, Y.; Ma, Y.; Teng, Y.; Xu, T.; Liang, S.; Zhu, Y. Synthesis of flower-like CuO nanostructures as a sensitive sensor for catalysis. *Sensors Actuators B Chem.* **2008**, *134*, 761–768. [CrossRef]
15. Anderson, S.A.; Root, T.W. Investigation of the effect of carbon monoxide on the oxidative carbonylation of methanol to dimethyl carbonate over Cu+X and Cu+ZSM-5 zeolites. *J. Mol. Catal. A Chem.* **2004**, *220*, 247–255. [CrossRef]
16. Shen, Y.; Meng, Q.; Huang, S.; Wang, S.; Gong, J.; Ma, X. Reaction mechanism of dimethyl carbonate synthesis on Cu/β zeolites: DFT and AIM investigations. *RSC Adv.* **2012**, *2*, 7109–7119. [CrossRef]
17. Zhang, G.; Liang, J.; Yin, J.; Yan, L.; Narkhede, N.; Zheng, H.; Li, Z. An efficient strategy to improve the catalytic activity of CuY for oxidative carbonylation of methanol: Modification of NaY by H4EDTA-NaOH sequential treatment. *Micropor. Mesopor. Mater.* **2020**, *307*, 110500. [CrossRef]
18. Wang, Y.; Liu, Z.; Tan, C.; Sun, H.; Li, Z. High catalytic activity of CuY catalysts prepared by high temperature anhydrous interaction for the oxidative carbonylation of methanol. *RSC Adv.* **2020**, *10*, 3293–3300. [CrossRef]
19. Zhang, Y.H.; Briggs, D.N.; De Smit, E.; Bell, A.T. Effects of zeolite structure and composition on the synthesis of dimethyl carbonate by oxidative carbonylation of methanol on Cu-exchanged Y, ZSM-5, and Mordenite. *J. Catal.* **2007**, *251*, 443–452. [CrossRef]
20. King, S.T. Reaction Mechanism of Oxidative Carbonylation of Methanol to Dimethyl Carbonate in Cu–Y Zeolite. *J. Catal.* **1996**, *161*, 530–538. [CrossRef]
21. Richter, M.; Fait, M.J.G.; Eckelt, R.; Schneider, M.; Radnik, J.; Heidemann, D.; Fricke, R. Gas-phase carbonylation of methanol to dimethyl carbonate on chloride-free Cu-precipitated zeolite Y at normal pressure. *J. Catal.* **2007**, *245*, 11–24. [CrossRef]
22. Zhang, Y.H.; Bell, A.T. The mechanism of dimethyl carbonate synthesis on Cu-exchanged zeolite Y. *J. Catal.* **2008**, *255*, 153–161. [CrossRef]
23. Huang, S.Y.; Wang, Y.; Wang, Z.Z.; Yan, B.; Wang, S.P.; Gong, J.L.; Ma, X.B. Cu-doped zeolites for catalytic oxidative carbonylation: The role of Brønsted acids. *Appl. Catal. A Gen.* **2012**, *417–418*, 236–242. [CrossRef]
24. Kieger, S.; Delahay, G.; Coq, B.; Neveu, B. Selective Catalytic Reduction of Nitric Oxide by Ammonia over Cu-FAU Catalysts in Oxygen-Rich Atmosphere. *J. Catal.* **1999**, *183*, 267–280. [CrossRef]
25. Richter, M.; Fait, M.J.G.; Eckelt, R.; Schreier, E.; Schneider, M.; Pohl, M.-M.; Fricke, R. Oxidative gas phase carbonylation of methanol to dimethyl carbonate over chloride-free Cu-impregnated zeolite Y catalysts at elevated pressure. *Appl. Catal. B Environ.* **2007**, *73*, 269–281. [CrossRef]
26. Engeldinger, J.; Domke, C.; Richter, M.; Bentrup, U. Elucidating the role of Cu species in the oxidative carbonylation of methanol to dimethyl carbonate on CuY: An in situ spectroscopic and catalytic study. *Appl. Catal. A Gen.* **2010**, *382*, 303–311. [CrossRef]
27. Engeldinger, J.; Richter, M.; Bentrup, U. Mechanistic investigations on dimethyl carbonate formation by oxidative carbonylation of methanol over a CuY zeolite: An operando SSITKA/DRIFTS/MS study. *Phys. Chem. Chem. Phys.* **2012**, *14*, 2183–2191. [CrossRef]
28. Zheng, H.-Y.; Wang, J.-Z.; Li, Z.; Yan, L.-F.; Wen, J.Z. Characterization and assessment of an enhanced CuY catalyst for oxidative carbonylation of methanol prepared by consecutive liquid-phase ion exchange and incipient wetness impregnation. *Fuel Process. Technol.* **2016**, *152*, 367–374. [CrossRef]
29. Anderson, S.A.; Root, T.W. Kinetic studies of carbonylation of methanol to dimethyl carbonate over Cu^+X zeolite catalyst. *J. Catal.* **2003**, *217*, 396–405. [CrossRef]
30. Zheng, X.B.; Bell, A.T. A Theoretical Investigation of Dimethyl Carbonate Synthesis on Cu^-Y Zeolite. *J. Phys. Chem. C* **2008**, *112*, 5043–5047. [CrossRef]
31. Zheng, H.; Qi, J.; Zhang, R.; Li, Z.; Wang, B.; Ma, X. Effect of environment around the active center Cu^+ species on the catalytic activity of CuY zeolites in dimethyl carbonate synthesis: A theoretical study. *Fuel Process. Technol.* **2014**, *128*, 310–318. [CrossRef]
32. Shen, Y.; Meng, Q.; Huang, S.; Gong, J.; Ma, X. DFT investigations for the reaction mechanism of dimethyl carbonate synthesis on Pd (ii)/β zeolites. *Phys. Chem. Chem. Phys.* **2013**, *15*, 13116–13127. [CrossRef] [PubMed]
33. Ren, J.; Wang, W.; Wang, D.; Zuo, Z.; Lin, J.; Li, Z. A theoretical investigation on the mechanism of dimethyl carbonate formation on Cu/AC catalyst. *Appl. Catal. A Gen.* **2014**, *472*, 47–52. [CrossRef]
34. Meng, Q.; Wang, Z.; Shen, Y.; Yan, B.; Wang, S.; Ma, X. DFT and DRIFTS studies of the oxidative carbonylation of methanol over γ-$Cu_2Cl(OH)_3$: The influence of Cl. *RSC Adv.* **2012**, *2*, 8752–8761. [CrossRef]

35. Zheng, H.; Zhang, R.; Li, Z. Theoretical Studies on the Interaction of CO and CH_3O on CuCl (111) Surface for Methanol Oxidative Carbonylation. *Chem. J. China Univ.* **2014**, *35*, 1926–1932.
36. Zhang, R.G.; Song, L.Z.; Wang, B.J.; Li, Z. A density functional theory investigation on the mechanism and kinetics of dimethyl carbonate formation on Cu_2O catalyst. *J. Comput. Chem.* **2012**, *33*, 1101–1110. [CrossRef]
37. Doornkamp, C.; Ponec, V. The universal character of the Mars and Van Krevelen mechanism. *J. Mol. Catal. A Chem.* **2000**, *162*, 19–32. [CrossRef]
38. Sherry, H.S. The Ion-Exchange Properties of Zeolites. I. Univalent Ion Exchange in Synthetic Faujasite. *J. Phys. Chem.* **1966**, *70*, 1158–1168. [CrossRef]
39. Zheng, H.; Narkhede, N.; Zhang, G.; Li, Z. Role of metal co-cations in improving CuY zeolite performance for DMC synthesis: A theoretical study. *Appl. Organomet. Chem.* **2020**, *34*, e5832. [CrossRef]
40. Berthomieu, D.; Ducéré, J.-M.; Goursot, A. A Theoretical Study of Cu(II) Sites in a Faujasite-Type Zeolite: Structures and Electron Paramagnetic Resonance Hyperfine Coupling Constants. *J. Phys. Chem. B* **2002**, *106*, 7483–7488. [CrossRef]
41. Drake, I.J.; Zhang, Y.H.; Briggs, D.; Lim, B.; Chau, T.; Bell, A.T. The Local Environment of Cu^+ in Cu^-Y Zeolite and Its Relationship to the Synthesis of Dimethyl Carbonate. *J. Phys. Chem. B* **2006**, *110*, 11654–11664. [CrossRef] [PubMed]
42. Zhang, R.; Li, J.; Wang, B. The effect of Si/Al ratios on the catalytic activity of CuY zeolites for DMC synthesis by oxidative carbonylation of methanol: A theoretical study. *RSC Adv.* **2013**, *3*, 12287–12298. [CrossRef]
43. Yang, G.; Wang, Y.; Zhou, D.; Liu, X.; Han, X.; Bao, X. Density functional theory calculations on various M/ZSM-5 zeolites: Interaction with probe molecule H_2O and relative hydrothermal stability predicted by binding energies. *J. Mol. Catal. A Chem.* **2005**, *237*, 36–44. [CrossRef]
44. Rejmak, P.; Sierka, M.; Sauer, J. Theoretical studies of Cu(i) sites in faujasite and their interaction with carbon monoxide. *Phys. Chem. Chem. Phys.* **2007**, *9*, 5446–5456. [CrossRef]
45. Campana, L.; Selloni, A.; Weber, J.; Goursot, A. Cation Siting and Dynamical Properties of Zeolite Offretite from First-Principles Molecular Dynamics. *J. Phys. Chem. B* **1997**, *101*, 9932–9939. [CrossRef]
46. Hill, J.-R.; Freeman, C.M.; Delley, B. Bridging Hydroxyl Groups in Faujasite: Periodic vs. Cluster Density Functional Calculations. *J. Phys. Chem. A* **1999**, *103*, 3772–3777. [CrossRef]
47. Kang, L.; Zhang, J.; Zhang, R.; Ling, L.; Wang, B. Insight into the formation mechanism and kinetics for the oxidative carbonylation of methanol to dimethyl carbonate over CuO catalyst: Effects of Cu valence state and solvent environment. *Mol. Catal.* **2018**, *449*, 38–48. [CrossRef]
48. Delley, B. From molecules to solids with the DMol3 approach. *J. Chem. Phys.* **2000**, *113*, 7756–7764. [CrossRef]
49. Perdew, J.P.; Burke, K.; Ernzerhof, M. Generalized gradient approximation made simple. *Phys. Rev. Lett.* **1996**, *77*, 3865–3868. [CrossRef] [PubMed]
50. Delley, B. An all-electron numerical method for solving the local density functional for polyatomic molecules. *J. Chem. Phys.* **1990**, *92*, 508–517. [CrossRef]

Disclaimer/Publisher's Note: The statements, opinions and data contained in all publications are solely those of the individual author(s) and contributor(s) and not of MDPI and/or the editor(s). MDPI and/or the editor(s) disclaim responsibility for any injury to people or property resulting from any ideas, methods, instructions or products referred to in the content.

Article

Study on the Formaldehyde Oxidation Reaction of Acid-Treated Manganese Dioxide Nanorod Catalysts

Yanqiu Li [1,2], Yuan Su [3], Yunfeng Yang [2,*], Ping Liu [1,*], Kan Zhang [1] and Keming Ji [1,*]

1 State Key Laboratory of Coal Conversion, Institute of Coal Chemistry, Chinese Academy of Sciences, Taiyuan 030001, China
2 State College of Chemistry and Chemical Engineering, North University of China, Taiyuan 030051, China
3 State China Sedinningbo Engineering Co., Ltd., Ningbo 315103, China
* Correspondence: yangyunfeng@nuc.edu.cn (Y.Y.); pingliu@sxicc.ac.cn (P.L.); jikeming@sxicc.ac.cn (K.J.)

Abstract: Formaldehyde is an important downstream chemical of syngas. Furniture and household products synthesized from formaldehyde will slowly decompose and release formaldehyde again during use, which seriously affects indoor air quality. In order to solve the indoor formaldehyde pollution problem, this paper took the catalytic oxidation of formaldehyde as the research object; prepared a series of low-cost, acid-treated manganese dioxide nanorod catalysts; and investigated the effect of the acid-treatment conditions on the catalysts' activity. It was found that the MnNR-0.3ac-6h catalyst with 0.3 mol/L sulfuric acid for 6 h had the best activity. The conversion rate of formaldehyde reached 98% at 150 °C and 90% at 25 °C at room temperature. During the reaction time of 144 h, the conversion rate of formaldehyde was about 90%, and the catalyst maintained a high activity. It was found that acid treatment could increase the number of oxygen vacancies on the surface of the catalysts and promote the production of reactive oxygen species. The amount of surface reactive oxygen species of the MnNR-0.3ac-6h catalyst was about 13% higher than that of the catalyst without acid treatment.

Keywords: syngas; formaldehyde; acid treatment; MnO_2; catalyst

Citation: Li, Y.; Su, Y.; Yang, Y.; Liu, P.; Zhang, K.; Ji, K. Study on the Formaldehyde Oxidation Reaction of Acid-Treated Manganese Dioxide Nanorod Catalysts. *Catalysts* **2022**, *12*, 1667. https://doi.org/10.3390/catal12121667

Academic Editor: Angelo Vaccari

Received: 25 October 2022
Accepted: 8 December 2022
Published: 18 December 2022

Publisher's Note: MDPI stays neutral with regard to jurisdictional claims in published maps and institutional affiliations.

Copyright: © 2022 by the authors. Licensee MDPI, Basel, Switzerland. This article is an open access article distributed under the terms and conditions of the Creative Commons Attribution (CC BY) license (https://creativecommons.org/licenses/by/4.0/).

1. Introduction

Formaldehyde is an important downstream product of syngas. Gaseous formaldehyde can be obtained via the partial oxidation or dehydrogenation of methanol from syngas through the methanol route. At present, 35% of the world's methanol production is used to meet the world's demand for formaldehyde, making formaldehyde the first product that is directly derived from methanol. In recent years, there have also been studies on the direct synthesis of formaldehyde using syngas, which bypasses the production of methanol and has a low reaction temperature, thus improving the conversion efficiency without producing CO_2 [1,2].

As an important chemical raw material, formaldehyde can be used to produce a variety of daily products such as processed wood, paint, cosmetics, resins, polymers, adhesives, etc. [3]. However, in artificial boards and other decoration materials that use formaldehyde as the raw material, paraformaldehyde depolymerizes and continuously releases [4], which seriously pollutes the indoor environment. Methods to eliminate indoor formaldehyde pollution include ventilation [5], plant purification [6], adsorption [7,8], plasma purification [9], catalytic oxidation [10], etc. Among these, the ventilation and air exchange method is simple and feasible but has limitations in application scenarios due to the influence of the house layout and climatic conditions. The removal process of the plant purification method is relatively passive and the processing speed is slow, so it is only suitable for the auxiliary means of formaldehyde purification [11]. The adsorption method is simple in principle and widely used, but it is difficult for it to play a long-term and stable purification role due to the inherent problem of adsorption saturation.

Plasma air-purification technology has a good formaldehyde purification effect and strong processing capacity, but the energy consumption of the process is high and there may be secondary pollution. The catalytic oxidation method, which uses oxygen in the air to react and generate H_2O and CO_2 with a good reaction stability [12], high treatment efficiency, mild reaction conditions, and no secondary pollution, has broad application prospects [13]. Formaldehyde oxidation catalysts can be divided into noble metal catalysts and non-noble-metal catalysts [14–16].

Among these, noble metal catalysts have specific excellent catalytic activity at low temperatures or even room temperature [17], but noble metals are expensive, which is not conducive to practical application [18]. Compared with noble metal catalysts, non-noble metals such as Mn, Co, and Ce are relatively abundant in reserves and have various electronic structures, reducibility, thermal stability, oxidative degradability, a low price, and other excellent characteristics that have attracted great attention in the fields of energy and environmental protection [19]. Formaldehyde oxidation of non-noble metal oxide catalysts is a hot research topic at present. As early as 2002, Sekinel [20] compared the catalytic oxidation performance of formaldehyde at a low temperature for a variety of non-noble metal oxides. The results showed that when ZnO, La_2O_3, and V_2O_5 were used as catalysts, the reaction activity was low and the conversion rate of the formaldehyde was less than 10%; while MnO_2 had a better catalytic performance and showed a conversion rate of the formaldehyde of more than 90%.

Mn is a commonly used transition metal element in catalytic reactions [21,22]. Mn atom has the outer electronic structure of $3d^54s^2$, and its oxide can form the [MnO_6] octahedron, which is used as the basic structural unit. The crystal structure of a one-dimensional tunnel, two-dimensional layer, or three-dimensional network is formed by common edges, common angular tops, or coplanes [23–25]. Under different conditions, crystalline manganese oxides of α, δ, γ, λ, β, and ε can be generated [21,25].

Zhang et al. [26] investigated the activity of four crystalline MnO_2 catalysts (α, δ, λ, and β) and found that the δ-MnO_2 catalyst could completely oxidize formaldehyde at 80 °C; while the α-, λ-, and β-MnO_2 catalysts had higher formaldehyde complete conversion temperatures of 125 °C, 200 °C, and 150 °C, respectively. This was due to the fact that the delta-MnO_2 layered structure had more active sites, which increased the adsorption and diffusion rate of formaldehyde. Other studies have shown that the (310) crystal surface of α-MnO_2 had the highest surface energy, which was conducive to the adsorption and activation of O_2 and H_2O, and could improve the oxidation activity of HCHO [27]; the (100) crystal surface of α-MnO_2 was conducive to the adsorption and activation of O_2; and the (001) crystal surface of layered δ-MnO_2 was conducive to the desorption of H_2O [28]. The pore structure of potassium manganese ore (0.26 nm) was similar to the dynamic diameter of the formaldehyde molecule (0.243 nm) and showed high catalytic activity in the reaction [29].

When manganese oxide has a special morphology, its unique microstructure and texture properties were also conducive to an improvement in catalytic performance [30–32]. Boyjoo et al. [33] prepared a hollow microglobular MnO_2 catalyst; subsequent studies showed that the catalyst had a large specific surface area, a highly dispersed mesoporous structure, and a large number of oxygen vacancies on the surface, which could promote the migration of oxygen species in the reaction process. Bai et al. [34] prepared a 3D mesoporous MnO_2 catalyst using the template method. The results showed that 3D-MnO_2 had a large specific surface area, that the Mn^{4+} on its (110) crystal plane provided the main active site for formaldehyde oxidation, and that it had good catalytic performance.

Acid treatment is an important way to improve the surface chemical properties of catalysts [35–37]. Wang et al. [37] found that after acid treatment of an MnO_2/$LaMnO_3$ catalyst, the number of oxygen vacancies and reactive oxygen species on the surface increased significantly, which promoted toluene adsorption and a catalytic reaction. Studies have found that acid treatment can affect the distribution and number of oxygen-containing functional groups on the surface of carbon nanotubes [35]. Cui et al. [36] found that sulfuric acid treatment made a TiO_2 nanoribbon surface rough and that the catalyst had an obvious mesoporous structure, which could promote the adsorption, diffusion, and transfer of reactants and provide more surface active sites. Gao et al. [38] found that the addition of dilute nitric acid in the forming process of an HZSM-5 molecular sieve increased the content of medium and strong acids and formed layered pores, which helped to reduce the reaction capacity of cracking, aromatization, and hydrogen transfer of the catalyst; improved the diffusion performance of the catalyst; and slowed the deposition rate of coke. In this study, MnO_2 nanorod catalysts were prepared using hydrothermal synthesis and treated with acid. By adjusting the acid concentration and reaction time during the acid treatment, a series of MnO_2 nanorod catalysts were obtained after acid treatment under different conditions. The structure and properties of the catalysts after acid treatment were explored through various representations, and the influence of the acid treatment on the catalytic oxidation activity of the formaldehyde was explained.

2. Results and Discussion

2.1. Microstructure of Catalysts

In order to study the crystal structure and phase of the catalysts treated with different concentrations of acid and different times, an XRD characterization was carried out, the results of which are shown in Figure 1. It can be seen that there were characteristic diffraction peaks of α-MnO_2 at 2θ = 12.8°, 18.1°, 28.8°, 37.5°, and 49.8°, which corresponded to the (110), (200), (310), (121), and (411) crystal planes, respectively. This indicated that the prepared MnO_2 nanorods were α-crystalline MnO_2. The (310) and (110) crystal surface energies of the α-MnO_2 were higher, which was conducive to the adsorption and activation of O_2 and H_2O and could improve the oxidation activity of HCHO [27,28]. It can be seen in Figure 1a that after treatment of the MnNR catalysts with different concentrations of acid, the increase in the acid concentration in the process of acid treatment did not change the characteristic diffraction peak position of the catalysts, which meant that the structure of the catalysts did not change during the treatment with different concentrations of acid. As shown in Figure 1b, the extension of acid treatment time did not change the structure of the catalysts.

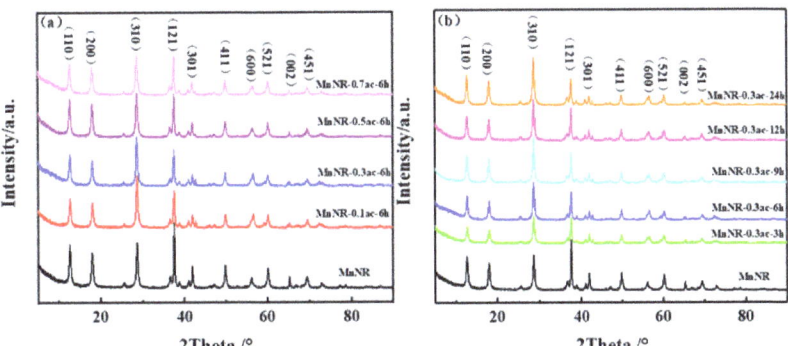

Figure 1. XRD patterns of catalysts. (**a**) XRD patterns of catalysts treated with different acid concentrations; (**b**) XRD patterns of catalysts with different acid treatment times.

The SEM characterization results for the manganese dioxide catalysts treated using different acid conditions are shown in Figure 2. Without acid treatment, as shown in Figure 2a, the MnNR catalyst process was obviously rod-like. After the treatment with different concentrations of acid, as shown in Figure 2b, the surface of the MnNR-0.1ac-6h catalyst prepared via the reaction of 0.1 mol/L sulfuric acid for 6 h was rough and weakly agglomerated, the particle size started to become large, and the rod-like structure could still be seen. As shown in Figure 2e, with the increase in acid concentration, the acid treatment changed the surface energy of the nanoparticles and increased the contact interface between the particles. The surface of the MnNR-0.7ac-6h catalyst prepared via the reaction of 0.7 mol/L sulfuric acid for 6 h was rougher with obvious agglomeration and no obvious rod-like structure. Figure 2f–i show the catalysts of MnO_2 nanorods treated with acid at different times. It can be seen that the MnNR-0.3ac-3h catalyst prepared via a 0.3 mol/L sulfuric acid reaction for 3 h showed weak agglomeration. With the gradual increase in the acid treatment time, the catalyst also gradually presented obvious nanoparticle agglomeration, the rod-like structure was no longer obvious with the extension of time, and the MnNR-0.3ac-24h catalyst did not show an obvious rod-like structure.

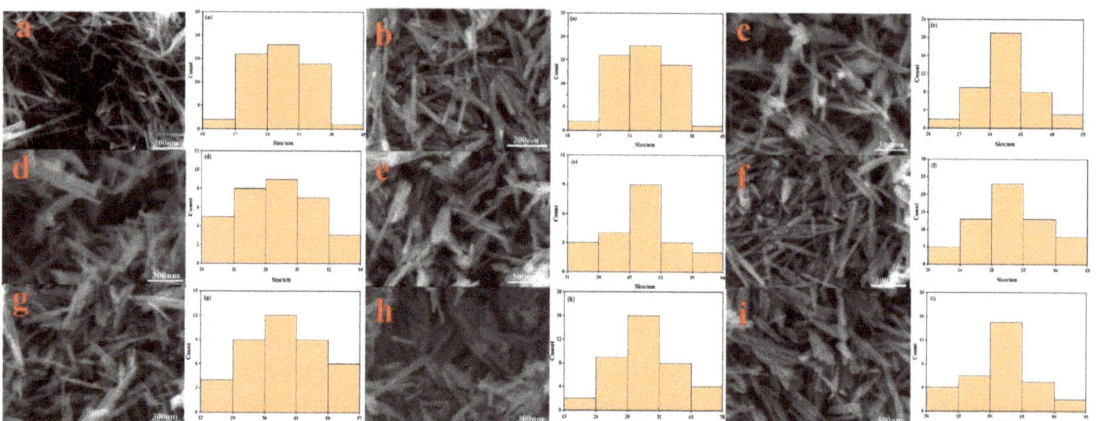

Figure 2. SEM images and particle size distributions of MnNR (**a**); MnNR-0.1ac-6h (**b**); MnNR-0.3ac-6h (**c**); MnNR-0.5ac-6h (**d**); MnNR-0.7ac-6h (**e**); MnNR-0.3ac-3h (**f**); MnNR-0.3ac-9h (**g**); MnNR-0.3ac-12h (**h**); MnNR-0.3ac-24h (**i**).

The specific surface area, pore size, and pore volume data of the MnNR series of catalysts are shown in Table 1. As can be seen in the table, the specific surface area of the MnNR catalyst without acid treatment was 47.6 m^2/g, the average pore size was 11.6 nm, and the average pore volume was 0.14 cm^3/g. When the MnNR catalyst was treated with different concentrations of acid, the specific surface area of the catalyst increased slightly, and the average pore size and average pore volume increased significantly, which may have been caused by the disappearance of a large number of micropores due to acid etching. The MnNR-0.3ac-6h catalyst prepared via the reaction of 0.3 mol/L sulfuric acid for 6 h had the largest specific surface area of 51.8 m^2/g, which meant that this catalyst could expose more active sites. The specific surface area of the MnNR catalyst gradually increased with time after acid treatment for different times. After 6 h of acid treatment, the specific surface area of the MnNR-0.3ac-6h catalyst was the largest at 51.8 m^2/g. With the extension of the acid treatment time, the specific surface area of the catalyst decreased. When combined with the SEM characterization results shown in Figure 2, it can be seen that as the concentration of acid treatment continued to increase and the time of acid treatment continued to increase, an agglomeration of the catalyst occurred, which may have be the reason for the slight decrease in the specific surface area.

Table 1. Physical properties of the as-prepared samples.

Sample	S_{BET} (m^2/g)	D_p (nm)	V_p (cm^3/g)
MnNR	47.6	11.6	0.14
MnNR-0.1ac-6h	49.7	18.2	0.20
MnNR-0.3ac-6h	51.8	20.2	0.21
MnNR-0.5ac-6h	49.6	20.3	0.24
MnNR-0.7ac-6h	48.0	20.6	0.25
MnNR-0.3ac-3h	48.2	17.1	0.20
MnNR 0.3ac-9h	50.9	20.9	0.20
MnNR-0.3ac-12h	48.4	21.3	0.22
MnNR-0.3ac-24h	45.5	22.2	0.22

2.2. Chemical Properties of the Catalysts

The reducibility of the catalysts was investigated using H$_2$-TPR characterization; the characterization results are shown in Figure 3. It can be seen in Figure 3a that the MnNR catalyst had a low-temperature reduction peak near 330 °C and a high-temperature reduction peak near 357 °C, which corresponded to the reduction of MnO$_2$ to Mn$_2$O$_3$ and of Mn$_2$O$_3$ to MnO, respectively [39,40]. After acid treatment, the high-temperature reduction peak of the MnNR-0.1ac-6h catalyst increased to 368 °C, while the low-temperature reduction peak did not change. With the increase in the acid concentration, the peak reduction temperature at a high temperature increased gradually, while the peak reduction temperature at a low temperature decreased. The high-temperature reduction peak of the MnNR-0.7ac-6h catalyst increased to 429 °C, and the low-temperature reduction peak was 316 °C. According to Figure 3b, after 3 h of acid treatment, the MnNR-0.3ac-3h catalyst had a low-temperature reduction peak near 330 °C and a high-temperature reduction peak near 357 °C. With the prolongation of the acid treatment process time, the peak reduction temperature for high-temperature reduction also increased, and the peak reduction temperature for low-temperature reduction decreased. The high-temperature reduction peak of the MnNR-0.3ac-24h catalyst increased to 381 °C, and the low-temperature reduction peak was 319 °C. The reduction peak temperature probably decreased because the Mn-O bond was more likely to break under the action of H$^+$ in the acid treatment process. A lower reduction temperature meant that the reduction was enhanced, the bond of the surface oxygen species was more easily broken, and the surface oxygen species was more easily activated, all of which were conducive to the enhancement of the activity. Lu et al. [41] believed that the reducibility of catalysts was related to oxygen–oxygen vacancies and that catalysts with a better reducibility could generate more oxygen vacancies [42]. Oxygen vacancies can promote oxygen activation, generate reactive oxygen species, and improve catalytic activity.

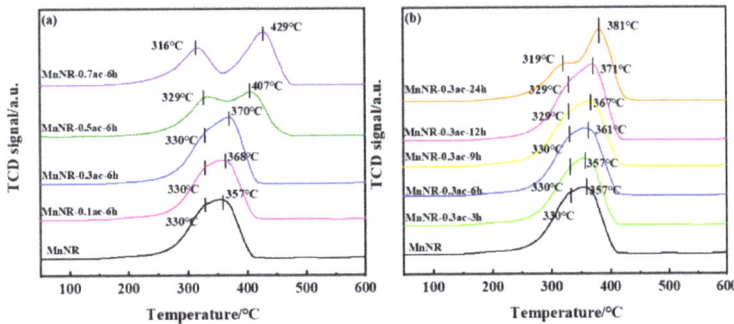

Figure 3. H$_2$-TPR profiles of various catalysts. (a) H$_2$-TPR profiles of catalysts treated with different acid concentrations; (b) H$_2$-TPR profiles of catalysts treated with different acid treatment times.

The O_2-TPD characterization results for the catalysts are shown in Figure 4. The O_2-TPD characterization could obtain the type of oxygen species on the surface of the catalyst. It was known from a literature survey [43,44] that the desorption order of oxygen species on the catalyst surface is: oxygen molecule (O_2) > oxygen molecule ion (O_2^-) > oxygen ion (O^-) > lattice oxygen (O^{2-}). Oxygen molecules can usually be adsorbed on the surface of a catalyst when the temperature is below 200 °C, while the desorption temperature of the reactive oxygen species O_2^- and O^- is usually between 200 and 400 °C. In addition, the lattice oxygen has the highest desorption temperature, which generally exceeds 400 °C [45,46].

It can be seen in Figure 4a that the MnNR catalyst had a weak desorption peak at 100 °C, which corresponded to the adsorption peak of oxygen molecules; the desorption peaks of reactive oxygen species O_2^- and O^- appeared at 300 °C; and the desorption peak of lattice oxygen appeared at 540 °C. After acid treatment, the desorption peak of the reactive oxygen species of the MnNR-0.7ac-6h catalyst also appeared at 300 °C. With the increase in the acid concentration, the desorption peak position of the reactive oxygen species shifted to the left; that is, the peak temperature decreased, and the lattice oxygen desorption peak of the catalyst gradually disappeared. The reactive oxygen desorption peak of the MnNR-0.7ac-6h catalyst appeared at 261 °C. It may have been that the strong acid reacted with part of the lattice oxygen to form oxygen vacancies, which led to the decrease in the lattice oxygen desorption peak. It can also be seen in Figure 4b that after acid treatment for different times, the desorption peak position of the reactive oxygen species of the catalysts also decreased with the increase in acid treatment time. The reactive oxygen species desorption peak of the MnNR-0.3ac-24h catalyst appeared at 270 °C. The desorption temperature of the oxygen was closely related to the activity of the catalysts. The lower the desorption temperature, the more likely the reactive oxygen species were to be generated. Reactive oxygen species can accelerate catalytic reactions [47], which is the most important factor that affects catalytic activity. In addition, Huang [43] found that oxygen vacancies could accelerate the generation of reactive oxygen species and that H_2-TPR characterization also proved that acid treatment could produce more oxygen vacancies in MnO_2 nanorod catalysts, so it could be predicted that acid treatment could improve the activity of the MnNR catalyst [41].

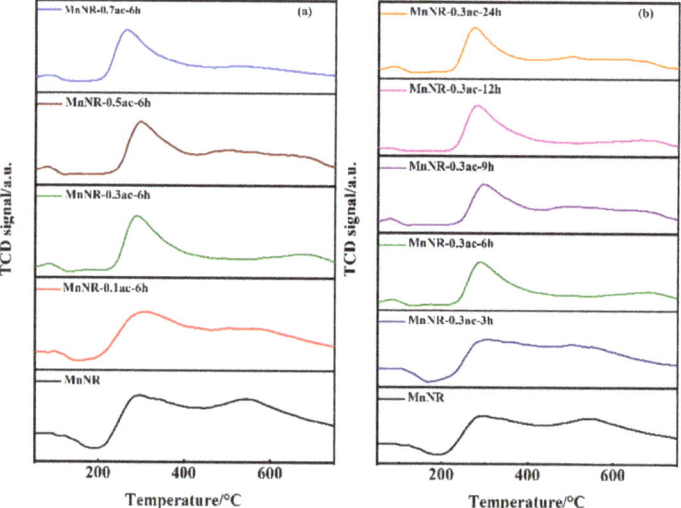

Figure 4. O_2-TPD profiles of various catalysts. (a) O_2-TPD profiles of catalysts treated with different acid concentrations; (b) O_2-TPD profiles of catalysts treated with different acid treatment times.

In order to explore the surface chemical composition and surface chemical valence of the MnNR catalysts under different treatment conditions, the high-resolution XPS spectra of the catalysts were obtained as shown in Figure 5. As can be seen in Figure 5a,b, all of the catalysts had obvious diffraction peaks for Mn and O elements, which meant that the catalysts contained manganese oxide compounds.

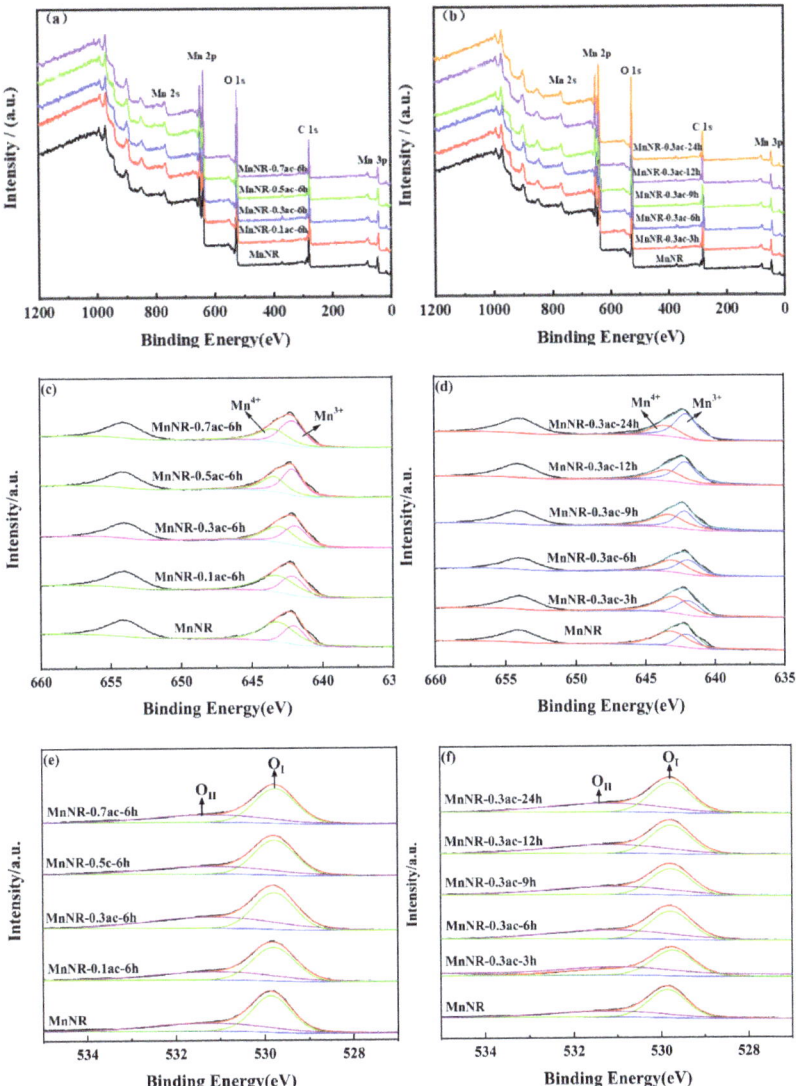

Figure 5. XPS spectra of catalysts for different acid treatment conditions and high-resolution XPS spectra of Mn 3s, Mn 2p, and O 1s. (**a**) high-resolution XPS spectra of catalysts treated with different acid concentrations of Mn 3s, Mn 2p, O 1s; (**b**) high-resolution XPS spectra of catalysts treated with different acid treatment times of Mn 3s, Mn 2p, O 1s; (**c**) XPS spectra of catalysts treated with different acid concentrations of Mn 2p; (**d**) XPS spectra of catalysts treated with different acid treatment times of Mn 2p; (**e**) XPS spectra of catalysts treated with different acid concentrations of O 1s; (**f**) XPS spectra of catalysts treated with different acid treatment times of O 1s.

Figure 5c,d reflect the Mn 2p spectra of the MnNR catalyst under different treatment conditions. It can be seen from the figure that the MnNR catalyst had two energy level peaks at 642 eV and 654 eV, which corresponded to Mn $2p_{3/2}$ and Mn $2p_{1/2}$, respectively. By using Xpspeak41 to divide the peaks of Mn $2p_{3/2}$, the Mn^{3+} and Mn^{4+} diffraction peaks appeared around 642 eV and 643 eV, respectively [48,49] and no Mn^{2+} diffraction peak appeared, which proved that there was no Mn^{2+} on the surface of the catalyst. The positions and proportions of the Mn^{3+} and Mn^{4+} peaks are shown in Table 2. As can be seen in the table, the MnNR catalyst's surface contained a small amount of Mn^{3+} (about 31%). The Mn^{3+} content on the surface of the MnNR-0.5ac-6h catalyst increased with the acid treatment; the Mn^{3+} content on the surface of the MnNR-0.5ac-6h catalyst was 56%. With the increase in the acid concentration, the Mn^{3+} and Mn^{4+} contents on the surface of the MnNR-0.7ac-6h catalyst increased continuously; the Mn^{3+} and Mn^{4+} contents were 57% and 43%, respectively. After acid treatment for different times, the Mn^{3+} content on the catalyst surface also increased while the Mn^{4+} content decreased. The contents of Mn^{3+} and Mn^{4+} on the surface of the MnNR-0.3ac-3h catalyst are 33% and 67%, respectively. The contents of Mn^{3+} and Mn^{4+} on the surface of the MnNR-0.3ac-24h catalyst were 61% and 39%, respectively. The increase in the Mn^{3+} content may have been due to the reaction of acid ions with lattice oxygen during the acid treatment, thereby leading to Mn electron transfer and the formation of oxygen vacancies. Oxygen vacancies were conducive to the adsorption and activation of oxygen molecules in the gas phase, which resulted in the generation of surface oxygen species and also promoted the stripping of reactive oxygen species at the adsorption site (oxygen vacancies) to enhance the activity of the catalysts.

Table 2. Chemical states of the surface elements determined using XPS.

Sample	Mn^{3+}	Mn^{4+}	O_I	O_{II}	Molar Ratio of Surface Elements		
					$O_{II}/O_I + O_{II}$ [a]	Mn^{3+} [a]	Mn^{4+} [a]
MnNR	641.8	643	529.9	531.1	35%	31%	69%
MnNR-0.1ac-6h	641.9	643.1	529.8	531.1	40%	35%	65%
MnNR-0.3ac-6h	641.9	643.1	529.8	531.2	48%	47%	53%
MnNR-0.5ac-6h	642	643.3	529.8	531.2	42%	56%	44%
MnNR-0.7ac-6h	642.1	643.4	529.7	531.3	41%	57%	43%
MnNR-0.3ac-3h	641.9	643	529.8	531.2	48%	33%	67%
MnNR-0.3ac-9h	641.9	643.1	529.8	531.2	48%	50%	50%
MnNR-0.3ac-12h	642	643.1	529.7	531.2	47%	52%	48%
MnNR-0.3ac-24h	642	643.5	529.7	531.2	45%	61%	39%

[a] Surface element molar ratio calculated according to the peak areas of the XPS.

The spectra of catalyst O 1s are shown in Figure 5e,f. It can be seen that the O 1s profile of the catalyst could be fitted into two subpeaks, which meant that there were two different types of oxygen species. According to the literature, the O_I diffraction peak around 529.8 eV corresponds to the lattice oxygen of the catalyst itself; and the O_{II} diffraction peak around 531.5 eV corresponds to O_2^-, O^-, the surface hydroxyl group, and other surface-adsorbed oxygen species. The positions and proportions of the O_{II} and O_I peaks in Table 2 showed that the O_I and O_{II} energy levels of the MnNR catalyst were 529.9 and 531.1 eV, respectively. After the acid treatment, the O_I level of the MnNR-0.7ac-6h catalyst was slightly decreased and the O_{II} level was slightly increased. This should be regarded as a negligible systematic displacement. It can be seen in Table 2 that the proportion of reactive oxygen species on the surface of the MnNR catalyst was 35%. After treatment at different concentrations and times, we found that the proportion of reactive oxygen species on the surface of the MnNR-0.3ac-6hcatalyst was the highest at 48%. The literature has proved that the surface adsorption of oxygen plays an important role in the process of HCHO oxidation at low temperatures. By combining the Mars–van Krevelen (MvK) mechanism and the previous results, it can be seen that surface oxygen groups could promote formaldehyde oxidation at room temperature. The MnNR-0.3ac-6h catalyst exhibited a better catalytic performance than the other catalysts.

2.3. Reaction Performance of Catalysts

The evaluation results of the formaldehyde catalytic oxidation of the MnO_2 nanorod catalysts treated with different concentrations of acid at 25–150 °C are shown in Figure 6. It can be seen that the formaldehyde conversion rate of the MnO_2 nanorod catalyst without acid treatment was only 68% at 25 °C. With the increase in the reaction temperature, the conversion rate of the formaldehyde increased; when the reaction temperature was 150 °C, the conversion rate of the formaldehyde increased to 92%. The catalytic activity of the MnO_2 nanorods increased after acid treatment; the conversion rate of formaldehyde of the MnNR-0.1ac-6h catalyst obtained via 0.1 mol/L sulfuric acid treatment increased to 97% at 150 °C. The formaldehyde conversion rate of the MnNR-0.3ac-6h catalyst reached 98% at 150 °C and 90% at 25 °C (room temperature). With the increase in the acid concentration, the catalyst activity decreased. The formaldehyde conversion rate of the MnNR-0.5ac-6h catalyst was 86% at 25 °C. The conversion rate of the MnNR-0.7ac-6h catalyst decreased to 92% at 150 °C, and the conversion rate of formaldehyde was as low as 78% at room temperature (25 °C).

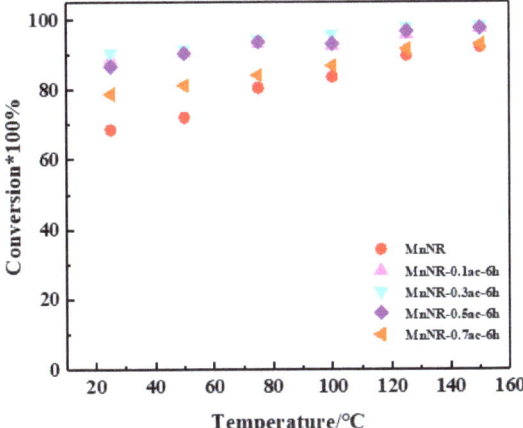

Figure 6. Experimental evaluation results of catalysts with different concentrations of acid treatment (GSHV = 3000 h^{-1}, 5 ± 1 ppm).

In addition, the effect of the acid treatment time on the catalyst activity was also evaluated via the formaldehyde reaction. The evaluation results of the formaldehyde catalytic oxidation of the catalysts at different temperatures are shown in Figure 7. It can be seen that when the concentration of sulfuric acid was 0.3 mol/L, the formaldehyde conversion rate of the MnNR-0.3ac-3h catalyst after three hours of reaction was 97% at 150 °C. With the extension of the reaction time, the formaldehyde conversion rate increased; the formaldehyde conversion rate of the MnNR-0.3ac-6h catalyst was 98% at 150 °C. The conversion rate of formaldehyde was still 90% at room temperature. With the increase in the acid treatment time, the formaldehyde conversion rate of the MnNR-0.3ac-9h catalyst decreased to 84% at 25 °C. When the acid reaction time continued to increase, the catalyst activity decreased; the formaldehyde conversion rate of the MnNR-0.3ac-24h catalyst decreased to 91% at 150 °C and was only 73% at 25 °C.

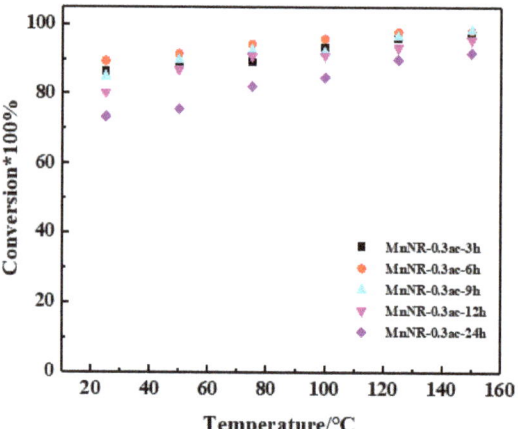

Figure 7. Experimental evaluation results of catalysts with different times of acid treatment (GSHV = 3000 h^{-1}, 5 ± 1 ppm).

Under the reaction condition of 50 °C and a formaldehyde concentration of 5 ppm, the effect of the space velocity on the conversion of the MnNR-0.3ac-6h catalyst was investigated; the results were shown in Figure 8. When the airspeed was 3000 h^{-1}, the conversion rate of the catalyst was 90%. With the increasing airspeed, the reaction conversion rate decreased gradually. When the airspeed was increased to 33,000 h^{-1}, the conversion rate of the reaction decreased to 70%, which was due to the reduced contact time between the catalyst and the formaldehyde.

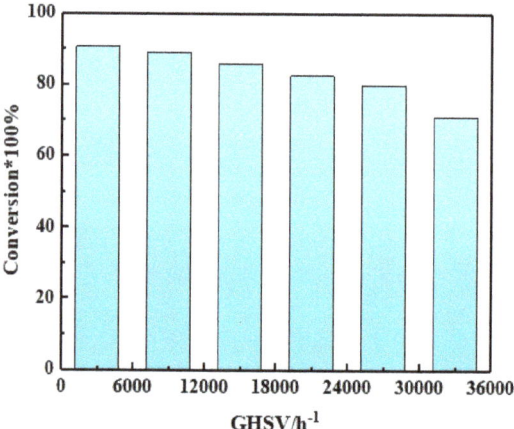

Figure 8. Experimental evaluation results of MnNR-0.3ac-6h catalysts with different GSHVs.

The catalytic stability of the MnNR-0.3ac-6h catalyst was evaluated at 50 °C and a formaldehyde concentration of 5 ppm; the results are shown in Figure 9. It can be seen in the figure that during the reaction time of 144 h, the conversion rate of formaldehyde remained around 90%, and the catalyst always maintained a high activity. When the reaction time exceeded 144 h, the catalyst activity decreased slightly until 168 h, and the MnNR-0.3ac-6h catalyst could still maintain a conversion rate of more than 87% under the condition of a formaldehyde concentration of 5 ppm. It can be seen that the MnNR-0.3ac-6h catalyst had a better stability in formaldehyde oxidation at a low temperature.

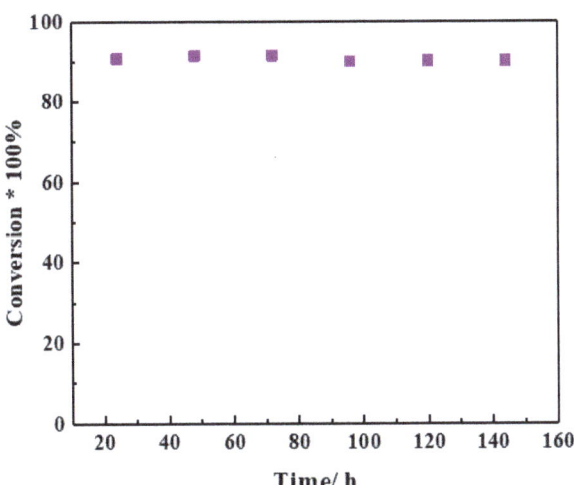

Figure 9. Stability test results of the MnNR-0.3ac-6h catalyst.

3. Experimental

3.1. Preparation of Catalysts

For the preparation of the α-MnO$_2$ nanorods, we added 3 mmol of KMnO$_4$ and 4.5 mmol of MnSO$_4$·H$_2$O to 30 mL of deionized water and stirred for one hour [41], then added 1.5 mL of sulfuric acid (98 wt.%), stirred the above solution at room temperature until completely dissolved, transferred it into PTFE lining, sealed it, and loaded it into a high-pressure reaction kettle. The reaction was conducted at 160 °C for 12 h. At the end of the reaction, the product was filtered and washed until neutral and was denoted as MnNR.

For the preparation of the catalysts for acid treatment with different concentrations, MnO$_2$ nanorods were placed in different concentrations of an H$_2$SO$_4$ solution and heated in an oil bath at 80 °C with reflux stirring. The ratio of the solid to liquid mass was kept at 1:60. Under the condition of maintaining the heating temperature as unchanged, the suspension was treated with sulfuric acid at 0.1, 0.3, 0.5, and 0.7 mol/L for 6 h; then the suspension was filtered and washed repeatedly with deionized water until the pH value of the eluent was about 7 and then dried at 105 °C. After drying, the powder was ground, pressed, crushed, and screened; the resulting products were respectively denoted as MnNR-0.1ac-6h, MnNR-0.3ac-6h, MnNR-0.5ac-6h, and MnNR-0.7ac-6h.

For the preparation of the catalysts for acid treatment at different times, MnO$_2$ nanorods were placed in 0.3 mol/L of the H$_2$SO$_4$ solution and heated in an oil bath at 80 °C with reflux stirring. The ratio of the solid to liquid mass was kept at 1:60. Under the condition of maintaining the heating temperature as unchanged, the reaction was carried out for 3 h, 9 h, 6 h, 12 h, and 24 h, respectively. The suspension was filtered and washed repeatedly with deionized water until the pH value of the eluent was about 7 and then dried at 105 °C. After drying, it was ground, pressed, crushed, and sieved; the resulting products were respectively denoted as MnNR-0.3ac-3h, MnNR-0.3ac-9h, MnNR-0.3ac-6h, MnNR-0.3ac-12h, and MnNR-0.3ac-24h.

3.2. Catalyst Characterization

The crystal structure of the sample was detected using an X-ray powder diffractometer (XRD, Bruker's D8 ADVANCE A2) using a Cu target (Kα-ray, λ = 1.5418 Å) at a large angle of 10~90°. The BET surface area of the sample was measured using the nitrogen adsorption and desorption isotherms at 77 K using a Tristar II (3020) mesoporous physical adsorption analyzer.

Before measurement, the samples were desorbed at 200 °C and 1.33 Pa for 6 h. The chemical valence of the catalyst surface was determined using X-ray photoelectron spectroscopy (XPS, AXIS ULTRA DLD model). The radiation source was Al Kα, and the modified binding energy was 284.8 eV with contaminated carbon C1s. Scanning electron microscopy (SEM) was carried out on a JSM-7900F (Japan Electronics, Japan, Tokyo) instrument. Firstly, a small amount of powder catalyst sample was coated on conductive adhesive and then the gold-sprayed treatment was performed. The temperature-programmed reduction (H_2-TPR) used H_2 as the reducing gas and was operated on an adsorption meter equipped with a TCD detector. The 100 mg samples were pretreated with N_2 at 300 °C for 60 min to remove the moisture on the catalyst surface and then cooled to room temperature (25 °C) under an N_2 atmosphere. After the baseline was stabilized in the H_2/N_2 atmosphere, the reduction treatment was carried out at 25–800 °C at a heating rate of 10 °C/min, and the data were detected and recorded by the TCD detector. For the oxygen-programmed temperature desorption (O_2-TPD), which used the same equipment as the H_2-TPR, 100 mg samples were pretreated with H_2 at 300 °C for 60 min then cooled to room temperature (25 °C) under a He atmosphere, adsorbed for 60 min under an O_2/He atmosphere, and then purged with He for 60 min. The desorption was performed under pure He (25–800 °C).

3.3. Catalyst Evaluation

The formaldehyde-catalyzed oxidation reaction was evaluated using a fixed-bed reactor under atmospheric pressure. A 20–40 mesh 2.0 g catalyst was weighed into the fixed-bed reactor, then the formaldehyde solution was sucked into the heating tube with a pump and heated and vaporized at 150 °C. The air carried away the formaldehyde vapor and part of the water vapor (a formaldehyde concentration of about 5 ± 1 ppm) through the heating tower into the fixed-bed reactor where the catalyst was located for the catalytic oxidation reaction. The raw material and formaldehyde after reaction were tested using a gas pump combined with a formaldehyde reaction tube.

$$\text{Conversion rate of formaldehyde (\%)} = (\text{Raw material formaldehyde concentration} - \text{product formaldehyde concentration})/\text{Raw material formaldehyde concentration} \times 100\% \quad (1)$$

4. Conclusions

The experimental results confirmed the reason why acid treatment can improve the catalytic activity and why the acid treatment time and acid concentration in the process of acid treatment will affect the catalytic activity. The MnNR-0.3ac-6h catalyst was the most active after being treated with 0.3 mol/L sulfuric acid for 6 h. The conversion rate of formaldehyde reached 98% at 150 °C and 90% at 25 °C. During the reaction time of 144 h, the conversion rate of formaldehyde was about 90%, and the catalyst maintained a high activity. It was found that acid treatment did not change the crystal shape of the catalysts. After acid treatment, the surface of the nanorods was obviously rough with obvious mesoporous characteristics, and the specific surface area was increased. The specific surface area of the MnNR-0.3ac-6h catalyst after acid treatment was increased by about 9% compared to that without acid treatment, and the increase in the specific surface area was conducive to the adsorption and diffusion of formaldehyde and oxygen. As the acid concentration became too high and the acid treatment time was prolonged, sulfuric acid reduced the surface energy of the nanoparticles, increased the contact interface between the nanoparticles, and caused the nanorods to easily agglomerate and the specific surface area to decrease. After acid treatment, MnO_2 nanorod catalysts had a good low-temperature reduction ability. The MnNR-0.3ac-6h catalyst had the best reducibility, which could promote the formation of oxygen vacancies and bond-breaking of oxygen species on the surface, and the oxygen molecules were more easily activated. After acid treatment, the oxygen desorption temperature of the catalysts decreased, and reactive oxygen species were more likely to be generated. The amount of reactive oxygen species on the surface of the MnNR-0.3ac-6h catalyst increased by about 13%. Acid treatment increased the content of Mn^{3+} on the surface of the catalysts, and the redox reaction between Mn^{4+} and Mn^{3+} promoted the activation of oxygen and the generation of reactive oxygen species, thereby promoting the improvement in the catalytic activity.

Author Contributions: Conceptualization, K.J.; methodology, Y.L.; software, Y.S.; validation, Y.S., P.L., K.Z., Y.Y. and K.J.; investigation, Y.S.; resources, K.Z.; data curation, Y.S.; writing—original draft preparation, Y.L.; writing—review and editing, Y.L.; visualization, Y.L. and Y.S.; supervision, P.L. and K.Z.; project administration, P.L. and K.Z., funding acquisition, P.L., K.Z. and K.J. All authors have read and agreed to the published version of the manuscript.

Funding: This research was funded by the Youth Foundation of Shanxi Province, grant number 202103021223460, the CAS Key Technology Talent Program, grant number E1YC916201 and the CAS Strategic Priority Research Program, grant number Y8YCZ11621.

Data Availability Statement: There are no other supplementary supporting data, excluding this project.

Acknowledgments: The authors acknowledge the financial support from the Youth Foundation of Shanxi Province, the CAS Key Technology Talent Program, and the CAS Strategic Priority Research Program.

Conflicts of Interest: The authors declare no conflict of interest.

References

1. Bahmanpour, A.-M.; Hoadley, A.; Mushrif, S.-H. Hydrogenation of Carbon Monoxide into Formaldehyde in Liquid Media. *ACS Sustain. Chem. Eng.* **2016**, *4*, 3970–3977. [CrossRef]
2. Kaithal, A.; Hölscher, M.; Leitner, W. Carbon monoxide and hydrogen (syngas) as a C1-building block for selective catalytic methylation. *Chem. Sci.* **2020**, *12*, 976–982. [CrossRef] [PubMed]
3. Heim, L.-E.; Konnerth, H.; Prechtl, M.-H.-G. The Prospecting Shortcut to an Old Molecule: Formaldehyde Synthesis at Low Temperature in Solution. *ChemSusChem* **2016**, *9*, 2905–2907. [CrossRef] [PubMed]
4. Jiang, C.-J.; Li, D.-D.; Zhang, P.-Y.; Li, J.-G.; Wang, J.; Yu, J.G. Formaldehyde and volatile organic compound (VOC) emissions from particleboard: Identification of odorous compounds and effects of heat treatment. *Build. Environ.* **2017**, *117*, 118–126. [CrossRef]
5. Gilbert, N.-L.; Guay, M.; Gauvin, D.; Dietz, R.-N.; Chan, C.-C.; Lévesque, B. Air change rate and concentration of formaldehyde in residential indoor air. *Atmos. Environ.* **2007**, *42*, 2424–2428. [CrossRef]
6. Giese, M.; Bauer-Doranth, U.; Langebartels, C.; Sandermann, H. Detoxification of Formaldehyde by the Spider Plant (*Chlorophytum comosum* L.) and by Soybean (*Glycine max* L.) Cell-Suspension Cultures. *Plant Physiol.* **1994**, *104*, 1301–1309. [CrossRef]
7. Suresh, S.; Bandosz, T.-J. Removal of formaldehyde on carbon-based materials: A review of the recent approaches and findings. *Carbon* **2018**, *137*, 207–221. [CrossRef]
8. Ye, J.-W.; Zhu, X.-F.; Cheng, B.; Yu, J.-G.; Jiang, C.-J. Few-Layered Graphene-like Boron Nitride: A Highly Efficient Adsorbent for Indoor Formaldehyde Removal. *Environ. Sci. Technol. Lett.* **2017**, *4*, 20–25. [CrossRef]
9. Liang, W.-J.; Li, J.; Li, J.-X.; Zhu, T.; Jin, Y.-Q. Formaldehyde removal from gas streams by means of $NaNO_2$ dielectric barrier discharge plasma. *J. Hazard. Mater.* **2009**, *175*, 1090–1095. [CrossRef]
10. Liao, Y.-C.; Xie, C.-S.; Liu, Y.; Chen, H.; Li, H.-Y.; Wu, J. Comparison on photocatalytic degradation of gaseous formaldehyde by TiO_2, ZnO and their composite. *Ceram. Int.* **2012**, *38*, 4437–4444. [CrossRef]
11. Kehr, E.; Riehl, G.; Hoferichter, E.; Roffael, E.; Dix, B. Feuchtebeständigkeit und Hydrolyseresistenz von Holz-zu-Holz-Bindungen in Spanplatten, hergestellt mit formaldehydarmen modifizierten Harnstoff-Formaldehydharzen unter Einsatz verschiedener Härtungsbeschleunigersysteme. *Holz als Roh-und Werkstoff* **1993**, *52*, 253–260. [CrossRef]
12. Duan, Y.-Y.; Song, S.-Q.; Cheng, B.; Yu, J.-G.; Jiang, C.-J. Effects of hierarchical structure on the performance of tin oxide-supported platinum catalyst for room-temperature formaldehyde oxidation. *Chin. J. Catal.* **2017**, *38*, 199–206. [CrossRef]
13. Ma, C.-Y.; Sun, S.-M.; Lu, H.; Hao, Z.-P.; Yang, C.-G.; Wang, B.; Chen, C.; Song, M.-Y. Remarkable MnO_2 structure-dependent H_2O promoting effect in HCHO oxidation at room temperature. *J. Hazard. Mater.* **2021**, *414*, 125542. [CrossRef] [PubMed]
14. Wang, S.; Wang, Y.; Wang, F.-G. Room temperature HCHO oxidation over the Pt/CeO_2 catalysts with different oxygen mobilities by changing ceria shapes. *Appl. Catal. A Gen.* **2022**, *630*, 118469. [CrossRef]
15. Wang, S.; Han, K.-H.; Deng, Z.-Y.; Wang, F.-G. CeO_2 Nanorods Decorated with Pt Nanoparticles as Catalysts for Oxidative Elimination of Formaldehyde. *ACS Appl. Nano Mater.* **2022**, *5*, 10036–10046. [CrossRef]
16. Wang, F.-G.; Wang, Y.; Han, K.-H.; Yu, H. Efficient elimination of formaldehyde over Pt/Fe_3O_4 catalyst at room temperature. *J. Environ. Chem. Eng.* **2020**, *8*, 104041. [CrossRef]
17. Dong, Y.-X.; Su, C.-G.; Liu, K.; Wang, H.-M.; Zheng, Z.; Zhao, W.; Lu, S.-H. The Catalytic Oxidation of Formaldehyde by FeO_x-MnO_2-CeO_2 Catalyst: Effect of Iron Modification. *Catalysts* **2021**, *11*, 555. [CrossRef]

18. Fan, Z.-Y.; Fang, W.-J.; Zhang, Z.-X.; Chen, M.-X.; Shangguan, W.-F. Highly active rod-like Co_3O_4 catalyst for the formaldehyde oxidation reaction. *Catal. Commun.* **2018**, *103*, 10–14. [CrossRef]
19. Cui, W.-Y.; Wang, C.; Wu, J.; Tan, N.-D. Research progress of formaldehyde oxidation over manganese oxide catalyst. *Fine Chem. Ind.* **2019**, *36*, 2353–2363. [CrossRef]
20. Sekine, Y. Oxidative decomposition of formaldehyde by metal oxides at room temperature. *Atmos. Environ.* **2002**, *36*, 5543–5547. [CrossRef]
21. Li, Y.-B.; Han, S.-H.; Zhang, L.-P.; Yu, Y.-F. Manganese-Based Catalysts for Indoor Volatile Organic Compounds Degradation with Low Energy Consumption and High Efficiency. *Trans. Tianjin Univ.* **2021**, *28*, 53–66. [CrossRef]
22. Li, S.-J.; Chen, L.-L.; Ma, Z.; Li, G.-S.; Zhang, D.-Q. Research Progress on Photocatalytic/Photoelectrocatalytic Oxidation of Nitrogen Oxides. *Trans. Tianjin Univ.* **2021**, *27*, 295–312. [CrossRef]
23. Li, S.-D.; Zheng, D.-S.; Guo, F. Research progress on catalytic oxidation of manganese dioxide. *Mod. Chem. Ind.* **2020**, *40*, 52–56. [CrossRef]
24. Tian, H.; He, J.-H. Research progress of formaldehyde oxidation catalyzed by manganese oxide. *Chem. Bull.* **2013**, *76*, 100–106. [CrossRef]
25. Robinson, D.-M.; Go, Y.-B.; Mui, M.; Gardner, G.; Zhang, Z.-G.; Mastrogiovanni, D.; Garfunkel, E.; Li, J.; Greenblatt, M.; Dismukes, G.C. Photochemical water oxidation by crystalline polymorphs of manganese oxides: Structural requirements for catalysis. *J. Am. Chem. Soc.* **2013**, *135*, 3494–3501. [CrossRef]
26. Zhang, J.-H.; Li, Y.-B.; Wang, L.; Zhang, C.-B.; He, H. Catalytic oxidation of formaldehyde over manganese oxides with different crystal structures. *Catal. Sci. Technol.* **2015**, *5*, 2305–2313. [CrossRef]
27. Rong, S.-P.; Zhang, P.-Y.; Liu, F.; Yang, Y.-J. Engineering Crystal Facet of α-MnO_2 Nanowire for Highly Efficient Catalytic Oxidation of Carcinogenic Airborne Formaldehyde. *ACS Catal.* **2018**, *8*, 3435–3446. [CrossRef]
28. Zhou, J.; Qin, L.-F.; Xiao, W.; Zeng, C.; Li, N.; Lv, T.; Zhu, H. Oriented growth of layered-MnO_2 nanosheets over α-MnO_2 nanotubes for enhanced room-temperature HCHO oxidation. *Appl. Catal. B Environ.* **2017**, *207*, 233–243. [CrossRef]
29. Chen, T.; Dou, H.-Y.; Li, X.-L.; Tang, X.-F.; Li, J.-H.; Hao, J.-M. Tunnel structure effect of manganese oxides in complete oxidation of formaldehyde. *Microporous Mesoporous Mater.* **2009**, *122*, 270–274. [CrossRef]
30. Liu, Z.; Zhang, X.-L.; Cai, J.; Zhan, Y.-K.; He, N.-D. Manganese based catalysts for catalytic oxidation of formaldehyde and their synergistic effects. *Adv. Chem.* **2019**, *31*, 311–321.
31. Bai, B.-Y.; Qiao, Q.; Li, J.-H.; Hao, J.-M. Progress in research on catalysts for catalytic oxidation of formaldehyde. *Chin. J. Catal.* **2016**, *37*, 102–122. [CrossRef]
32. Wei, Y.-Z.; Chen, M.-Y.; Ren, X.-Y.; Wang, Q.-F.; Han, J.-F.; Wu, W.-Z.; Yang, X.-G.; Wang, S.; Yu, J.-H. One-Pot Three-Dimensional Printing Robust Self-Supporting MnO_x/Cu-SSZ-13 Zeolite Monolithic Catalysts for NH_3-SCR. *CCS Chem.* **2022**, *4*, 1708–1719. [CrossRef]
33. Boyjoo, Y.; Rochard, G.; Giraudon, J.-M.; Liu, J.; Lamonier, J.-F. Mesoporous MnO_2 hollow spheres for enhanced catalytic oxidation of formaldehyde. *Sustain. Mater. Technol.* **2018**, *20*, e00091. [CrossRef]
34. Bai, B.-Y.; Qiao, Q.; Li, J.-H.; Hao, J.-M. Synthesis of three-dimensional ordered mesoporous MnO_2 and its catalytic performance in formaldehyde oxidation. *Chin. J. Catal.* **2016**, *37*, 27–31. [CrossRef]
35. Wepasnick, K.-A.; Smith, B.-A.; Schrote, K.-E. Surface and structural characterization of multi-walled carbon nanotubes following different oxidative treatments. *Carbon* **2011**, *49*, 24–36. [CrossRef]
36. Cui, W.-Y.; Xue, D.; Yuan, X.-L.; Zheng, B.; Jia, M.-J.; Zhang, W.-X. Acid-treated TiO_2 nanobelt supported platinum nanoparticles for the catalytic oxidation of formaldehyde at ambient conditions. *Appl. Surf. Sci.* **2017**, *411*, 105–112. [CrossRef]
37. Wang, S.-H.; Liu, Q.; Zhao, Z.-Q.; Fan, C.; Chen, X.-P.; Xu, G.; Wu, M.-H.; Chen, J.-J.; Li, J.-H. Enhanced Low-Temperature Activity of Toluene Oxidation over the Rod-like MnO_2/$LaMnO_3$ Perovskites with Alkaline Hydrothermal and Acid-Etching Treatment. *Ind. Eng. Chem. Res.* **2020**, *59*, 6556–6564. [CrossRef]
38. Gao, J.-H.; Zhou, H.; Zhang, F.-C.; Ji, K.-M.; Liu, P.; Liu, Z.-H.; Zhang, K. Effect of Preparation Method on the Catalytic Performance of HZSM-5 Zeolite Catalysts in the MTH Reaction. *Materials* **2022**, *15*, 2206. [CrossRef]
39. Fang, R.-M.; Feng, Q.-Y.; Huang, H.-B.; Ji, J.; He, M.; Zhan, Y.-J.; Liu, B.-Y.; Leung, D.-Y.-C. Effect of K + ions on efficient room-temperature degradation of formaldehyde over MnO_2 catalysts. *Catal. Today* **2019**, *327*, 154–160. [CrossRef]
40. Sun, D.; Wageh, S.; Al-Ghamdi, A.-A.; Le, Y.; Yu, J.-G.; Jiang, C.-J. Pt/C@MnO_2 composite hierarchical hollow microspheres for catalytic formaldehyde decomposition at room temperature. *Appl. Surf. Sci.* **2018**, *466*, 301–308. [CrossRef]
41. Lu, S.-H.; Wang, X.; Zhu, Q.-Y.; Chen, C.-C.; Zhou, X.-F.; Huang, F.-L.; Li, K.-L.; He, L.-L.; Liu, Y.-X.; Pang, F.-J. Ag–K/MnO_2 nanorods as highly efficient catalysts for formaldehyde oxidation at low temperature. *RSC Adv.* **2018**, *8*, 14221–14228. [CrossRef] [PubMed]
42. Li, D.-D.; Yang, G.-L.; Li, P.-L.; Wang, J.-L.; Zhang, P.-Y. Promotion of formaldehyde oxidation over Ag catalyst by Fe doped MnO_x support at room temperature. *Catal. Today* **2016**, *277*, 257–265. [CrossRef]
43. Huang, H.; Dai, Q.-G.; Wang, X.-Y. Morphology effect of Ru/CeO_2 catalysts for the catalytic combustion of chlorobenzene. *Appl. Catal. B Environ.* **2014**, *158–159*, 96–105. [CrossRef]
44. Cai, T.; Huang, H.; Deng, W.; Dai, Q.-G.; Liu, W.; Wang, X.-Y. Catalytic combustion of 1,2-dichlorobenzene at low temperature over Mn-modified Co_3O_4 catalysts. *Appl. Catal. B Environ.* **2015**, *166–167*, 393–405. [CrossRef]

45. Xue, L.; Zhang, C.-B.; He, H.; Teraoka, Y. Catalytic decomposition of N_2O over CeO_2 promoted Co_3O_4 spinel catalyst. *Appl. Catal. B Environ.* **2007**, *75*, 167–174. [CrossRef]
46. Ma, C.-Y.; Wang, D.-H.; Xue, W.-J. Investigation of Formaldehyde Oxidation over Co_3O_4-CeO_2 and Au/ Co_3O_4-CeO_2 Catalysts at Room Temperature: Effective Removal and Determination of Reaction Mechanism. *Environ. Sci. Technol. ES&T* **2011**, *45*, 3628–3634. [CrossRef]
47. Jiang, T.-T.; Wang, X.; Chen, J.-Z.; Mai, Y.-L.; Liao, B.; Hu, W. Hierarchical Ni/Co-LDHs catalyst for catalytic oxidation of indoor formaldehyde at ambient temperature. *J. Mater. Sci. Mater. Electron.* **2020**, *31*, 3500–3509. [CrossRef]
48. Wang, M.; Zhang, L.-X.; Huang, W.-M.; Xiu, T.-P.; Zhuang, C.-G.; Shi, J.-L. The catalytic oxidation removal of low-concentration HCHO at high space velocity by partially crystallized mesoporous MnO_x. *Chem. Eng. J.* **2017**, *320*, 667–676. [CrossRef]
49. Han, Z.; Wang, C.; Zou, X.; Chen, T.; Dong, S.; Zhao, Y.; Xie, J.; Liu, H. Diatomite-supported birnessite-type MnO_2 catalytic oxidation of formaldehyde: Preparation, performance and mechanism. *Appl. Surf. Sci.* **2020**, *502*, 144201. [CrossRef]

Article

Efficient Pyrolysis of Low-Density Polyethylene for Regulatable Oil and Gas Products by ZSM-5, HY and MCM-41 Catalysts

Ting Liu [1], Yincui Li [1], Yifan Zhou [2], Shengnan Deng [1] and Huawei Zhang [1,*]

[1] School of Environmental and Municipal Engineering, Qingdao University of Technology, Qingdao 266033, China
[2] College of Chemical and Biological Engineering, Shandong University of Science and Technology, Qingdao 266590, China
* Correspondence: sdkdzhw@163.com; Tel.: +86-0532-85071133

Abstract: In this research, catalytic cracking of low-density polyethylene (LDPE) has been carried out in the presence of three kinds of typical molecular sieves, including ZSM-5, HY and MCM-41, respectively. The effects of different catalysts on the composition and quantity of pyrolysis products consisting of gas, oil and solid material were systematically investigated and summarized. Specially, the three kinds of catalysts were added into LDPE for pyrolysis to obtain regulatable oil and gas products (H_2, CH_4 and a mixture of C_2–C_4^+ gaseous hydrocarbons). These catalysts were characterized with BET, NH_3-TPD, SEM and TEM. The results show that the addition of MCM-41 improved the oil yield, indicating that the secondary cracking of intermediate species in primary pyrolysis decreased with the case of the catalyst. The highest selectivity of MCM-41 to liquid oil (78.4% at 650 °C) may be attributed to its moderate total acidity and relatively high BET surface area. The ZSM-5 and HY were found to produce a great amount of gas products (61.4% and 67.1% at 650 °C). In particular, the aromatic yield of oil production reached the maximum (65.9% at 500 °C) when the ZSM-5 was used. Accordingly, with the three kinds of catalysts, a new environment-friendly and efficient recovery approach may be developed to obtain regulatable and valuable products by pyrolysis of LDPE-type plastic wastes.

Keywords: LDPE; ZSM-5; HY; MCM-41; catalytic pyrolysis

1. Introduction

Global demand for plastics is growing rapidly due to their widespread applications in many fields [1–3]. Plastic production has increased 20-fold over the past half century and is expected to exceed 500 million tons by 2050 [4]. There are many kinds of plastics, while polyethylene (PE) ranks first with 32%. Low-density polyethylene (LDPE) is one of the most widely applied plastic [5]. LDPE has a high degree of short- and long-chain branching, which prevents the chains from entering the crystal structure [4,6]. The environment and global ecosystems are negatively affected by the excessive use, improper management and disposal of plastics [3]. Therefore, how to achieve a clean and efficient utilization of waste plastics with high value, for example, regulatable oil and gas products, has become an urgent problem to be solved.

Pyrolysis is considered an emerging recycling technology that has attracted wide attention. It is the process of breaking polymer molecular chains and converting them into liquid oil, char and gases at a high temperature (300–900 °C) in an inert atmosphere [7,8]. Given the feasibility of regulatable gas, recycling waste plastics using pyrolysis is considered a promising treatment [9–11]. However, the factors influencing the pyrolysis process, such as temperature, pressure, catalyst type, heating mode, etc., are complex and there is still a great difference in the yield of gas products, restraining the application. At the same time, pyrolysis suffers from a large variety of reaction products, high pyrolysis temperatures and low yields of valuable chemicals. However, the introduction of catalysts can decrease

the reaction temperature as well as the activation energy of pyrolysis and change the way of plastic pyrolysis to achieve the selective collection of target products. At the same time, catalytic pyrolysis can promote the fracture of long-chain molecules to achieve light pyrolysis products [12] and reduce the viscosity of the liquid phase of pyrolysis, obtaining valuable chemical production.

There are various catalysts applied in the process of plastic pyrolysis, but the most widely employed catalysts are ZSM-5, Y-zeolite, FCC(fluid catalytic cracking) and MCM-41 [7]. Especially zeolites have found widespread applications in plastic cracking because of their structural advantages such as diverse skeleton structure, highly ordered pores, sufficient acid sites, large specific surface area as well as great stability [7,11].

ZSM-5 has been generally employed in the thermal cracking of waste plastic and gas adsorption-separation industry because of its strong acidity and shape selectivity. Ding et al. [13] and Du et al. [14] found that ZSM-5 is a kind of crystalline aluminosilicate material with a unique two-dimensional pore structure. The pores intersect each other with a diameter of 0.55 nm, which is conducive to the generation of hydrocarbons with a carbon number of less than 10. It also has excellent thermal stability and hydrothermal stability, strong acid resistance and anti-carbon deposition, adjustable acidity, great shape selectivity, isomerization capacity and other catalytic properties. Wei et al. [15] found that HY zeolite shows good catalytic performance with the advantages of regular pore structure, high stability as well as reactivity. At the same time, Ding et al. [16] found that HY is used as a catalyst for co-pyrolysis with LDPE, increasing from 23.5% to 80.4% as the ratio of HY to LDPE rose from 0 to 1:5. It is known that oil production and quality achieve the best balance at the HY to LDPE ratio of 1:10. Zhang et al. [2] reported that MCM-41 is a mesoporous material with a high surface area, which can enhance the yield of hydrocarbons and the quality of pyrolytic oil. Chi et al. [17] found MCM-41 has a unique advantage because its larger pore size makes macromolecular catalysis, adsorption and separation possible, reducing the diffusion resistance of molecules in the channel. Furthermore, MCM-41 has a high specific surface (about 1000 m^2/g), which provides adequate surface sites for adsorption and catalytic reactions of active ingredients. It also gets relatively fewer coke products. However, there have been few studies on the co-pyrolysis of different molecular sieves with LDPE [10,18–20] and the systematic analysis of the catalytic mechanism has not been perfected, lacking systematic analysis and summary of the pyrolysis characteristics of different molecular sieves and waste plastics.

In this paper, we aimed to make clean and efficient utilization of waste plastics with high value, obtaining regulatable gaseous products or liquids. Using ZSM-5, HY and MCM-41 as catalysts, catalytic pyrolysis of LDPE was performed in a fixed-bed reactor to achieve the three-phase products. The effect of pyrolysis temperature and type of catalysts on the product yield was explored. Furthermore, the characteristic and distribution of the pyrolysis products catalyzed by three kinds of catalysts were compared to obtain the interaction path and scheme of the catalytic pyrolysis. It provides the theoretical basis for the clean application of waste plastics, selectivity of valuable chemicals and selection of catalysts.

2. Results and Discussion

2.1. BET Results

Table 1 shows the structural properties of the three kinds of catalysts. MCM-41 has the largest BET-specific surface area caused by mesopores, facilitating multiple contacts of plastics with catalytic active centers and favoring the passage of large pyrolysis products (like olefins and aromatics) [21,22]. Compared to ZSM-5 and HY, MCM-41 obtains a larger average pore size (3.653 nm), specific surface area (962 m^2/g) and the total pore volume (0.718 cm^3/g), which is mainly manifested in the catalytic activity of MCM-41. ZSM-5 obtains the minimum average pore size of 0.411 nm, allowing the heavy chemicals to crack further.

Table 1. Textural properties of different catalysts.

Catalysts	BET Surface Area (m²/g)	Total Pore Volume (cm³/g)	Average Pore Diameter (nm)
ZSM-5	361	0.206	0.411
HY	701	0.392	0.811
MCM-41	962	0.718	3.653

2.2. Acid Properties of Zeolites

Figure 1 shows the NH_3-TPD results of the three kinds of catalysts. The weak, medium and strong acid sites of the catalysts correspond to the characteristic peaks at 155 °C, 275 °C and 505 °C, respectively. The results present that most of the acid sites of ZSM-5 are as the same as that of the HY, achieving uniform acid strength, while the MCM-41 zeolite-based catalyst with a SiO_2/Al_2O_3 ratio of 30 has low acid strength and no strong acid. The acidity shown in Table 2 further confirms the results obtained.

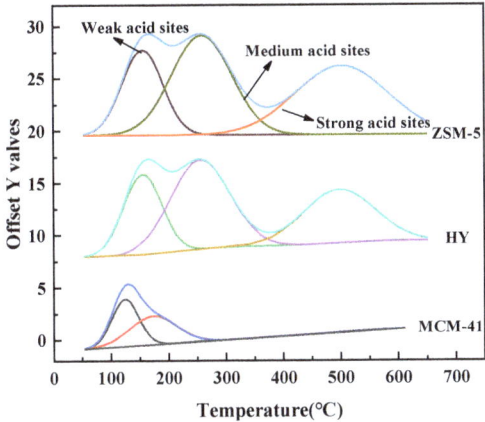

Figure 1. NH_3 adsorption/desorption isotherm distribution of samples.

Table 2. Acidity distribution of three kinds of catalysts.

Catalysts	Acid Content (µmol/g)			
	Weak Acidity	Medium Acidity	Strong Acidity	Total Acidity
ZSM-5	368	838	549	1755
HY	366	748	652	1766
MCM-41	138	221	-	359

In addition, the acid distribution of the catalyst was estimated by Gaussian fitting. As presented in Table 2, HY has the highest total acid contents and a strong acid site, which were 1766 µmol/g and 652 µmol/g, respectively. The MCM-41 has the lowest total acid content (359 µmol/g). The acidity of the catalysts has a great influence on the catalytic performance of the final product of plastic pyrolysis, which is covered in detail in Sections 2.4–2.6.

2.3. SEM and TEM Results

The SEM and TEM images of ZSM-5, MCM-41 and HY are exhibited in Figure 2. The ZSM-5 is constituted by clear quadrangular prism-like crystallites [23], which is consistent with the study reported by Haswin Kaur Gurdeep Singh et al. [24], with sizes ranging from 250–450 nm (Figure 2(a1)). And the image for ZSM-5 shows a relatively irregular sheet structure (Figure 2(a2)). It is worth noting that, in the presence of MCM-41, various particles are uniformly distributed on the surface of the carrier (Figure 2(b2)), ranging from 20 nm to 90 nm, which might be conducive to its relatively higher specific surface area among three kinds of catalysts. As seen from the SEM image (Figure 2(c1)), the hierarchical HY zeolite retains its intact crystal structures [25,26], ranging from 50 nm to 80 nm and the regular sheet structure is observed in the HY (Figure 2(c2)).

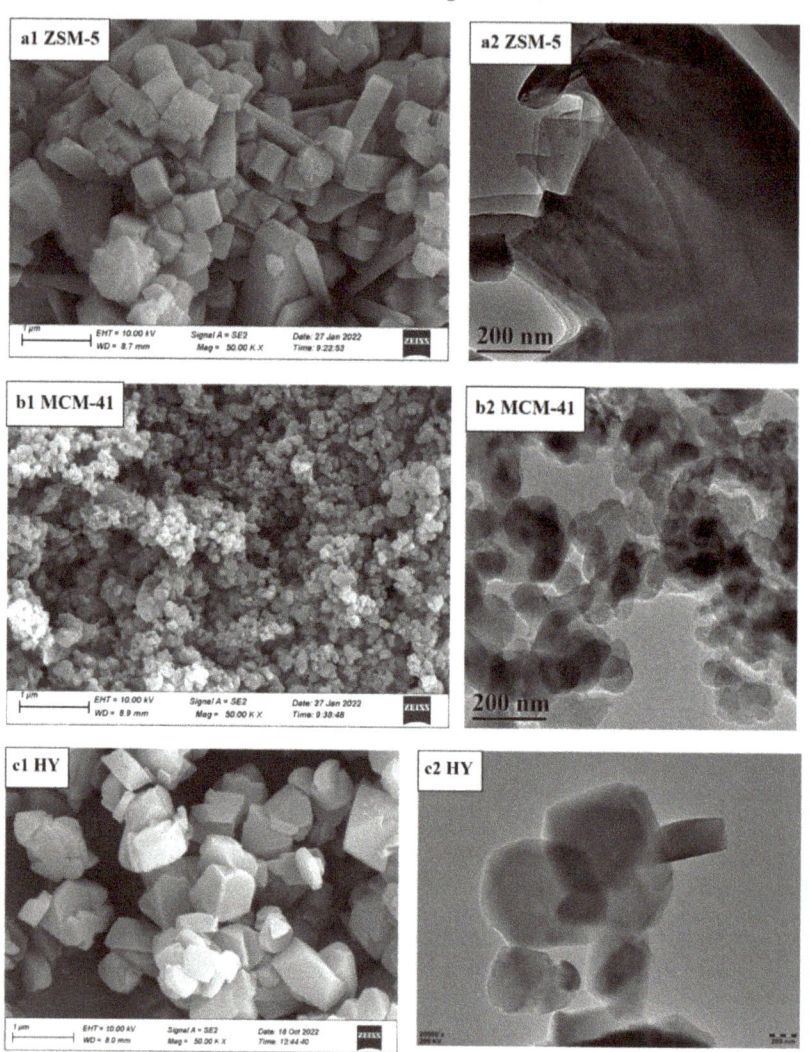

Figure 2. The SEM and TEM images of three kinds of catalysts. (**a1,b1,c1**): SEM; (**a2,b2,c2**): TEM.

2.4. Effects of Temperature on Gas–Liquid–Solid Three-Phase Yield of LDPE Pyrolysis

As depicted in Figure 3, the performance on catalytic conversion of LDPE over different catalysts from 450 °C to 650 °C was contrasted. It is obvious that the yield distribution was significantly affected by pyrolysis temperature. With the increase of pyrolysis temperature, the total yield of gas and oil enhanced largely while the yield of solid decreased greatly, which may be due to the decomposition and secondary reaction of LDPE pyrolysis volatiles [27]. As the temperature continued to rise, it provided more heat to the polymer, weakening the chain structure and causing more polymer chains to break [28] and the trend is consistent with most research on polymer pyrolysis [29,30]. As can be seen from Figure 3a, the liquid phase yield was lower at 450 °C (31.3%) and at 500 °C (55.6%), with the increase in temperature, the conversion of the polymer improved [28], so the liquid yield rose to 82.0% at 550 °C. This is due to further increases in temperature causing further cracking of the oligomer to form smaller hydrocarbons in the form of gaseous compounds, while liquid production does not change significantly.

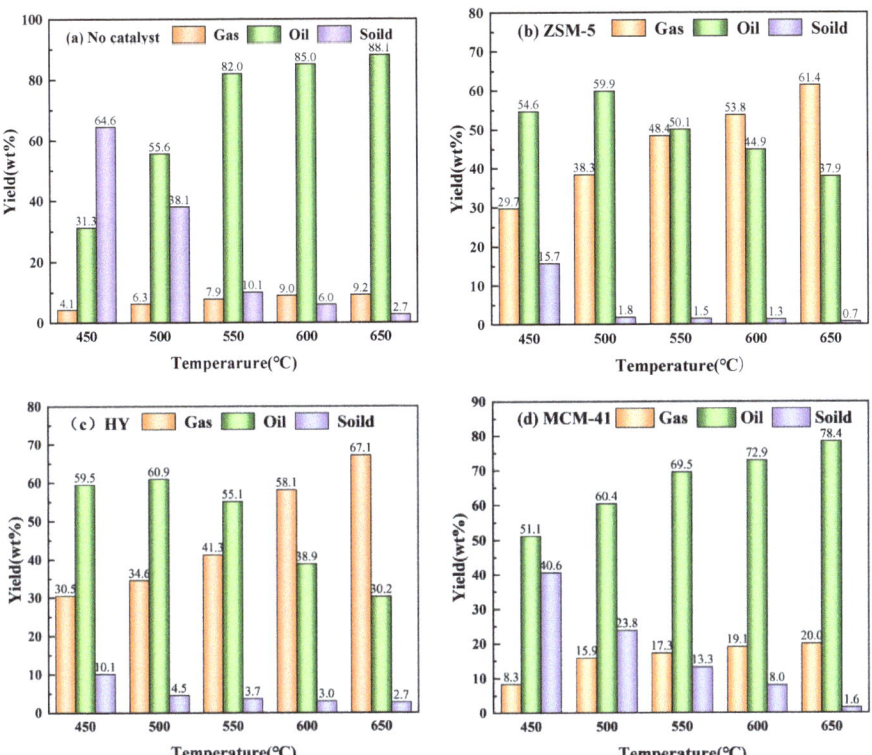

Figure 3. Three-phase yield diagram of LDPE (**a**): no catalyst; (**b**): ZSM-5 catalyst; (**c**): HY catalyst; (**d**): MCM-41 catalyst.

Compared to the case without catalysts, the pyrolysis gas yield of LDPE increased, indicating that the catalyst has a moderate acidity, leading to obvious secondary cracking of liquid oil [21,28]. As illustrated in Figure 3a, the liquid phase yield of non-catalytic pyrolysis at 500 °C occupied about 55.6%, which is similar to that of catalytic pyrolysis at 450 °C in Figure 3b. It can be concluded that the catalysts could significantly decrease the reaction temperature.

Furthermore, it can be observed that the ZSM-5 and HY catalysts resulted in much higher gas yields with increasing temperature while the MCM-41 obtained more oil yield. As seen from Figure 3b,c, there was no significant difference in gas yield between ZSM-5 and HY, both of which had higher gas yield than MCM-41. The yield of gas on the ZSM-5 catalyst increased from 29.7% to 61.4% as the temperature increased from 450 °C to 650 °C; however, the yield of oil on MCM-41 increased from 51.1% to 78.4%.

For ZSM-5, the main cause of such phenomenon comes from the function of the acidic sites and framework structure of ZSM-5 [31]. It was found that the interaction during catalytic pyrolysis could promote the formation of light molecular gases from chain-breaking volatiles [32]. At the same time, ZSM-5 has a smaller pore size and a larger intracrystalline pore structure, allowing further cracking of heavy chemicals. Since the initial decomposition sample on the outer surface of the ZSM-5 can diffuse into the inner cavity of the ZSM-5, further decomposition into gaseous products resulted in very high gas yields [33]. Compared with MCM-41, the HY catalyst clearly provided a higher gas yield as a result of the strong acid sites and high acid density of the HY zeolite, which provided higher cracking activity than MCM-41 with only weak acid sites [34]. Additionally, the difference in gas yield among the three kinds of catalysts is due to the difference in carbon deposition [20,34,35], resulting in the difference in strong acid sites.

For MCM-41, it exhibited the greatest amounts of oil yield (78.4%) and the lowest amounts of gas yield (20.0%) at 650 °C. HY showed the second-highest oil yield (60.9%) and ZSM-5 presented a slightly lower oil yield (59.9%). This manifested that, in the case of these catalysts, secondary cracking was slightly enhanced and the difference in pyrolysis yields was largely as a result of the differences in acidity and structural properties discussed earlier. In addition, MCM-41 with uniform morphology was easy to produce the pyrolysis product with similar carbon distribution, leading to more oil products produced by MCM-41 than other catalysts [21].

2.5. Effects of Catalyst on the Composition and Quality of Gaseous Products

The gaseous product composition for non-catalytic and catalytic experiments from 450 °C to 650 °C are depicted in Figure 4. The pyrolysis gas consists of H_2, CH_4 and C_2–C_4^+ gaseous hydrocarbon mixtures. The contents of H_2, CH_4, C_2 and C_3 increased by 2.35%, 1.45%, 19.21% and 19.72% as the temperature rose from 450 °C to 650 °C, while C_4^+ gas reduced by 35.59%, which was mainly due to the C_4^+ gaseous products being further cleaved to CH_4 and other small molecule gases as temperature increased [27,28].

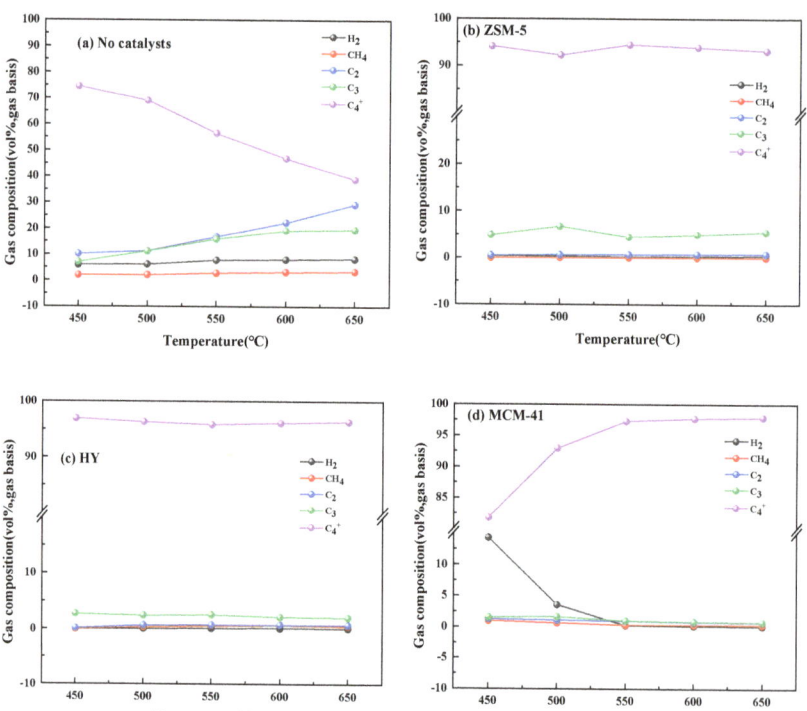

Figure 4. Comparison of gas–phase composition during LDPE pyrolysis with three kinds of catalysts (**a**) no catalysts; (**b**) ZSM-5; (**c**) HY; (**d**) MCM−41.

In the non-catalytic run, a considerable amount of C_2 and C_3 were observed. After adding the catalysts into pyrolysis, the number of C_4^+ gaseous products raised significantly. It is interesting that all the gaseous products of the catalyzed reactions showed a percentage of C_4^+ around 95% and MCM-41 at 500 °C. Additionally, the wider pore size distribution observed in HY and MCM-41 zeolite resulted in the diffusion of the reactant and product, which was more favorable for macromolecular hydrocarbons to enter the pore size of the molecular sieve to react with the recombinant gas [36]. As illustrated in Figure 4d, at temperatures of 450–500 °C, the gas phase products of MCM-41 had a higher content of H_2 and lower content of C_4^+ compared to both ZSM-5 and HY catalysts. This is because MCM-41 has a larger pore size and higher selectivity to heavier components. However, due to its weaker acidity, fewer active sites, lower catalytic activity and selectivity at lower temperatures, the content of hydrogen and methane was higher. However, as temperatures rose further, rapid product formation did not allow more cracking gas to occur in the reactor, so the heavier hydrocarbon component of the product increased, resulting in the formation of heavy hydrocarbons and low hydrogen [37]. Therefore, the main component of the gas phase product was C_4^+. The difference in the pyrolysis yield can be directly related to changes in the structural and acid properties of the catalysts.

2.6. Effects of Catalyst on Oil Distribution and Quantity

The carbon number distribution of oil is exhibited in Figure 5 and the major constituents of the oil product as well as relative content are presented in the Supplementary Material. As shown in Figure 5, in the experiment without catalysts, the oil products comprised four comparable fractions (<C_{11}, C_{12}–C_{18}, C_{19}–C_{30} and aromatics), indicating that the carbon number distribution was relatively uniform compared to the catalytic experiment. As the temperature increased from 500 °C to 650 °C, the yield of light hydrocarbon fractions improved and the yield of heavy hydrocarbon fractions decreased in the non-catalytic run (Figure 5a). The liquid fractions are mainly composed of linear paraffins (C_{10}~C_{30}) and produced almost no aromatic hydrocarbons [3] (Tables S1 and S2). The pyrolysis of LDPE was the result of its characteristic long-carbon chain structure, converting the feedstock into wax rather than liquid oil [3].

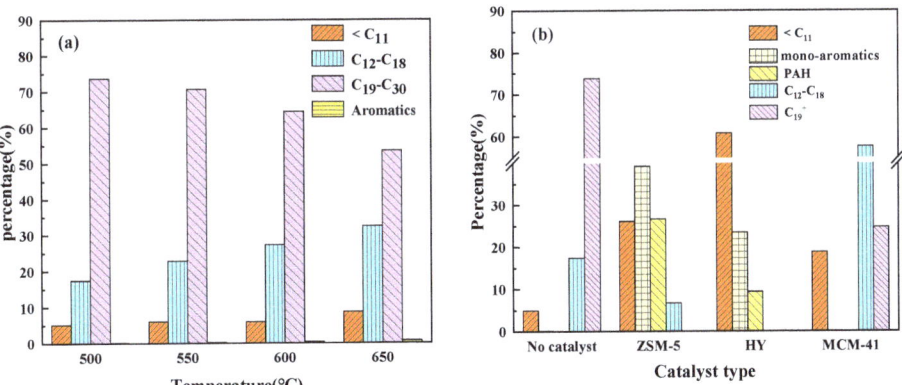

Figure 5. Distribution of the oil products in terms of carbon number (a) Liquid product composition of LDPE at different temperatures; (b) Liquid product composition of LDPE at different catalysts.

Moreover, it is obvious that the representative of diesel products is C_{12}–C_{18} hydrocarbons and the MCM-41 catalyst has a potential application value in the use of plastic waste to produce diesel. Among the three kinds of catalysts, ZSM-5 presented the especially high selectivity for the aromatics and low selectivity for the C_{12}–C_{18} fraction. Because ZSM-5 exhibited the second-highest total acid site, it is not difficult to infer that high diesel production was due, in part, to the mild acidity as well as excessive cracking of hydrocarbons.

Compared to the pyrolysis of LDPE without catalysts, the contents of mono-aromatics and polycyclic aromatic hydrocarbons (PAHs) in ZSM-5 catalytic pyrolysis were higher by 65.9% (Figure 5b). The pore size and structure of ZSM-5 played an important role in the formation of aromatic hydrocarbons, owing to its shape selectivity [20,38]. It is well-known that ZSM-5 has acid sites, a suitable pore size and shape selectivity, which is beneficial for the formation of aromatics [39] and the conversion of aliphatics to aromatic production using the Diels–Alder reaction [40]. It is worth noting that MCM-41 displayed the lowest selectivity for the aromatic compounds due to its weak aromatizing ability [19].

Furthermore, the results showed that the alkane content of heavy hydrocarbon fractions obtained during the catalytic process was more than the olefin content (Tables S3–S5). This is consistent with the conclusion in the study of Liu et al. [41], which may be caused by alkylation of the primary intermediate.

2.7. Effect of Catalysts on Some Reaction Pathways

It is concluded that the pyrolysis of LDPE followed the random-chain-breaking mechanism and the catalytic thermal decomposition of LDPE underwent the carbocation theory [4,8,33]. The catalytic effect of the catalysts was primarily due to their acidity during pyrolysis. The carbonate ion theory was based on the acid sites of the catalyst [33].

Thermal cracking of LDPE is often related to the free-radical mechanism [28]. The thermal pyrolysis of LDPE was partly caused by the homolysis of C–C bonds in the polymer chain under thermal action. As shown in Figure 6, LDPE formed free radical fragments of different lengths through random fracture (Step 1 and Step 2). Then the terminal free radical fragments generated alkenes through a hydrogen transfer reaction and further bond-breaking reaction (Step 3) and alkanes and hydrogen gas were further generated through a bimolecular reaction (Step 4).

As shown in Figure 7, the scheme of catalyst participation in LPDE pyrolysis reaction pathways is proposed. When the ZSM-5 was introduced into the pyrolysis process, the smaller pore size and larger pore structure of ZSM-5 allowed the initial decomposition sample on the outer surface to diffuse into the interior of ZSM-5, favoring the recombinant fraction to further decompose into gaseous products, resulting in higher gas production. Additionally, there was a higher aromatic content in the liquid phase of the ZSM-5 catalyst, probably due to its high acidity and shape selectivity. The acid sites could facilitate the formation of aromatics by catalyzing hydrogen transfer reactions and the Diels–Alder reactions [42] and the gas–liquid products underwent further aromatization. The hydrogen transfer reaction is considered to be the main source of aromatics and alkanes. Dehydrogenation active sites can promote the Diels–Alder reactions and cyclization intermediates. In addition, the heavy aromatic hydrocarbons were more easily decomposed into light aromatic hydrocarbons by hydrogenation than monocyclic and bicyclic aromatic hydrocarbons.

The wider pore size distribution observed in HY zeolite contributed to the diffusion of the reactant and product [36]. Therefore, the main component of the gas phase product was C_4^+, while the content of recombinant fraction (C_{19}^+) in the liquid phase product was seldom. Due to the strong acid site and high acid density of HY, it provided more pyrolysis activity leading to a higher gas yield. Bimolecular pyrolysis and hydrogen transfer reaction on HY can produce a large number of incondensable gas and a high yield of alkenes and alkanes below C_{11}.

When the MCM-41 was added to the pyrolysis, its mesoporous structure could greatly promote the accessibility of macromolecules to zeolite, reduce residence time, inhibit the secondary reaction and thus improve the yield of liquid [25]. Meanwhile, the mesoporous catalyst has a large pore volume and pore size, which was conducive to the free diffusion of the primary thermal decomposition molecules of LDPE and was easy to be converted into liquid products [43]. Therefore, compared with ZSM-5 and HY, MCM-41 had a greater promotion effect on crude oil fractions, which was less favorable for cracking and aromatization reactions, resulting in a higher content of C_{12}–C_{18}. The lower proportion of aromatics obtained by MCM-41 may be related to its lower acid strength, lower catalytic activity and poor selectivity.

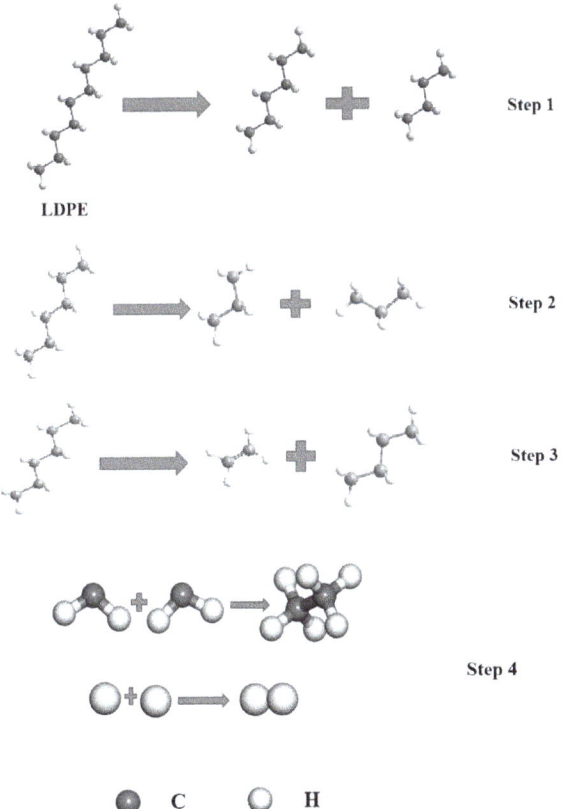

Figure 6. A plausible reaction pathway of LDPE pyrolysis without catalysts.

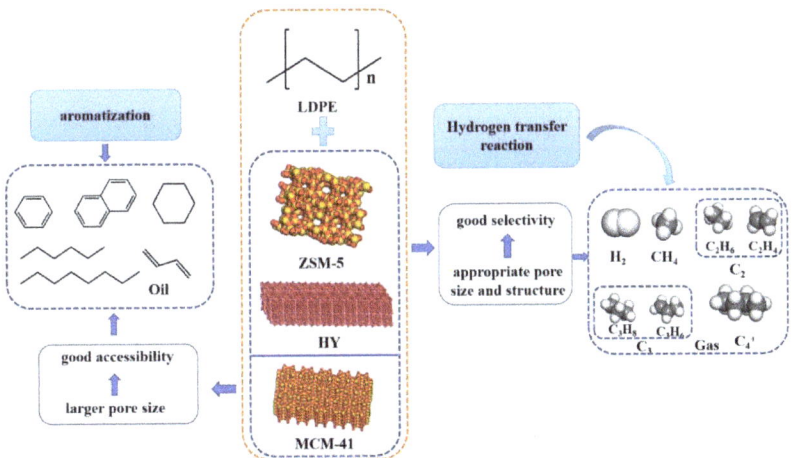

Figure 7. The scheme of catalyst participation in LPDE pyrolysis reaction pathways.

3. Materials and Methods

3.1. Materials

Powdered LDPE (100 mesh, Zhongyanshan Petrochemical Co., Ltd., Beijing, China) was commercially available. ZSM-5 powder with a SiO_2/Al_2O_3 ratio of 20, HY powder with a SiO_2/Al_2O_3 ratio of four and MCM-41 powder a with SiO_2/Al_2O_3 ratio of 30 were obtained from Zhongyanshan Petrochemical Co., Ltd. All of the molecular sieve catalysts were about 60–100 mesh in size and were calcined in air at 550 °C for 3 h before pyrolysis experiments.

3.2. Experimental Setup

The pyrolysis experiment of waste plastics was performed in a fixed bed reactor, as pictured in Figure 8. Briefly, the device was composed of an electric heating tube furnace, a temperature-controlled system, a quartz reactor (ID = 50 mm, L = 440 mm), as well as a cooling system. In the typical pyrolysis run, the catalyst and the plastic sample were mixed at a mass ratio of 1:2 (the proportion of catalyst to plastic remained constant throughout the experiment). Each plastic sample weighed approximately 5 g, the exact catalyst and plastic sample mixture were placed in the quartz tube between two sections of quartz wool. The tubular furnace was first blown with nitrogen (100 mL/min) for about 30 min to remove air and then the system was heated at a rate of 15 °C/min to the desired temperature (450, 500, 550, 600 or 650 °C) and held for 0 min.

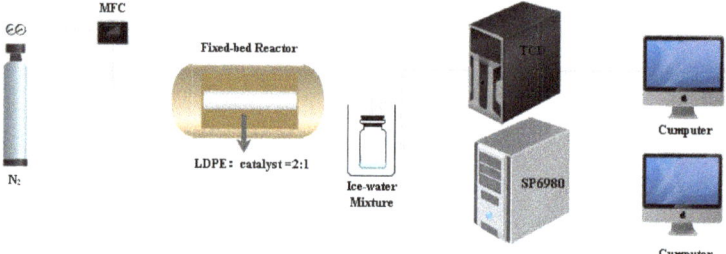

Figure 8. The pyrolysis experimental setup.

The gas and oil vapor generated from pyrolysis was blown into a condenser, which was cooled by the ice–water mixture. The condensate oil products were collected in a glass bottle and the mass of liquid oil was judged by the weight difference before and after the glass bottle. During the pyrolysis process, a solid mass was obtained by calculating the weight difference of the quartz tube before and after the reaction. The gas products were collected by gas bags and the mass was calculated by the difference. The equations involved were defined as follows:

$$Y_{p1} = \frac{m_1}{m_0} \times 100\% \tag{1}$$

$$Y_{p2} = \frac{m_2}{m_0} \times 100\% \tag{2}$$

$$Y_{p3} = \frac{m_0 - m_1 - m_2}{m_0} \times 100\% \tag{3}$$

where Y_{P1}, Y_{P2}, Y_{P3} were the yields of oil, solid products and gas after pyrolysis, respectively. m_0 was the mass of the LDPE sample, m_1 and m_2 were defined as the mass of the liquid oil and solid product after pyrolysis.

Additionally, to ensure the accuracy of the experimental data, all experiments were repeated three times.

3.3. Characterization

The physicochemical properties of the three kinds of catalysts were determined by scanning electron microscopy (SEM), N_2 adsorption–desorption isotherms, temperature-programmed desorption of ammonia (NH_3-TPD) and high-resolution transmission electron microscopy (HRTEM). The detailed characterization methods of the samples are presented in the Supplementary Material.

3.4. Product Analysis

The determination of H_2 and CH_4 in cracking gas was done using GC-TCD (Ruihong, SP-6800A, Zaozhuang, China), analysis of hydrocarbon gases such as CH_4 and C_2^+ was done using GC-FID (Fuli, SP-6890, Nanjing, China). Each gas sample was measured three times to obtain the average. An analysis of pyrolysis oil composition was done using GC-MS (Agilent, 6890-5973, Santa Clara, CA, USA) with the HP-5MS capillary column (30 m × 250 μm × 0.25 μm). The operating parameters of GC-MS were described below: 60 °C for 3 min; 60 to 240 °C for 2 min at 12 °C/min; 240 to 300 °C for 10 min at 6 °C/min. The split ratio was kept at 100:1.

4. Conclusions

In this study, three different catalysts of ZSM-5, HY and MCM-41 were added, respectively, into non-catalytic pyrolysis of LDPE for regulatable oil and gas products. On the basis of analyzing the structure and characterization of the catalysts, the distribution of the pyrolysis products and the reaction mechanism of LDPE on different catalysts were discussed. The NH_3-TPD and BET characterizations of these catalysts exhibited the differences in pore size as well as acidity and their unique structural characteristics. The results of NH_3-TPD and BET presented that MCM-41 had the lowest acid strength and the largest pore size. The morphologies of the different catalysts were characterized by SEM and TEM. In the presence of MCM-41, a uniformly distributed granular structure could be observed. Because of the proper combination of acidity and structural properties, MCM-41 has been observed to produce a great deal of oil products, while ZSM-5 and HY were found to produce a great amount of gas products. Specially, ZSM-5 showed the greatest amounts of the aromatic products. This facilitates the selection of catalysts for cleaning applications of waste plastics and targeted access to valuable chemicals.

Supplementary Materials: The following supporting information can be downloaded at: https://www.mdpi.com/article/10.3390/catal13020382/s1. Table S1. Liquid product composition of LDPE at different temperatures. Table S2. Liquid product composition of LDPE at different temperatures. Table S3. Liquid phase GC-MS table of catalytic pyrolysis of LDPE by ZSM-5 molecular sieve at 500 °C. Table S4. Liquid phase GC-MS table of catalytic pyrolysis of LDPE by HY molecular sieve at 500 °C. Table S5. Liquid phase GC-MS table of catalytic pyrolysis of LDPE by MCM-41 molecular sieve at 500 °C.

Author Contributions: Resources, S.D.; data curation, Y.Z.; writing: original draft preparation, Y.L.; writing: review and editing, T.L.; project administration, H.Z. All authors have read and agreed to the published version of the manuscript.

Funding: This research was funded by the National Natural Science Foundation of China (22078168, NO. 52272086), Huawei Zhang, School of Environmental and Municipal Engineering, Qingdao University of Technology, Qingdao 266033, China.

Acknowledgments: Financial support was sponsored by the National Natural Science Foundation of China (22078168, 52272086).

Conflicts of Interest: The authors declare no conflict of interest.

References

1. Aboul-Enein, A.A.; Awadallah, A.E. Production of nanostructured carbon materials using Fe–Mo/MgO catalysts via mild catalytic pyrolysis of polyethylene waste. *Chem. Eng. J.* **2018**, *354*, 802–816. [CrossRef]
2. Zhang, Y.; Huang, J.; Williams, P.T. Fe–Ni–MCM-41 Catalysts for Hydrogen-Rich Syngas Production from Waste Plastics by Pyrolysis–Catalytic Steam Reforming. *Energy Fuels* **2017**, *31*, 8497–8504. [CrossRef]
3. Maqsood, T.; Dai, J.; Zhang, Y.; Guang, M.; Li, B. Pyrolysis of plastic species: A review of resources and products. *J. Anal. Appl. Pyrolysis* **2021**, *159*, 105295. [CrossRef]
4. Peng, Y.; Wang, Y.; Ke, L.; Dai, L.; Wu, Q.; Cobb, K.; Zeng, Y.; Zou, R.; Liu, Y.; Ruan, R. A review on catalytic pyrolysis of plastic wastes to high-value products. *Energy Convers. Manag.* **2022**, *254*, 115243. [CrossRef]
5. Tao, L.; Ma, X.; Ye, L.; Jia, J.; Wang, L.; Ma, P.; Liu, J. Interactions of lignin and LDPE during catalytic co-pyrolysis: Thermal behavior and kinetics study by TG-FTIR. *J. Anal. Appl. Pyrolysis* **2021**, *158*, 105267. [CrossRef]
6. Sogancioglu, M.; Yel, E.; Ahmetli, G. Pyrolysis of waste high density polyethylene (HDPE) and low density polyethylene (LDPE) plastics and production of epoxy composites with their pyrolysis chars. *J. Clean. Prod.* **2017**, *165*, 369–381. [CrossRef]
7. Miandad, R.; Rehan, M.; Barakat, M.A.; Aburiazaiza, A.S.; Khan, H.; Ismail, I.M.I.; Dhavamani, J.; Gardy, J.; Hassanpour, A.; Nizami, A.-S. Catalytic Pyrolysis of Plastic Waste: Moving Toward Pyrolysis Based Biorefineries. *Front. Energy Res.* **2019**, *7*, 27. [CrossRef]
8. Miandad, R.; Barakat, M.A.; Aburiazaiza, A.S.; Rehan, M.; Nizami, A.S. Catalytic pyrolysis of plastic waste: A review. *Process Saf. Environ. Prot.* **2016**, *102*, 822–838. [CrossRef]
9. Li, D.; Lei, S.; Wang, P.; Zhong, L.; Ma, W.; Chen, G. Study on the pyrolysis behaviors of mixed waste plastics. *Renew. Energy* **2021**, *173*, 662–674. [CrossRef]
10. De Souza, M.J.B.; Silva, T.H.A.; Ribeiro, T.R.S.; Da Silva, A.O.S.; Pedrosa, A.M.G. Thermal and catalytic pyrolysis of polyvinyl chloride using micro/mesoporous ZSM-35/MCM-41 catalysts. *J. Therm. Anal. Calorim.* **2019**, *140*, 167–175. [CrossRef]
11. Anuar Sharuddin, S.D.; Abnisa, F.; Wan Daud, W.M.A.; Aroua, M.K. A review on pyrolysis of plastic wastes. *Energy Convers. Manag.* **2016**, *115*, 308–326. [CrossRef]
12. Nishu; Li, Y.; Liu, R. Catalytic pyrolysis of lignin over ZSM-5, alkali, and metal modified ZSM-5 at different temperatures to produce hydrocarbons. *J. Energy Inst.* **2022**, *101*, 111–121. [CrossRef]
13. Ding, Y.L.; Wang, H.Q.; Xiang, M.; Yu, P.; Li, R.Q.; Ke, Q.P. The Effect of Ni-ZSM-5 Catalysts on Catalytic Pyrolysis and Hydro-Pyrolysis of Biomass. *Front. Chem.* **2020**, *8*, 790. [CrossRef] [PubMed]
14. Du, S.; Gamliel, D.P.; Valla, J.A.; Bollas, G.M. The effect of ZSM-5 catalyst support in catalytic pyrolysis of biomass and compounds abundant in pyrolysis bio-oils. *J. Anal. Appl. Pyrolysis* **2016**, *122*, 7–12. [CrossRef]
15. Wei, B.; Jin, L.; Wang, D.; Hu, H. Catalytic upgrading of lignite pyrolysis volatiles over modified HY zeolites. *Fuel* **2020**, *259*, 116234. [CrossRef]
16. Ding, K.; Liu, S.; Huang, Y.; Liu, S.; Zhou, N.; Peng, P.; Wang, Y.; Chen, P.; Ruan, R. Catalytic microwave-assisted pyrolysis of plastic waste over NiO and HY for gasoline-range hydrocarbons production. *Energy Convers. Manag.* **2019**, *196*, 1316–1325. [CrossRef]
17. Chi, Y.; Xue, J.; Zhuo, J.; Zhang, D.; Liu, M.; Yao, Q. Catalytic co-pyrolysis of cellulose and polypropylene over all-silica mesoporous catalyst MCM-41 and Al-MCM-41. *Sci. Total Environ.* **2018**, *633*, 1105–1113. [CrossRef]
18. Yu, L.; Farinmade, A.; Ajumobi, O.; Su, Y.; John, V.T.; Valla, J.A. MCM-41/ZSM-5 composite particles for the catalytic fast pyrolysis of biomass. *Appl. Catal. A Gen.* **2020**, *602*, 117727. [CrossRef]
19. Li, X.; Dong, L.; Zhang, J.; Hu, C.; Zhang, X.; Cai, Y.; Shao, S. In-situ catalytic upgrading of biomass-derived vapors using HZSM-5 and MCM-41: Effects of mixing ratios on bio-oil preparation. *J. Energy Inst.* **2019**, *92*, 136–143. [CrossRef]
20. Kim, Y.-M.; Han, T.U.; Kim, S.; Jae, J.; Jeon, J.-K.; Jung, S.-C.; Park, Y.-K. Catalytic co-pyrolysis of epoxy-printed circuit board and plastics over HZSM-5 and HY. *J. Clean. Prod.* **2017**, *168*, 366–374. [CrossRef]
21. Li, K.; Lei, J.; Yuan, G.; Weerachanchai, P.; Wang, J.-Y.; Zhao, J.; Yang, Y. Fe-, Ti-, Zr- and Al-pillared clays for efficient catalytic pyrolysis of mixed plastics. *Chem. Eng. J.* **2017**, *317*, 800–809. [CrossRef]
22. Kelkar, S.; Saffron, C.M.; Andreassi, K.; Li, Z.; Murkute, A.; Miller, D.J.; Pinnavaia, T.J.; Kriegel, R.M. A survey of catalysts for aromatics from fast pyrolysis of biomass. *Appl. Catal. B Environ.* **2015**, *174–175*, 85–95. [CrossRef]
23. Abdalla, A.; Arudra, P.; Al-Khattaf, S.S. Catalytic cracking of 1-butene to propylene using modified H-ZSM-5 catalyst: A comparative study of surface modification and core-shell synthesis. *Appl. Catal. A Gen.* **2017**, *533*, 109–120. [CrossRef]
24. Gurdeep Singh, H.K.; Yusup, S.; Quitain, A.T.; Abdullah, B.; Ameen, M.; Sasaki, M.; Kida, T.; Cheah, K.W. Biogasoline production from linoleic acid via catalytic cracking over nickel and copper-doped ZSM-5 catalysts. *Environ. Res.* **2020**, *186*, 109616. [CrossRef]
25. Li, W.; Zheng, J.; Luo, Y.; Tu, C.; Zhang, Y.; Da, Z. Hierarchical Zeolite Y with Full Crystallinity: Formation Mechanism and Catalytic Cracking Performance. *Energy Fuels* **2017**, *31*, 3804–3811. [CrossRef]
26. Zhao, J.; Wang, G.; Qin, L.; Li, H.; Chen, Y.; Liu, B. Synthesis and catalytic cracking performance of mesoporous zeolite Y. *Catal. Commun.* **2016**, *73*, 98–102. [CrossRef]
27. Zhang, S.; Yang, M.; Shao, J.; Yang, H.; Zeng, K.; Chen, Y.; Luo, J.; Agblevor, F.A.; Chen, H. The conversion of biomass to light olefins on Fe-modified ZSM-5 catalyst: Effect of pyrolysis parameters. *Sci. Total Environ.* **2018**, *628*, 350–357. [CrossRef]
28. Wong, S.L.; Ngadi, N.; Abdullah, T.A.T.; Inuwa, I.M. Conversion of low density polyethylene (LDPE) over ZSM-5 zeolite to liquid fuel. *Fuel* **2017**, *192*, 71–82. [CrossRef]

29. Roozbehani, B.; Sakaki, S.A.; Shishesaz, M.; Abdollahkhani, N.; Hamedifar, S. Taguchi method approach on catalytic degradation of polyethylene and polypropylene into gasoline. *Clean Technol. Environ. Policy* **2015**, *17*, 1873–1882. [CrossRef]
30. Mo, Y.; Zhao, L.; Wang, Z.; Chen, C.L.; Tan, G.Y.; Wang, J.Y. Enhanced styrene recovery from waste polystyrene pyrolysis using response surface methodology coupled with Box-Behnken design. *Waste Manag.* **2014**, *34*, 763–769. [CrossRef] [PubMed]
31. Zhao, J.-P.; Cao, J.-P.; Wei, F.; Zhao, X.-Y.; Feng, X.-B.; Huang, X.; Zhao, M.; Wei, X.-Y. Sulfation-acidified HZSM-5 catalyst for in-situ catalytic conversion of lignite pyrolysis volatiles to light aromatics. *Fuel* **2019**, *255*, 115784. [CrossRef]
32. Tang, Z.; Chen, W.; Hu, J.; Li, S.; Chen, Y.; Yang, H.; Chen, H. Co-pyrolysis of microalgae with low-density polyethylene (LDPE) for deoxygenation and denitrification. *Bioresour. Technol.* **2020**, *311*, 123502. [CrossRef]
33. Song, J.; Sima, J.; Pan, Y.; Lou, F.; Du, X.; Zhu, C.; Huang, Q. Dielectric Barrier Discharge Plasma Synergistic Catalytic Pyrolysis of Waste Polyethylene into Aromatics-Enriched Oil. *ACS Sustain. Chem. Eng.* **2021**, *9*, 11448–11457. [CrossRef]
34. Namchot, W.; Jitkarnka, S. Catalytic pyrolysis of waste tire using HY/MCM-41 core-shell composite. *J. Anal. Appl. Pyrolysis* **2016**, *121*, 297–306. [CrossRef]
35. Wei, B.; Yang, H.; Hu, H.; Wang, D.; Jin, L. Enhanced production of light tar from integrated process of in-situ catalytic upgrading lignite tar and methane dry reforming over Ni/mesoporous Y. *Fuel* **2020**, *279*, 118533. [CrossRef]
36. Liu, Y.; Yan, L.; Bai, Y.; Li, F. Catalytic upgrading of volatile from coal pyrolysis over faujasite zeolites. *J. Anal. Appl. Pyrolysis* **2018**, *132*, 184–189. [CrossRef]
37. Singh, R.K.; Ruj, B. Time and temperature depended fuel gas generation from pyrolysis of real world municipal plastic waste. *Fuel* **2016**, *174*, 164–171. [CrossRef]
38. Lok, C.M.; Van Doorn, J.; Aranda Almansa, G. Promoted ZSM-5 catalysts for the production of bio-aromatics, a review. *Renew. Sustain. Energy Rev.* **2019**, *113*, 109248. [CrossRef]
39. Sun, L.; Zhang, X.; Chen, L.; Zhao, B.; Yang, S.; Xie, X. Comparision of catalytic fast pyrolysis of biomass to aromatic hydrocarbons over ZSM-5 and Fe/ZSM-5 catalysts. *J. Anal. Appl. Pyrolysis* **2016**, *121*, 342–346. [CrossRef]
40. Xu, D.; Yang, S.; Su, Y.; Shi, L.; Zhang, S.; Xiong, Y. Simultaneous production of aromatics-rich bio-oil and carbon nanomaterials from catalytic co-pyrolysis of biomass/plastic wastes and in-line catalytic upgrading of pyrolysis gas. *Waste Manag.* **2021**, *121*, 95–104. [CrossRef]
41. Liu, W.-W.; Hu, C.-W.; Yang, Y.; Tong, D.-M.; Zhu, L.-F.; Zhang, R.-N.; Zhao, B.-H. Study on the effect of metal types in (Me)-Al-MCM-41 on the mesoporous structure and catalytic behavior during the vapor-catalyzed co-pyrolysis of pubescens and LDPE. *Appl. Catal. B Environ.* **2013**, *129*, 202–213. [CrossRef]
42. Sun, K.; Themelis, N.J.; Bourtsalas, A.C.; Huang, Q. Selective production of aromatics from waste plastic pyrolysis by using sewage sludge derived char catalyst. *J. Clean. Prod.* **2020**, *268*, 122038. [CrossRef]
43. Casoni, A.I.; Nievas, M.L.; Moyano, E.L.; Álvarez, M.; Diez, A.; Dennehy, M.; Volpe, M.A. Catalytic pyrolysis of cellulose using MCM-41 type catalysts. *Appl. Catal. A Gen.* **2016**, *514*, 235–240. [CrossRef]

Disclaimer/Publisher's Note: The statements, opinions and data contained in all publications are solely those of the individual author(s) and contributor(s) and not of MDPI and/or the editor(s). MDPI and/or the editor(s) disclaim responsibility for any injury to people or property resulting from any ideas, methods, instructions or products referred to in the content.

Article

The Promoting Effect of Ti on the Catalytic Performance of V-Ti-HMS Catalysts in the Selective Oxidation of Methanol

Ley Boon Sim [1,†], Kek Seong Kim [1,†], Jile Fu [1,*] and Binghui Chen [1,2,*]

[1] School of Energy and Chemical Engineering, Xiamen University Malaysia, Jalan Sunsuria, Bandar Sunsuria, Sepang 43900, Malaysia
[2] National Engineering Laboratory for Green Productions of Alcohols-Ethers-Esters, Department of Chemical and Biochemical Engineering, College of Chemistry and Chemical Engineering, Xiamen University, Xiamen 361005, China
* Correspondence: jile.fu@xmu.edu.my (J.F.); chenbh@xmu.edu.cn (B.C.)
† These authors contributed equally to this work.

Abstract: The effects of Ti modification on the structural properties and catalytic performance of vanadia on hexagonal mesoporous silica (V-HMS) catalysts are studied for selective methanol-to-dimethoxymethane oxidation. Characterizations including N_2 adsorption–desorption (S_{BET}), X-ray diffraction (XRD), UV-Vis diffuse reflectance spectroscopy (DRS UV-Vis), Micro-Raman spectroscopy, FTIR spectroscopy, and H_2 temperature-programmed reduction (H_2-TPR) were carried out to investigate the property and structure of the catalysts. The results show that Ti can be successfully incorporated into the HMS framework in a wide range of Si/Ti ratios from 50 to 10. Ti modification can effectively improve the distribution of vanadium species and thus enhance the overall redox properties and catalytic performance of the catalysts. The catalytic activity of the V-Ti-HMS catalysts with the Si/Ti ratio of 30 is approximately two times higher than that of V-HMS catalysts with comparable selectivity. The enhanced activity exhibited by the V-Ti-HMS catalyst is attributed to the improved dispersion and reducibility of vanadium oxides.

Keywords: mesoporous titanosilicate; hexagonal mesoporous silica (HMS); vanadium catalyst; methanol-selective oxidation

1. Introduction

Methanol synthesized from syngas is an important building block for the synthesis of many important chemicals, e.g., dimethoxymethane, which is in high demand in various fields, including the pharmaceutical, cosmetics, and petroleum industries, due to its excellent chemical stability, desirable solubility, and low toxicity. The industrial production of dimethoxymethane from methanol involves a two-step, formaldehyde-mediated acidic route, which requires high energy and additional maintenance expenses [1,2]. Therefore, the design of catalysts for one-pot methanol-to-dimethoxymethane processes is of great potential. Continuous endeavors have been dedicated to improving the efficiency of this process [3–5].

Metal oxides comprising metals of high oxidation states generally show high activity in selective oxidation reactions, which can be correlated to the redox pair of the oxides [6]. Among these metal oxides, vanadium oxide stands out for its wide applications in catalytic reactions, including the selective oxidation of o-xylene to phthalic anhydride [7–9], the oxidative dehydrogenation of alkanes [10–12], and the selective oxidation of methanol to dimethoxymethane [13–16]. The performance of vanadium oxide is dramatically influenced by its dispersion degree. Vanadium oxides must be well-dispersed on supports due to the non-selective behavior exhibited by their bulk forms [17–20]. Among the studied supports (e.g., TiO_2, Al_2O_3, ZrO_2, CeO_2, and SiO_2), the use of TiO_2 results in the best catalytic activity due to the synergistic metal-support interaction, which greatly improves the dispersion of vanadium oxide [15,21–24]. However, the low surface area and poor thermal stability of TiO$_2$ limit its practical application.

SiO$_2$ possesses a high surface area and excellent thermal stability, but pure SiO$_2$ is unsuitable to support vanadium oxides. It is surmised that calcination subjects silica-supported vanadium oxides to agglomeration [25]. Hence, the incorporation of Ti species is suggested to improve the dispersion of vanadium oxides on silica supports and thus result in high catalytic activity. Compared with pure titania or silica supports, titanosilicates have high surface areas that greatly accommodate the dispersion of active components [20,26–28]. Additionally, titanosilicates have high thermal stability up to 800 °C and can stabilize the highly dispersed amorphous vanadium oxides (up to 20 wt%) under a calcination temperature of 500 °C [29].

Many efforts have been made towards the fabrication of titanosilicate-based catalysts. The conventional methods involve the coating of Ti species on the silica surface via the impregnation method, the grafting method, or the atomic layer deposition method [30–35]. The disadvantages of these methods are the complicated procedures and calcination steps that easily contribute to the formation of bulk Ti species, which often block the paths toward the active components [36]. Therefore, the direct integration of Ti species into the silica framework is a desirable alternative to the traditional coating-based methods. Although Ti species have been directly incorporated into mesoporous silicas such as SBA-15 [37–39], HMS [40–42], MCM-41 [43–45], and MCM-48 [46–48], there are few reports on such heterogeneous supports involving vanadium oxide catalysts [12,19], much less in terms of selective methanol-to-dimethoxymethane oxidation applications.

Herein, we prepared the hexagonal mesoporous silica (HMS) and Ti-modified HMS (Ti-HMS), which were used as supports for vanadium oxide to selectively oxidize methanol to dimethoxymethane. Characterizations including X-ray diffraction (XRD), UV-vis diffuse reflectance spectroscopy (DRS), N$_2$ adsorption–desorption, Raman spectroscopy, Fourier transform infrared (FT-IR) spectroscopy, and H$_2$–TPR were performed to reveal the effects of Ti species in promoting the structural properties and catalytic performance of V$_2$O$_5$–HMS catalysts.

2. Results and Discussion

2.1. Physicochemical Properties of Supports and Catalysts

The small-angle XRD patterns of Ti-HMS and HMS supports are shown in Figure 1. It can be seen that all the Ti-HMS samples exhibit a small-angle peak at approximately 1–2°, which is attributed to the d$_{100}$ diffraction [41,49]. This observation indicates the preservation of the HMS's mesoporous structures after the incorporation of Ti species. With the increasing amount of Ti species, the d$_{100}$ peak shifts to lower angles with increased intensity, suggesting the occurrences of lattice expansion, scattering-domain dimension reduction, and pore diameter enlargement [50]. As the Si/Ti ratio reaches 10, the d$_{100}$ peak becomes poorly resolved, which may be due to the partial destruction of the HMS framework. For all the Ti-incorporated samples, no characteristic peak of TiO$_2$ is observed (Figure 1b), indicating the high dispersion of Ti on the HMS surface and/or in the HMS framework.

Figure 2 shows the wide-angle XRD patterns of the supported catalysts in the range of 10–90°. All the supported catalysts present similar well-resolved peaks belonging to crystalline V$_2$O$_5$, suggesting the formation of bulk V$_2$O$_5$ and/or V$_2$O$_5$ nanoparticles on the surface of the supports. It is noteworthy that the V$_2$O$_5$ in the 28V-Ti-HMS(30) catalyst presents less intensive XRD peaks. The relative crystallinity of V$_2$O$_5$ for 28V-HMS and 28V-HMS(50) is the highest (14%), followed by 28V-HMS(20) and 28V-HMS(10). 28V-HMS(30) has the lowest relative crystallinity, which is only 8%. This could be due to the better-dispersed V$_2$O$_5$ in the 28V-Ti-HMS(30) catalyst compared with the other catalysts.

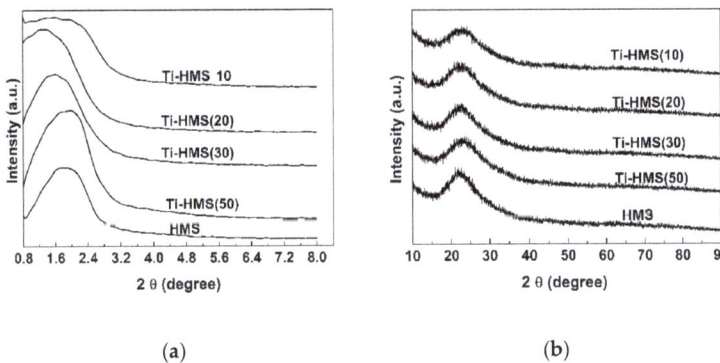

Figure 1. XRD patterns of HMS and Ti-HMS. (**a**) Low-angle. (**b**) Wide-angle.

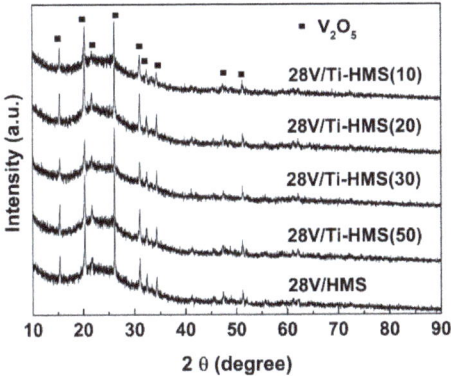

Figure 2. XRD patterns of catalysts with different supports.

Nitrogen physisorption is a practical technique used to study the textural properties of mesoporous materials. The BET specific surface area (S_{BET}) of supports and supported catalysts are summarized in Table 1. The pure HMS demonstrates the highest surface area. After the incorporation of Ti species, the S_{BET} of the supports decreases, while the cumulative pore volume (V_P) and pore diameter increase. This is consistent with the results of the XRD analysis, i.e., the peak position of Ti-HMS slightly shifts to lower angles due to the increasing interplanar distance. The deposition of vanadium species on the supports significantly decreased the S_{BET}, which is commonly observed for impregnated catalysts. Such surface-area loss is ascribed to the partial destruction of the support framework and/or the blocking of catalyst pores by bulk oxide or oxide nanoparticles.

2.2. Confirmation of the Framework Ti

The IR spectra of HMS and Ti-HMS with various Si/Ti ratios in the region of 400–1400 cm^{-1} are shown in Figure 3. All the samples exhibit a symmetric stretching vibration band at 807 cm^{-1} and asymmetric vibration band at 1090 cm^{-1} for the tetrahedral SiO_4^{4-} species. Meanwhile, the pure HMS shows a weak peak at 960 cm^{-1}, which indicates the Si-OH stretching vibration of free silanol groups in the amorphous region [51,52]. For Ti-containing materials, the 960 cm^{-1} band marginally shifts to a lower wavenumber, which has been reported in published studies [53,54]. Additionally, the bulk/crystalline TiO_2 (anatase and/or rutile) with a representative broad band at 600–650 cm^{-1} was not observed. These results suggest that Ti species have been successfully embedded into the HMS framework.

Table 1. Textural properties and Eg of the materials.

Sample	S_{BET} [a] (m^2/g)	V_P [b] (cm^3/g)	D_{ads} [c] (nm)	E_g [d] (eV)
HMS	764	0.67	3.71	-
Ti-HMS(50)	730	0.97	4.46	-
Ti-HMS(30)	662	0.93	5.22	-
Ti-HMS(20)	662	0.90	4.78	-
Ti-HMS(10)	661	0.59	3.80	-
28V-HMS	382	0.45	4.68	2.47
28V-Ti-HMS(50)	364	0.43	5.66	2.59
28V-Ti-HMS(30)	329	0.38	5.22	2.73
28V-Ti-HMS(20)	345	0.44	5.24	2.57
28V-Ti-HMS(10)	343	0.39	4.84	2.55

[a] Specific surface area calculated by the BET method. [b] Total pore volume determined at P/P0 = 0.99. [c] BJH adsorption average pore diameter. [d] Energy of adsorption edge.

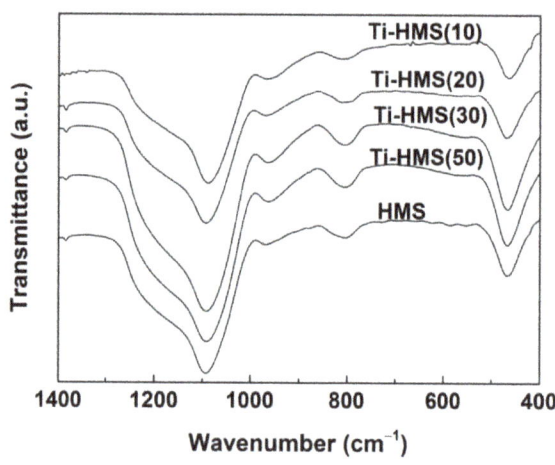

Figure 3. Infrared spectra of HMS and Ti-HMS samples with different Si/Ti ratios.

The incorporation of Ti species into the HMS framework was further verified by UV-vis DRS. This technique has been extensively employed for the characterization of the nature and coordination of Ti^{4+} ions in titanosilicates [55]. The corresponding spectra of the Ti-substituted supports are displayed in Figure 4, along with pure HMS and anatase (bulk). The band at ~210 nm is assigned to the titanium-to-silica charge transfer, which is used to confirm the Ti^{4+} framework sites. The broad absorption band at ~325 nm denotes the presence of bulk TiO$_2$ (anatase) [56]. It is apparent from Figure 4 that the HMS shows no adsorption band. The incorporation of Ti species introduces a peak centered at 210 nm with gradually strengthening intensity with increased Ti content. Despite the high loading of Ti for Ti-HMS(10), no visible peak at 325 nm is observed. Such a phenomenon strongly demonstrates that most Ti species successfully occupy the isolated site positions in the HMS framework in a wide range of Si/Ti ratios from 50 to 10. However, it is noticeable that the position of the 210 nm band shifts to a higher wavelength when the Si/Ti ratio is 10.

2.3. Effect of Ti on the Vanadia Dispersion

Figure 5 presents the UV-vis DRS spectra of the supported catalysts. The abrupt variation of all the spectra at 350 nm is due to the change of light source during detection. All the spectra consist of several ligand-to-metal charge transfer absorption bands characterizing the V$_2$O$_5$ constituted siliceous surface. The d-d absorption bands in the region of 495–660 nm representing the vanadium (+IV) are not detected, which confirms the fact that all the vanadium species were successfully oxidized to vanadium (+V). The region of 170–290 nm is attributed to the presence of coordinated tetrahedral vanadium species, whereas the 290–410 nm range features the O_h point group vanadium species (square-pyramidal or octahedral) [57]. These results indicate that the well-dispersed and bulk-like vanadium species coexisted in all the catalysts. Therefore, to further understand the vanadium's state, the absorption-edge energies (E_g) of the spectra are obtained to retrieve information regarding the type and number of ligands neighboring the core metal ion of the first coordination sphere, as there is a linear relationship between E_g and the number of bridging V-O-V bonds in bulk vanadium species [19,58].

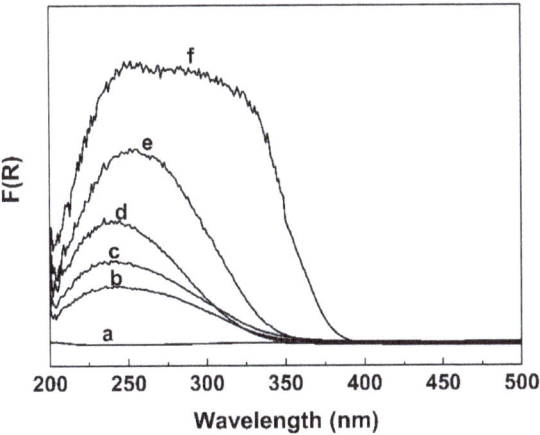

Figure 4. UV-vis spectra of calcined Ti-containing materials (**a**) HMS, (**b**) Ti-HMS(50), (**c**) Ti-HMS(30), (**d**) Ti-HMS(20), (**e**) Ti-HMS(10), and (**f**) anatase.

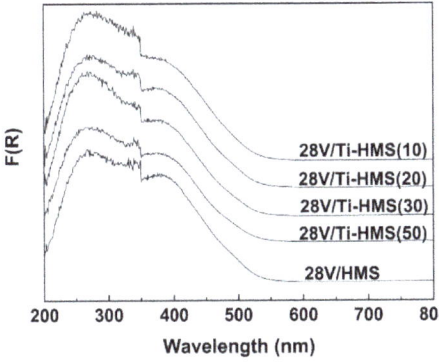

Figure 5. UV-vis spectra of catalysts with different supports.

The E_g of the supported catalysts are tabulated in Table 1, and the spectra of V-HMS and V-Ti-HMS in the form of $[F(R\infty)h\nu]^2$-against-E_g plot are depicted in Figure 6. Tian et al. [59] have demonstrated the use of E_g values to qualitatively monitor vanadium species polymerization. It has been verified that a high E_g value reflects the high dispersion of vanadium species. As shown in Table 1, Ti-HMS-supported V_2O_5 catalysts have higher E_g values than pure-HMS-supported catalysts, wherein Ti-HMS(30) exhibits the highest E_g value. Our results thus suggest the better dispersion of vanadium species supported on Ti-HMS catalysts than those on pure HMS catalysts.

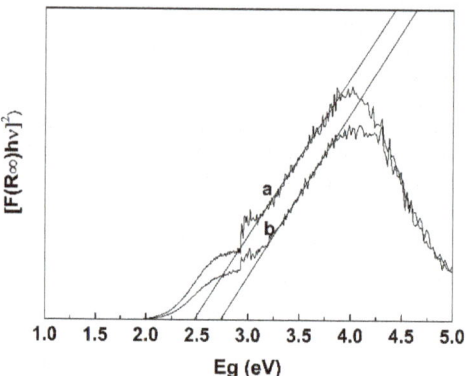

Figure 6. UV-vis spectra of catalysts with different supports. (a) 28V/HMS. (b) 28V/Ti-HMS (30).

The Raman spectra of the samples in the region of 200 to 1200 cm^{-1} are presented in Figure 7. The distinctive Raman characteristics of crystalline V_2O_5 are the bands centered at 282, 301, 405, 478, 525, 700, and 995 cm^{-1}, which appear for all samples [31] and thus agree well with the XRD results as shown in Figure 2 where well-resolved V_2O_5 patterns are observed. It is notable that a new band at 1035 cm^{-1}, which is identified as the terminal mono-oxo V=O bond of isolated surface vanadium species bonded to the surface silica species (O=V(OSi)$_3$) [60], is only observed for the Ti-modified catalysts. This finding further confirms that the dispersion of V_2O_5 over the Ti-modified catalysts, especially for V-Ti-HMS(30), is better than that over the Ti-free catalysts. This result is consistent with the outcomes of UV-vis DRS studies.

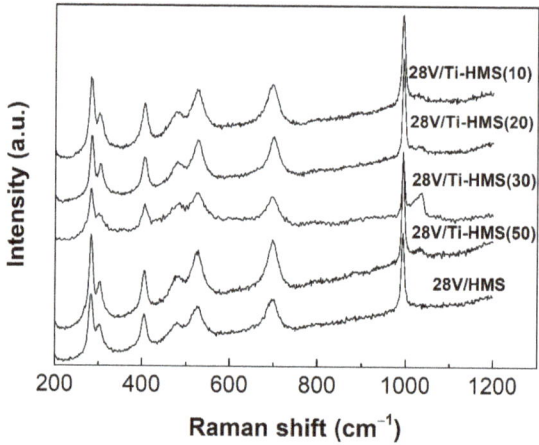

Figure 7. Raman spectra of vanadia-supported catalysts with different supports.

2.4. Redox Property and Catalytic Performance of Catalysts

The TPR profiles of different catalysts are presented in Figure 8. Since no reduction peak was observed for Ti-HMS and HMS, the corresponding TPR files are not shown here. As surmised, the addition of Ti species can improve the redox properties of catalysts. The main H_2 consumption peak of V-HMS catalysts emerges at ~890 K, with a shoulder occurring at ~860 K. After the incorporation of Ti species into the HMS framework, the main reduction peaks shift to lower temperatures (~865 K). Further increasing the amount of Ti, a slight increase in the reduction temperature is observed for V-Ti-HMS(10), which may be attributed to the destruction of HMS. Besselmann et al. [61] associated the change in redox properties with the different types of vanadium species. The crystalline and polymeric vanadium species are more challenging to reduce than their monomeric counterparts. The resultant spectra of both UV-vis DRS and Raman spectroscopy revealed the improved dispersion of V_2O_5 over the Ti-HMS supports compared with that over the pure HMS supports. Therefore, it is reasonable to believe that modification using Ti species improves the dispersion of V_2O_5 and thus enhances the reducibility of catalysts.

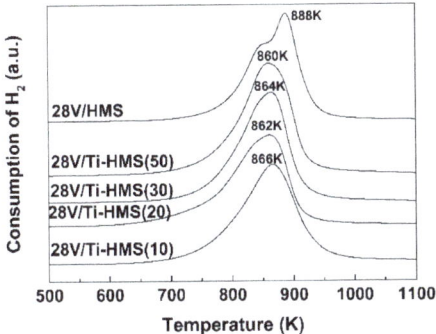

Figure 8. Temperature programmed reduction (TPR) profiles of catalysts with different supports.

Table 2 presents the catalytic performance of various catalysts at a low temperature (413 K). The V-HMS (without Ti) shows much lower catalytic activity with merely 6% of methanol conversion. It is evident that Ti modification can enhance catalysts' methanol conversion to a value higher than 10% without a significant loss in selectivity. Among the Ti-modified catalysts, V-Ti-HMS(30) exhibits the best catalytic performance with the methanol-to-dimethoxymethane conversion and selectivity of 15% and 87%, respectively. The improvement in catalytic performance is not related to the specific surface area since all the samples have similar specific surface area. Well-dispersed VO_4 oxides (isolated and polymeric surface species) are considered to be the catalytically active sites for the selective oxidation of methanol [62]. Additionally, the selectivity for DMM is mainly determined by the acidity of the catalysts [15]. According to the characterization results above, the incorporation of Ti significantly improved the dispersion of vanadia, which had minor effect on the acidity. The V-Ti-HMS(30) with the best V_2O_5 dispersion exhibits the best catalytic behaviors. Hence, the high dispersion of V_2O_5 is likely the key reason for the high activity observed for the Ti-modified catalysts. With the improvement in the dispersion of V_2O_5, the catalysts become more reducible, which has been confirmed by the TPR results. Based on the Mars-van Krevelen mechanism [63], the lattice oxygen of vanadia participates in oxidation. With the improved reducibility, both the oxidation of methanol and the re-oxidation of catalysts becomes easier. Therefore, the excellent dispersibility and high reducibility are most likely the major contributions to the improved catalytic performance of the V-Ti-HMS catalysts. Additionally, the reusability of V-Ti-HMS(30) was also tested for 50 h. The results in Figure 9 indicate the high stability of the catalysts.

Table 2. Performance of catalysts in the methanol oxidation reaction.

Sample	Conversion (%)	Selectivity (%)				
		DMM	MF	FA	DME	CO_x
28V-HMS	6	89	2	6	3	0
28V-Ti-HMS(50)	12	85	7	6	2	0
28V-Ti-HMS(30)	15	87	4	7	2	0
28V-Ti-HMS(20)	11	84	5	8	3	0
28V-Ti-HMS(10)	8	81	5	12	2	0

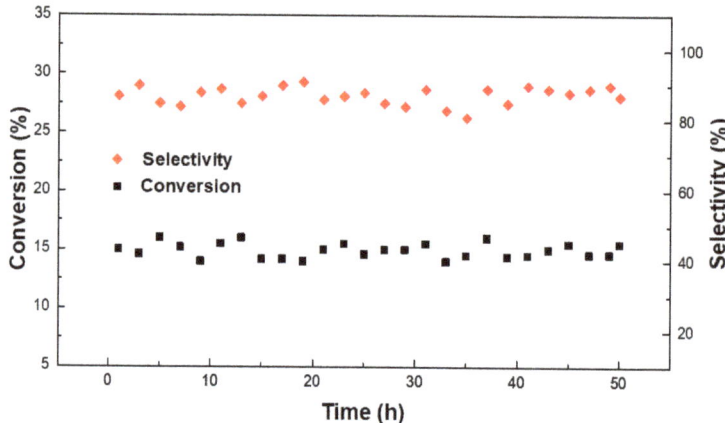

Figure 9. Conversion of methanol and selectivity of DMM using V-Ti-HMS(30) in the methanol oxidation.

3. Materials and Methods

3.1. Synthesis of the Supports

The hexagonal mesoporous silica (HMS) and hexagonal mesoporous titanosilicate (Ti-HMS) were synthesized under ambient conditions using dodecylamine ($C_{12}H_{25}NH_2$, Aldrich 98%) as a surfactant [56]. Generally, different ratios of tetraethyl orthosilicate (TEOS) to titanium tetrabutoxide (TBOT) were added to the solution of dodecylamine (DDA) in water and ethanol under vigorous stirring. The molar composition of the reaction mixture is xTi: 1.0 Si: 0.2 DDA: 9.0 EtOH: 160 H_2O. The reaction mixture was aged under ambient temperature under vigorous stirring for 24 h to obtain crystalline products. The Si/Ti molar ratios (denoted as y) used for the preparation of Ti-HMS(y) were 50, 30, 20, and 10. Pure HMS samples without Ti content were also prepared accordingly. All samples were filtered, washed thoroughly with water, dried under ambient temperature, and calcined at 923 K for 4 h.

3.2. Preparation of the Catalysts

The vanadium oxide species were introduced onto the support using the conventional wet impregnation method. NH_4VO_3 was dissolved in the oxalic acid solution and then impregnated onto the HMS or Ti-HMS supports. After drying at 383 K for 12 h and calcined at 673 K for 3 h, zV-HMS or zV-Ti-HMS(y) catalysts were obtained, where a and b are the Si/Ti molar ratio and loading of V_2O_5 (z wt%), respectively. The powdered vanadium catalysts were dissolved in a solution with 5% aqua regia and 30% hydrogen peroxide for 10 min. Then the percentage of V_2O_5 in the obtained catalysts was confirmed by inductively coupled plasma optical emission spectroscopy (ICP-OES) in the range of 27.5 wt% to 28.6 wt%. This is consistent with the estimated loading (28 wt%).

3.3. Characterization Methods

Powder XRD analysis was conducted using a Philips PANalytical X'pert Pro diffractometer equipped with the Cu Kα radiation at λ = 0.1542 nm, 40 kV, and 30 mA. The nitrogen adsorption–desorption isotherms were obtained using the Tristar 3000 equipment. Prior to the experiments, the samples were degassed at 543 K under vacuum for 5 h. The volume of the adsorbed N_2 was normalized to the standard temperature and pressure. The specific surface area (S_{BET}) was calculated by applying the BET equation in the relative pressure range of $0.05 < P/P_0 < 0.30$. The average pore diameter was calculated by applying the Barret-Joyner-Halenda (BJH) method to the adsorption branches of the N_2 isotherms. The cumulative pore volume was obtained from the isotherms at $P/P_0 = 0.99$.

DR UV-vis spectra were collected using a Varian Cary 5000 UV-vis-NIR spectrometer equipped with diffuse-reflectance accessories in the range of 200–1000 nm, and dehydrated $BaSO_4$ was used as the reference. All the samples were dehydrated at 473 K overnight and then stored in a sealed serum bottle to prevent the hydration of vanadium species. The Fourier transform infrared (FTIR) spectra were recorded in the range 400–1400 cm^{-1} using 32 scans and a resolution of 4 cm^{-1} on the VERTEX 70 of spectroscopic grade potassium bromide (KBr). Raman spectra were obtained using the Renishaw inVia Raman System equipped with a CCD detector and a Leica DMLM microscope. The excitation was provided by a 532 nm Ar^+ ion laser. The range and resolution were 100–1200 and 0.5 cm^{-1}, respectively.

H_2-TPR measurements were carried out in a continuous mode using a quartz microreactor (3.5 mm in diameter). A sample of ~50 mg was contacted with an H_2:Ar mixture (5% volume of H_2 in Ar) at a total flow rate of 20 ml min^{-1}. The sample was heated at a rate of 10 K min^{-1} from room temperature to 1073 K. The hydrogen consumption was monitored using a thermal conductivity detector (TCD). The reducing gas was first passed through the reference arm of the TCD before entering the reactor. The exit flow of the reactor was directed through a trap filled with $Mg(ClO_4)_2$ for dehydration, and then the flow proceeded to the second arm of the TCD.

3.4. Catalytic Activity Measurements

The selective oxidation of methanol was carried out at the atmospheric pressure in a fixed-bed microreactor (glass) with an inner diameter of 8 mm. Methanol was introduced into the reaction zone by bubbling O_2/N_2 (1/4) through a glass saturator filled with methanol (99.9%) maintained at 288 K. In each test, 0.2 g of catalyst was loaded, and the gas hourly space velocity (GHSV) was 11,400 ml g^{-1} h^{-1}. The feed composition was maintained as 1 methanol:1 O_2:4 N_2 (v/v). Methanol, dimethoxymethane, formaldehyde, and other organic compounds were analyzed using a GC equipped with an FID detector connected to Porapak N columns. CO and CO_2 were detected using a TDX-01 column. The gas lines were kept at 393 K to prevent the condensation of reactants and products.

4. Conclusions

In this study, Ti was successfully incorporated into the HMS framework with a wide range of Si/Ti ratios from 10 to 50. Compared with pure HMS, the Ti-HMS-supported vanadium oxide catalysts exhibited higher activities with a similar selectivity in the selective oxidation of methanol to dimethoxymethane. It was revealed that the incorporation of Ti rendered the well-dispersed oxides on the catalyst surface and hence increased the catalysts' reducibility. Excellent dispersibility and high reducibility are most likely the major contributions to the improved performance of the V-Ti-HMS catalysts.

Author Contributions: Conceptualization, J.F. and B.C.; methodology, L.B.S.; validation, L.B.S.; formal analysis, L.B.S. and J.F.; investigation, L.B.S. and K.S.K.; resources, B.C.; data curation, K.S.K.; writing—original draft preparation, L.B.S. and K.S.K.; writing—review and editing, J.F. and B.C.; visualization, L.B.S.; supervision, J.F. and B.C.; project administration, J.F.; funding acquisition, J.F. and B.C. All authors have read and agreed to the published version of the manuscript.

Funding: This research was funded by the Xiamen University Malaysia Campus, XMUMRF/2021-C8/IENG/0040.

Acknowledgments: The authors would like to thank the support from Xiamen University Malaysia Campus.

Conflicts of Interest: The authors declare no conflict of interest.

References

1. Ahmad, W.; Chan, F.L.; Chaffee, A.L.; Wang, H.; Hoadley, A.; Tanksale, A. Dimethoxymethane Production via Catalytic Hydrogenation of Carbon Monoxide in Methanol Media. *ACS Sustain. Chem. Eng.* **2020**, *8*, 2081–2092. [CrossRef]
2. Ftouni, K.; Lakiss, L.; Thomas, S.; Daturi, M.; Fernandez, C.; Bazin, P.; El Fallah, J.; El-Roz, M. TiO_2/Zeolite Bifunctional (Photo)Catalysts for a Selective Conversion of Methanol to Dimethoxymethane: On the Role of Brønsted Acidity. *J. Phys. Chem. C* **2018**, *122*, 29359–29367. [CrossRef]

3. Peláez, R.; Marín, P.; Ordóñez, S. Effect of formaldehyde precursor and water inhibition in dimethoxymethane synthesis from methanol over acidic ion exchange resins: Mechanism and kinetics. *Biofuel Bioprod. Biorefin.* **2021**, *15*, 1696–1708. [CrossRef]
4. To, A.T.; Wilke, T.J.; Nelson, E.; Nash, C.P.; Bartling, A.; Wegener, E.C.; Unocic, K.A.; Habas, S.E.; Foust, T.D.; Ruddy, D.A. Dehydrogenative Coupling of Methanol for the Gas-Phase, One-Step Synthesis of Dimethoxymethane over Supported Copper Catalysts. *ACS Sustain. Chem. Eng.* **2020**, *8*, 12151–12160. [CrossRef]
5. Sun, R.; Delidovich, I.; Palkovits, R. Dimethoxymethane as a Cleaner Synthetic Fuel: Synthetic Methods, Catalysts, and Reaction Mechanism. *ACS Catal.* **2019**, *9*, 1298–1318. [CrossRef]
6. Reddy, K.P.; Choi, H.; Kim, D.; Choi, M.; Ryoo, R.; Park, J.Y. The facet effect of ceria nanoparticles on platinum dispersion and catalytic activity of methanol partial oxidation. *Chem. Commun.* **2021**, *57*, 7382–7385. [CrossRef] [PubMed]
7. Akbari, A.; Alavi, S.M. The effect of cesium and antimony promoters on the performance of Ti-phosphate-supported vanadium(V) oxide catalysts in selective oxidation of o-xylene to phthalic anhydride. *Chem. Eng. Res. Des.* **2015**, *102*, 286–296. [CrossRef]
8. Wellmann, A.; Grazia, L.; Bermejo-Deval, R.; Sanchez-Sanchez, M.; Lercher, J.A. Effect of promoters on o-xylene oxidation pathways reveals nature of selective sites on TiO$_2$ supported vanadia. *J. Catal.* **2022**, *408*, 330–338. [CrossRef]
9. Dias, C.R.; Portela, M.F.; Bañares, M.A.; Galán-Fereres, M.; López-Granados, M.; Peña, M.A.; Fierro, J.L.G. Selective oxidation of o-xylene over ternary V-Ti-Si catalysts. *Appl. Catal. A Gen.* **2002**, *224*, 141–151. [CrossRef]
10. Jiang, X.; Zhang, X.; Purdy, S.C.; He, Y.; Huang, Z.; You, R.; Wei, Z.; Meyer, H.M.; Yang, J.; Pan, Y.; et al. Multiple Promotional Effects of Vanadium Oxide on Boron Nitride for Oxidative Dehydrogenation of Propane. *JACS Au* **2022**, *2*, 1096–1104. [CrossRef]
11. Kazerooni, H.; Towfighi Darian, J.; Mortazavi, Y.; Khadadadi, A.A.; Asadi, R. Titania-Supported Vanadium Oxide Synthesis by Atomic Layer Deposition and Its Application for Low-Temperature Oxidative Dehydrogenation of Propane. *Catal. Lett.* **2020**, *150*, 2807–2822. [CrossRef]
12. Wang, C.; Chen, J.-G.; Xing, T.; Liu, Z.-T.; Liu, Z.-W.; Jiang, J.; Lu, J. Vanadium Oxide Supported on Titanosilicates for the Oxidative Dehydrogenation of n-Butane. *Ind. Eng. Chem. Res.* **2015**, *54*, 3602–3610. [CrossRef]
13. Chen, S.; Ma, X. The role of oxygen species in the selective oxidation of methanol to dimethoxymethane over VOx/TS-1 catalyst. *J. Ind. Eng. Chem.* **2017**, *45*, 296–300. [CrossRef]
14. Wang, T.; Meng, Y.; Zeng, L.; Gong, J. Selective oxidation of methanol to dimethoxymethane over V$_2$O$_5$/TiO$_2$–Al$_2$O$_3$ catalysts. *Sci. Bull.* **2015**, *60*, 1009–1018. [CrossRef]
15. Lu, X.; Qin, Z.; Dong, M.; Zhu, H.; Wang, G.; Zhao, Y.; Fan, W.; Wang, J. Selective oxidation of methanol to dimethoxymethane over acid-modified V$_2$O$_5$/TiO$_2$ catalysts. *Fuel* **2011**, *90*, 1335–1339. [CrossRef]
16. Fan, Z.; Guo, H.; Fang, K.; Sun, Y. Efficient V$_2$O$_5$/TiO$_2$ composite catalysts for dimethoxymethane synthesis from methanol selective oxidation. *RSC Adv.* **2015**, *5*, 24795–24802. [CrossRef]
17. Mamedov, E.A.; Cortés Corberán, V. Oxidative dehydrogenation of lower alkanes on vanadium oxide-based catalysts. The present state of the art and outlooks. *Appl. Catal. A Gen.* **1995**, *127*, 1–40. [CrossRef]
18. Khodakov, A.; Yang, J.; Su, S.; Iglesia, E.; Bell, A.T. Structure and properties of vanadium oxide-zirconia catalysts for propane oxidative dehydrogenation. *J. Catal.* **1998**, *177*, 343–351. [CrossRef]
19. Setnička, M.; Čičmanec, P.; Bulánek, R.; Zukal, A.; Pastva, J. Hexagonal mesoporous titanosilicates as support for vanadium oxide—Promising catalysts for the oxidative dehydrogenation of n-butane. *Catal. Today* **2013**, *204*, 132–139. [CrossRef]
20. Santacesaria, E.; Cozzolino, M.; Di Serio, M.; Venezia, A.M.; Tesser, R. Vanadium based catalysts prepared by grafting: Preparation, properties and performances in the ODH of butane. *Appl. Catal. A Gen.* **2004**, *270*, 177–192. [CrossRef]
21. Meng, Y.; Wang, T.; Chen, S.; Zhao, Y.; Ma, X.; Gong, J. Selective oxidation of methanol to dimethoxymethane on V$_2$O$_5$–MoO$_3$/γ-Al$_2$O$_3$ catalysts. *Appl. Catal. B* **2014**, *160–161*, 161–172. [CrossRef]
22. Kaichev, V.V.; Popova, G.Y.; Chesalov, Y.A.; Saraev, A.A.; Andrushkevich, T.V.; Bukhtiyarov, V.I. Active component of supported vanadium catalysts in the selective oxidation of methanol. *Kinet. Catal.* **2016**, *57*, 82–94. [CrossRef]
23. Kropp, T.; Paier, J.; Sauer, J. Support Effect in Oxide Catalysis: Methanol Oxidation on Vanadia/Ceria. *J. Am. Chem. Soc.* **2014**, *136*, 14616–14625. [CrossRef]
24. Yoon, S.; Oh, K.; Liu, F.; Seo, J.H.; Somorjai, G.A.; Lee, J.H.; An, K. Specific metal–support interactions between nanoparticle layers for catalysts with enhanced methanol oxidation activity. *ACS Catal.* **2018**, *8*, 5391–5398. [CrossRef]
25. Jaegers, N.R.; Wan, C.; Hu, M.Y.; Vasiliu, M.; Dixon, D.A.; Walter, E.; Wachs, I.E.; Wang, Y.; Hu, J.Z. Investigation of Silica-Supported Vanadium Oxide Catalysts by High-Field 51V Magic-Angle Spinning NMR. *J. Phys. Chem. C* **2017**, *121*, 6246–6254. [CrossRef]
26. Castillo, R.; Koch, B.; Ruiz, P.; Delmon, B. Influence of the Amount of Titania on the Texture and Structure of Titania Supported on Silica. *J. Catal.* **1996**, *161*, 524–529. [CrossRef]
27. Gao, X.; Wachs, I.E. Molecular Engineering of Supported Vanadium Oxide Catalysts Through Support Modification. *Top. Catal.* **2002**, *18*, 243–250. [CrossRef]
28. Gao, X.; Bare, S.R.; Fierro, J.L.G.; Wachs, I.E. Structural Characteristics and Reactivity/Reducibility Properties of Dispersed and Bilayered V$_2$O$_5$/TiO$_2$/SiO$_2$ Catalysts. *J. Phys. Chem. B* **1999**, *103*, 618–629. [CrossRef]
29. Mahipal Reddy, B.; Mehdi, S.; Padmanabha Reddy, E. Dispersion and thermal stability of vanadium oxide catalysts supported on titania-silica mixed oxide. *Catal. Lett.* **1993**, *20*, 317–327. [CrossRef]
30. Chakraborty, S.; Nayak, S.C.; Deo, G. TiO$_2$/SiO$_2$ supported vanadia catalysts for the ODH of propane. *Catal. Today* **2015**, *254*, 62–71. [CrossRef]

31. Shee, D.; Mitra, B.; Chary, K.V.R.; Deo, G. Characterization and reactivity of vanadium oxide supported on TiO$_2$-SiO$_2$ mixed oxide support. *Mol. Catal.* **2018**, *451*, 228–237. [CrossRef]
32. Hamilton, N.; Wolfram, T.; Tzolova Müller, G.; Hävecker, M.; Kröhnert, J.; Carrero, C.; Schomäcker, R.; Trunschke, A.; Schlögl, R. Topology of silica supported vanadium–titanium oxide catalysts for oxidative dehydrogenation of propane. *Catal. Sci. Technol.* **2012**, *2*, 1346–1359. [CrossRef]
33. Kwak, J.H.; Herrera, J.E.; Hu, J.Z.; Wang, Y.; Peden, C.H.F. A new class of highly dispersed VO$_x$ catalysts on mesoporous silica: Synthesis, characterization, and catalytic activity in the partial oxidation of ethanol. *Appl. Catal. A Gen.* **2006**, *300*, 109–119. [CrossRef]
34. Keränen, J.; Carniti, P.; Gervasini, A.; Iiskola, E.; Auroux, A.; Niinistö, L. Preparation by atomic layer deposition and characterization of active sites in nanodispersed vanadia/titania/silica catalysts. *Catal. Today* **2004**, *91–92*, 67–71. [CrossRef]
35. Keränen, J.; Auroux, A.; Ek, S.; Niinistö, L. Preparation, characterization and activity testing of vanadia catalysts deposited onto silica and alumina supports by atomic layer deposition. *Appl. Catal. A Gen.* **2002**, *228*, 213–225. [CrossRef]
36. Bérubé, F.; Kleitz, F.; Kaliaguine, S. Surface properties and epoxidation catalytic activity of Ti-SBA15 prepared by direct synthesis. *J. Mater. Sci.* **2009**, *44*, 6727–6735. [CrossRef]
37. Chen, Y.; Huang, Y.; Xiu, J.; Han, X.; Bao, X. Direct synthesis, characterization and catalytic activity of titanium-substituted SBA-15 mesoporous molecular sieves. *Appl. Catal. A Gen.* **2004**, *273*, 185–191. [CrossRef]
38. Devi, P.; Das, U.; Dalai, A.K. Production of glycerol carbonate using a novel Ti-SBA-15 catalyst. *Chem. Eng. J.* **2018**, *346*, 477–488. [CrossRef]
39. Bérubé, F.; Nohair, B.; Kleitz, F.; Kaliaguine, S. Controlled Postgrafting of Titanium Chelates for Improved Synthesis of Ti-SBA-15 Epoxidation Catalysts. *Chem. Mater.* **2010**, *22*, 1988–2000. [CrossRef]
40. Zepeda, T.A.; Pawelec, B.; Fierro, J.L.G.; Halachev, T. Removal of refractory S-containing compounds from liquid fuels on novel bifunctional CoMo/HMS catalysts modified with Ti. *Appl. Catal. B* **2007**, *71*, 223–236. [CrossRef]
41. Düzenli, D.; Sahin, Ö.; Kazıcı, H.Ç.; Aktaş, N.; Kivrak, H. Synthesis and characterization of novel Ti doped hexagonal mesoporous silica catalyst for nonenzymatic hydrogen peroxide oxidation. *Microporous Mesoporous Mater.* **2018**, *257*, 92–98. [CrossRef]
42. Zepeda, T.A.; Infantes-Molina, A.; de Leon, J.N.D.; Obeso-Estrella, R.; Fuentes, S.; Alonso-Nuñez, G.; Pawelec, B. Synthesis and characterization of Ga-modified Ti-HMS oxide materials with varying Ga content. *J. Mol. Catal. A Chem.* **2015**, *397*, 26–35. [CrossRef]
43. Wang, S.; Shi, Y.; Ma, X.; Gong, J. Tuning Porosity of Ti-MCM-41: Implication for Shape Selective Catalysis. *ACS Appl. Mater. Interfaces* **2011**, *3*, 2154–2160. [CrossRef] [PubMed]
44. Peng, R.; Zhao, D.; Dimitrijevic, N.M.; Rajh, T.; Koodali, R.T. Room Temperature Synthesis of Ti–MCM-48 and Ti–MCM-41 Mesoporous Materials and Their Performance on Photocatalytic Splitting of Water. *J. Phys. Chem. C* **2012**, *116*, 1605–1613. [CrossRef]
45. Song, H.; Wang, J.; Wang, Z.; Song, H.; Li, F.; Jin, Z. Effect of titanium content on dibenzothiophene HDS performance over Ni$_2$P/Ti-MCM-41 catalyst. *J. Catal.* **2014**, *311*, 257–265. [CrossRef]
46. Zhao, D.; Budhi, S.; Rodriguez, A.; Koodali, R.T. Rapid and facile synthesis of Ti-MCM-48 mesoporous material and the photocatalytic performance for hydrogen evolution. *Int. J. Hydrog. Energy* **2010**, *35*, 5276–5283. [CrossRef]
47. Peng, R.; Wu, C.-M.; Baltrusaitis, J.; Dimitrijevic, N.M.; Rajh, T.; Koodali, R.T. Ultra-stable CdS incorporated Ti-MCM-48 mesoporous materials for efficient photocatalytic decomposition of water under visible light illumination. *Chem. Commun.* **2013**, *49*, 3221–3223. [CrossRef]
48. Huang, F.; Hao, H.; Sheng, W.; Dong, X.; Lang, X. Embedding an organic dye into Ti-MCM-48 for direct photocatalytic selective aerobic oxidation of sulfides driven by green light. *Chem. Eng. J.* **2022**, *432*, 134285. [CrossRef]
49. Tanev, P.T.; Pinnavaia, T.J. A Neutral Templating Route to Mesoporous Molecular Sieves. *Science* **1995**, *267*, 865–867. [CrossRef]
50. Zepeda, T.A.; Fierro, J.L.G.; Pawelec, B.; Nava, R.; Klimova, T.; Fuentes, G.A.; Halachev, T. Synthesis and Characterization of Ti-HMS and CoMo/Ti-HMS Oxide Materials with Varying Ti Content. *Chem. Mater.* **2005**, *17*, 4062–4073. [CrossRef]
51. Davis, R.J.; Liu, Z. Titania−Silica: A Model Binary Oxide Catalyst System. *Chem. Mater.* **1997**, *9*, 2311–2324. [CrossRef]
52. Schraml-Marth, M.; Walther, K.L.; Wokaun, A.; Handy, B.E.; Baiker, A. Porous silica gels and TiO$_2$/SiO$_2$ mixed oxides prepared via the sol-gel process: Characterization by spectroscopic techniques. *J. Non-Cryst. Solids* **1992**, *143*, 93–111. [CrossRef]
53. Boahene, P.E.; Soni, K.; Dalai, A.K.; Adjaye, J. Hydrotreating of coker light gas oil on Ti-modified HMS supports using Ni/HPMo catalysts. *Appl. Catal. B* **2011**, *101*, 294–305. [CrossRef]
54. Soni, K.; Boahene, P.E.; Chandra Mouli, K.; Dalai, A.K.; Adjaye, J. Hydrotreating of coker light gas oil on Ti-HMS supported heteropolytungstic acid catalysts. *Appl. Catal. A Gen.* **2011**, *398*, 27–36. [CrossRef]
55. Boccuti, M.R.; Rao, K.M.; Zecchina, A.; Leofanti, G.; Petrini, G. Spectroscopic Characterization of Silicalite and Titanium-Silicalite. In *Studies in Surface Science and Catalysis*; Morterra, C., Zecchina, A., Costa, G., Eds.; Elsevier: Amsterdam, The Netherlands, 1989; Volume 48, pp. 133–144.
56. Zhang, W.; Fröba, M.; Wang, J.; Tanev, P.T.; Wong, J.; Pinnavaia, T.J. Mesoporous Titanosilicate Molecular Sieves Prepared at Ambient Temperature by Electrostatic (S$^+$I$^-$, S$^+$X$^-$I$^+$) and Neutral (S°I°) Assembly Pathways: A Comparison of Physical Properties and Catalytic Activity for Peroxide Oxidations. *J. Am. Chem. Soc.* **1996**, *118*, 9164–9171. [CrossRef]
57. Bulánek, R.; Čapek, L.; Setnička, M.; Čičmanec, P. DR UV–vis Study of the Supported Vanadium Oxide Catalysts. *J. Phys. Chem. C* **2011**, *115*, 12430–12438. [CrossRef]

58. Gao, X.; Wachs, I.E. Investigation of Surface Structures of Supported Vanadium Oxide Catalysts by UV−vis−NIR Diffuse Reflectance Spectroscopy. *J. Phys. Chem. B* **2000**, *104*, 1261–1268. [CrossRef]
59. Tian, H.; Ross, E.I.; Wachs, I.E. Quantitative Determination of the Speciation of Surface Vanadium Oxides and Their Catalytic Activity. *J. Phys. Chem. B* **2006**, *110*, 9593–9600. [CrossRef]
60. Magg, N.; Immaraporn, B.; Giorgi, J.B.; Schroeder, T.; Bäumer, M.; Döbler, J.; Wu, Z.; Kondratenko, E.; Cherian, M.; Baerns, M.; et al. Vibrational spectra of alumina- and silica-supported vanadia revisited: An experimental and theoretical model catalyst study. *J. Catal.* **2004**, *226*, 88–100. [CrossRef]
61. Besselmann, S.; Freitag, C.; Hinrichsen, O.; Muhler, M. Temperature-programmed reduction and oxidation experiments with V_2O_5/TiO_2 catalysts. *Phys. Chem. Chem. Phys.* **2001**, *3*, 4633–4638. [CrossRef]
62. Kim, T.; Wachs, I.E. CH3OH oxidation over well-defined supported V_2O_5/Al_2O_3 catalysts: Influence of vanadium oxide loading and surface vanadium–oxygen functionalities. *J. Catal.* **2008**, *255*, 197–205. [CrossRef]
63. Kaichev, V.V.; Popova, G.Y.; Chesalov, Y.A.; Saraev, A.A.; Zemlyanov, D.Y.; Beloshapkin, S.A.; Knop-Gericke, A.; Schlögl, R.; Andrushkevich, T.V.; Bukhtiyarov, V.I. Selective oxidation of methanol to form dimethoxymethane and methyl formate over a monolayer V_2O_5/TiO_2 catalyst. *J. Cata.* **2014**, *311*, 59–70. [CrossRef]

MDPI
St. Alban-Anlage 66
4052 Basel
Switzerland
www.mdpi.com

Catalysts Editorial Office
E-mail: catalysts@mdpi.com
www.mdpi.com/journal/catalysts

Disclaimer/Publisher's Note: The statements, opinions and data contained in all publications are solely those of the individual author(s) and contributor(s) and not of MDPI and/or the editor(s). MDPI and/or the editor(s) disclaim responsibility for any injury to people or property resulting from any ideas, methods, instructions or products referred to in the content.

www.ingramcontent.com/pod-product-compliance
Lightning Source LLC
LaVergne TN
LVHW070639100526
838202LV00013B/838